# FUNDAMENTAL MECHANICS OF FLUIDS

# Also Available from McGraw-Hill

## Schaum's Outline Series in Mechanical Engineering

Most outlines include basic theory, definitions, and hundreds of example problems solved in step-by-step detail, and supplementary problems with answers.

Titles on the current list include:

*Acoustics*
*Basic Equations of Engineering Science*
*Continuum Mechanics*
*Engineering Economics*
*Engineering Mechanics, 4th edition*
*Engineering Thermodynamics*
*Fluid Dynamics, 2d edition*
*Fluid Mechanics & Hydraulics, 2d edition*
*Heat Transfer*
*Lagrangian Dynamics*

*Machine Design*
*Mathematical Handbook of Formulas*
   *& Tables*
*Mechanical Vibrations*
*Operations Research*
*Statics & Mechanics of Materials*
*Statics & Strength of Materials*
*Strength of Materials, 2d edition*
*Theoretical Mechanics*
*Thermodynamics with Chemical*
   *Applications, 2d edition*

## Schaum's Solved Problems Books

Each title in this series is a complete and expert source of solved problems containing thousands of problems with worked out solutions.

Related titles on the current list include:

*3000 Solved Problems in Calculus*
*2500 Solved Problems in Differential*
  *Equations*
*2500 Solved Problems in Fluid Mechanics*
  *& Hydraulics*
*1000 Solved Problems in in Heat Transfer*
*3000 Solved Problems in Linear Algebra*

*2000 Solved Problems in Mechanical*
  *Engineering Thermodynamics*
*2000 Solved Problems in Numerical Analysis*
*700  Solved Problems in Vector Mechanics*
  *for Engineers: Dynamics*
*800  Solved Problems in Vector Mechanics*
  *for Engineers: Statics*

Available at most college bookstores, or for a complete list of titles and prices, write to:   Schaum Division
         McGraw-Hill, Inc.
         Princeton Road, S-1
         Hightstown, NJ 08520

# FUNDAMENTAL
# MECHANICS
# OF FLUIDS

## Second Edition

**I. G. Currie**

*Professor*
*Department of Mechanical Engineering*
*University of Toronto*

Boston, Massachusetts   Burr Ridge, Illinois
Dubuque, Iowa   Madison, Wisconsin   New York, New York
San Francisco, California   St. Louis, Missouri

## McGraw-Hill

*A Division of The McGraw-Hill Companies*

This book was set in Times Roman by Science Typographers, Inc.
The editors were John J. Corrigan and John M. Morriss;
the production supervisor was Richard A. Ausburn.
The cover was designed by Carla Bauer.
Project supervision was done by Science Typographers, Inc.

QA
901
C8
1993

This book is printed on acid-free paper.

**FUNDAMENTAL MECHANICS OF FLUIDS**

5 6 7 8 9 10 11 12 13 14 BKMBKM 9 9 8

ISBN 0-07-015000-1


**Library of Congress Cataloging-in-Publication Data**

Currie, Iain G.
    Fundamental mechanics of fluids / I. G. Currie. —2nd ed.
        p.        cm.
    Includes bibliographical references and index.
    ISBN 0-07-015000-1
    1. Fluid mechanics.    I. Title.
QA901.C8        1993
532–dc20                                    92-40110

**I. G. Currie** is a Professor of Mechanical Engineering at the University of Toronto, where he has taught and carried out research in fluid mechanics since 1966. He is a registered Professional Engineer in the Province of Ontario, a Member of the American Society of Mechanical Engineers, a Fellow of the Canadian Society for Mechanical Engineering, and a Member of Sigma Xi. Subsequent to receiving his Bachelor's degree from the University of Strathclyde, Professor Currie received a Masters degree from the University of British Columbia and a Doctorate from the California Institute of Technology.

To my wife CATHIE,
our daughter KAREN, and
our sons DAVID and BRIAN

# CONTENTS

**ix**

# Part III  Viscous Flows of Incompressible Fluids

# Part IV  Compressible Flow of Inviscid Fluids

# PREFACE TO THE SECOND EDITION

It has been most gratifying to receive, over the years, letters from students, instructors, and professors dealing with the contents of the original edition of the book. Errors have been pointed out, and many useful and constructive suggestions have been received. In addition, many complimentary remarks have been received. All of these communications have been most valuable, and they are all appreciated.

The continued acceptance of the book over a period of almost two decades is probably due to three factors which relate to the nature of the contents. First, the material is broad-based, covering a wide range of topics in an introductory manner. Second, the material content involves classical results, rather than attempting to include the most recent advances in the subject. Third, the material is presented in a sufficiently explicit manner that students do not have to spend hours filling in the missing steps. These characteristics are consistent with the stated objective of supplying a teaching text at the first advanced level of the subject. They also tend to maintain relevancy of the book as a text over a long period of time.

In view of the factors pointed out above, preparation of the second edition posed some strong temptations. Since none of the topics is covered in great depth, there is always the temptation of adding more material to existing chapters. The question of which material to extend is a personal matter, and any additional material is as liable to irk the reader as it is to endear him or her. Then there is the temptation of adding whole new chapters of material which were not covered at all in the original edition. Whole books have been written on topics which are omitted from this text, but many of these topics cannot be given a brief, meaningful introduction. Turbulence is an example of this type of topic.

After many discussion with many people, and after much personal reflection, it was decided to add one new chapter. This chapter appears in Part III of the book, and it deals with buoyancy-driven flows. This is the main new feature of the book with respect to the addition of material. The existing material has been reviewed and revised, and the main addition to the existing chapters has been to the problems at the end of each chapter. Increasing the number of problems, and the need for a solutions manual for instructors, are the only features which most users agreed upon. There are now over 100 problems in the book.

In addition to the many users who took the time and the trouble to write to me about the book, the author would like to thank the students at the University of Toronto for their continuous feedback relating to the contents. The following reviewers deserve special recognition for providing many valuable comments and suggestions. H. L. Moses, Virginia Polytechnic Institute and State University; D. N. Riahi, University of Illinois–Urbana, Champaign; Richard Salant, Georgia Institute of Technology; Stephen G. Schwarz, Tulane University; S. Thangam, Stevens Institute of Technology; Albert Tong, University of Texas–Arlington; and Houston G. Wood III, University of Virginia. Thanks are also due to Ernesto Morala for taking the photographs which now form Plate 1 of the book.

*I. G. Currie*

# PREFACE TO THE FIRST EDITION

This book covers the fundamental mechanics of fluids as it is treated at the senior level or in first graduate courses. Many excellent books exist which treat special areas of fluid mechanics such as ideal-fluid flow or boundary-layer theory. However, there are very few books indeed at this level which sacrifice an in-depth study of one of these special areas of fluid mechanics for a briefer treatment of a broader area of the fundamentals of fluid mechanics. This situation exists despite the fact that many institutions of higher learning offer a broad, fundamental course in a wide spectrum of their students before offering more advanced specialized courses to those who are specializing in fluid mechanics. Recognition of this situation is the prime motivation for introducing this book.

The book is divided into four parts, each of which contains three chapters. Part I is entitled "Governing Equations," and it deals with the derivation of the basic conservation laws, flow kinematics, and some basic theorems of fluid mechanics. Part II is entitled "Ideal-Fluid Flow," and it covers two-dimensional potential flows, three-dimensional potential flows, and surface waves. Part III deals with "Viscous Flows of Incompressible Fluids," and it contains chapters on exact solutions, low-Reynolds-number approximations, and boundary-layer theory. The final part of the book is entitled "Compressible Flow of Inviscid Fluids," and this part contains chapters which deal with shock waves, one-dimensional flows, and multi-dimensional flows. Appendixes are also included which summarize vectors, tensors, the governing equations in the common coordinate systems, complex variables, and thermodynamics.

The treatment of the material is such as to emphasize the phenomena which are associated with the various properties of fluids while providing techniques for solving specific classes of fluid-flow problems. The treatment is not geared to any one discipline, and it may readily be studied by physicists and chemists as well as by engineers from various branches. Since the book is intended for teaching purposes, phrases such as "it can be shown that" and similar clichés which cause many hours of effort for many students have been avoided. In order to aid the teaching process, several problems are included at the end of each of the twelve chapters. These problems serve to illustrate points which are brought out in the text and to extend the material covered in the text.

Most of the material contained in this book can be covered in about 50 lecture hours. For more extensive courses the material contained here may be completely covered and even augmented. Parts II, III, and IV are essentially independent so that they may be interchanged or any one or more of them may be omitted. This permits a high degree of teaching flexibility, and permits the instructor to include or substitute material which is not covered in the text. Such additional material may include free convection, density stratification, hydrodynamic stability, and turbulence with applications to pollution, meteorology, etc. These topics are not included here, not because they do not involve fundamentals, but rather because the author set up a priority of what he considers to be the basic fundamentals.

Many people are to be thanked for their direct or indirect contributions to this text. The author had the privilege of taking lectures from F. E. Marble, C. B. Millikan, and P. G. Saffman. Some of the style and methods of these great scholars is probably evident on some of the following pages. The National Research Council of Canada are due thanks for supplying the photographs which appear in this book. My colleagues at the University of Toronto have been a constant source of encouragement, and the staff of the Department of Mechanical Engineering provided excellent typing and drafting services. Finally, sincere appreciation is extended to the many students who have taken my lectures at the University of Toronto and who have pointed out errors and deficiencies in the material content of the draft of this text.

*I. G. Currie*

# PART
# I

# GOVERNING
# EQUATIONS

In this first part of the book a sufficient set of equations will be derived, based on physical laws and postulates, governing the dependent variables of a fluid which is moving. The dependent variables are the fluid-velocity components, pressure, density, temperature, and internal energy, or some similar set of variables. The equations governing these variables will be derived from the principles of mass, momentum, and energy conservation and from equations of state. Having established a sufficient set of governing equations, some purely kinematical aspects of fluid flow are discussed, at which time the concept of vorticity is introduced. The final section of this part of the book introduces certain relationships which can be derived from the governing equations under certain simplifying conditions. These relationships may be used in conjunction with the basic governing equations or as alternatives to them.

1

Taken as a whole, this part of the book establishes the mathematical equations which result from invoking certain physical laws which are postulated to be valid for a moving fluid. These equations may assume different forms, depending upon which variables are chosen and upon which simplifying assumptions are made. The remaining parts of the book are devoted to solving these governing equations for different classes of fluid flows and thereby explaining quantitatively some of the phenomena which are observed in fluid flow.

# CHAPTER

# 1

# BASIC CONSERVATION LAWS

The essential purpose of this chapter is to derive the set of equations which results from invoking the physical laws of conservation of mass, momentum, and energy. In order to realize this objective, it is necessary to discuss certain preliminary topics. The first topic of discussion is the two basic ways in which the conservation equations may be derived, the statistical method and the continuum method. Having selected the basic method to be used in deriving the equations, one is then faced with the choice of reference frame to be employed, eulerian or lagrangian. Next, a general theorem, called Reynolds' transport theorem, is derived, since this theorem relates derivatives in the lagrangian framework to derivatives in the eulerian framework.

Having established the basic method to be employed and the tools to be used, the basic conservation laws are then derived. The conservation of mass yields the so-called continuity equation. The conservation of momentum leads ultimately to the Navier-Stokes equations, while the conservation of thermal energy leads to the energy equation. The derivation is followed by a discussion of the set of equations so obtained, and finally a summary of the basic conservation laws is given.

## 1.1 STATISTICAL AND CONTINUUM METHODS

There are basically two ways of deriving the equations which govern the motion of a fluid. One of these methods approaches the question from the molecular point of view. That is, this method treats the fluid as consisting of molecules whose motion is governed by the laws of dynamics. The macroscopic phenomena are assumed to arise from the molecular motion of the molecules, and the theory attempts to predict the macroscopic behavior of the fluid from the laws of mechanics and probability theory. For a fluid which is in a state not too far removed from equilibrium, this approach yields the equations of mass, momentum, and energy conservation. The molecular approach also yields expressions for the transport coefficients, such as the coefficient of viscosity and the thermal conductivity, in terms of molecular quantities such as the forces acting between molecules or molecular diameters. The theory is well developed for light gases, but it is incomplete for polyatomic gas molecules and for liquids.

The alternative method which is used to derive the equations which govern the motion of a fluid uses the continuum concept. In the continuum approach, individual molecules are ignored and it is assumed that the fluid consists of continuous matter. At each point of this continuous fluid there is supposed to be a unique value of the velocity, pressure, density, and other so-called "field variables." The continuous matter is then required to obey the conservation laws of mass, momentum, and energy, which give rise to a set of differential equations governing the field variables. The solution to these differential equations then defines the variation of each field variable with space and time which corresponds to the mean value of the molecular magnitude of that field variable at each corresponding position and time.

The statistical method is rather elegant, and it may be used to treat gas flows in situations where the continuum concept is no longer valid. However, as was mentioned before, the theory is incomplete for dense gases and for liquids. The continuum approach requires that the mean free path of the molecules be very small compared with the smallest physical-length scale of the flow field (such as the diameter of a cylinder or other body about which the fluid is flowing). Only in this way can meaningful averages over the molecules at a "point" be made and the molecular structure of the fluid be ignored. However, if this condition is satisfied, there is no distinction amongst light gases, dense gases, or even liquids—the results apply equally to all. Since the vast majority of phenomena encountered in fluid mechanics fall well within the continuum domain and may involve liquids as well as gases, the continuum method will be used in this book. With this background, the meaning and validity of the continuum concept will now be explored in some detail. The field variables such as the density $\rho$ and the velocity vector $\mathbf{u}$ will in general be functions of the spatial coordinates and time. In symbolic form this is written as $\rho = \rho(\mathbf{x}, t)$ and $\mathbf{u} = \mathbf{u}(\mathbf{x}, t)$, where $\mathbf{x}$ is the position vector whose cartesian coordinates are $x$, $y$, and $z$. At any particular point in space these continuum variables are defined in terms of the properties of the various molecules which occupy a small volume in the neighborhood of that point.

Consider a small volume of fluid $\Delta V$ containing a large number of molecules. Let $\Delta m$ and $\mathbf{v}$ be the mass and velocity of any individual molecule contained within the volume $\Delta V$ as indicated in Fig. 1.1. The density and the

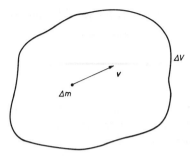

**FIGURE 1.1**
An individual molecule in a small volume $\Delta V$ having a mass $\Delta m$ and a velocity **v**.

velocity at a point in the continuum are then defined by the following limits:

$$\rho = \lim_{\Delta V \to \epsilon} \left( \frac{\sum \Delta m}{\Delta V} \right)$$

$$\mathbf{u} = \lim_{\Delta V \to \epsilon} \left( \frac{\sum \mathbf{v} \Delta m}{\sum \Delta m} \right)$$

where $\epsilon$ is a volume which is sufficiently small that $\epsilon^{1/3}$ is small compared with the smallest significant length scale in the flow field but is sufficiently large that it contains a large number of molecules. The summations in the above expressions are taken over all the molecules contained within the volume $\Delta V$. The other field variables may be defined in terms of the molecular properties in an analogous way.

A sufficient condition, though not a necessary condition, for the continuum approach to be valid is

$$\frac{1}{n} \ll \epsilon \ll L^3$$

where $n$ is the number of molecules per unit volume and $L$ is the smallest significant length scale in the flow field, which is usually called the *macroscopic length scale*. The characteristic *microscopic length scale* is the mean free path between collisions of the molecules. Then the above condition states that the continuum concept will certainly be valid if some volume $\epsilon$ can be found which is much larger than the volume occupied by a single molecule of the fluid but which is much smaller than the cube of the smallest macroscopic length scale (such as cylinder diameter). Since a cube of gas, at normal temperature and pressure, whose side is 2 micrometers contains about $2 \times 10^8$ molecules and the corresponding figure for a liquid is about $2 \times 10^{11}$ molecules, the continuum condition is readily met in the vast majority of flow situations encountered in physics and engineering. It may be expected to break down in situations where the smallest macroscopic length scale approaches microscopic dimensions, such as in the structure of a shock wave, and where the microscopic length scale approaches macroscopic dimensions, such as when a rocket passes through the edge of the atmosphere.

## 1.2 EULERIAN AND LAGRANGIAN COORDINATES

Having selected the continuum approach as the method which will be used to derive the basic conservation laws, one is next faced with a choice of reference frames in which to formulate the conservation laws. There are two basic coordinate systems which may be employed, these being eulerian and lagrangian coordinates.

In the eulerian framework the independent variables are the spatial coordinates $x$, $y$, and $z$ and time $t$. This is the familiar framework in which most problems are solved. In order to derive the basic conservation equations in this framework, attention is focused on the fluid which passes through a control volume which is fixed in space. The fluid inside the control volume at any instant in time will consist of different fluid particles from that which was there at some previous instant in time. If the principles of conservation of mass, momentum, and energy are applied to the fluid which passes through the control volume, the basic conservation equations are obtained in eulerian coordinates.

In the lagrangian approach, attention is fixed on a particular mass of fluid as it flows. Suppose we could color a small portion of the fluid without changing its density. Then in the lagrangian framework we follow this colored portion as it flows and changes its shape, but we are always considering the same particles of fluid. The principles of mass, momentum, and energy conservation are then applied to this particular element of fluid as it flows, resulting in a set of conservation equations in lagrangian coordinates. In this reference frame $x$, $y$, $z$, and $t$ are no longer independent variables, since if it is known that our colored portion of fluid passed through the coordinates $x_0$, $y_0$, and $z_0$ at some time $t_0$, then its position at some later time may be calculated if the velocity components $u$, $v$, and $w$ are known. That is, as soon as a time interval $(t - t_0)$ is specified, the velocity components uniquely determine the coordinate changes $(x - x_0)$, $(y - y_0)$, and $(z - z_0)$ so that $x$, $y$, $z$, and $t$ are no longer independent. The independent variables in the lagrangian system are $x_0$, $y_0$, $z_0$, and $t$, where $x_0$, $y_0$, and $z_0$ are the coordinates which a specified fluid element passed through at time $t_0$. That is, the coordinates $x_0$, $y_0$, and $z_0$ identify which fluid element is being considered, and the time $t$ identifies its instantaneous location.

The choice of which coordinate system to employ is largely a matter of taste. It is probably more convincing to apply the conservation laws to a control volume which always consists of the same fluid particles rather than one through which different fluid particles pass. This is particularly true when invoking the law of conservation of energy, which consists of applying the first law of thermodynamics, since the same fluid particles are more readily justified as a thermodynamic system. For this reason, the lagrangian coordinate system will be used to derive the basic conservation equations. Although the lagrangian system will be used to derive the basic equations, the eulerian system is the preferred one for solving the majority of problems. In the next section the relation between the different derivatives will be established.

## 1.3 MATERIAL DERIVATIVE

Let $\alpha$ be any field variable such as the density or temperature of the fluid. From the eulerian viewpoint, $\alpha$ may be considered to be a function of the independent variables $x$, $y$, $z$, and $t$. But if a specific fluid element is observed for a short period of time $\delta t$ as it flows, its position will change by amounts $\delta x$, $\delta y$, and $\delta z$ while its value of $\alpha$ will change by an amount $\delta \alpha$. That is, if the fluid element is observed in the lagrangian framework, the independent variables are $x_0$, $y_0$, $z_0$, and $t$, where $x_0$, $y_0$, and $z_0$ are initial coordinates for the fluid element. Thus $x$, $y$, and $z$ are no longer independent variables but are functions of $t$ as defined by the trajectory of the element. During the time $\delta t$ the change in $\alpha$ may be calculated from differential calculus to be

$$\frac{\partial \alpha}{\partial t}\delta t + \frac{\partial \alpha}{\partial x}\delta x + \frac{\partial \alpha}{\partial y}\delta y + \frac{\partial \alpha}{\partial z}\delta z$$

Equating the above change in $\alpha$ to the observed change $\delta \alpha$ in the lagrangian framework and dividing throughout by $\delta t$ gives

$$\frac{\delta \alpha}{\delta t} = \frac{\partial \alpha}{\partial t} + \frac{\delta x}{\delta t}\frac{\partial \alpha}{\partial x} + \frac{\delta y}{\delta t}\frac{\partial \alpha}{\partial y} + \frac{\delta z}{\delta t}\frac{\partial \alpha}{\partial z}$$

The left-hand side of this expression represents the total change in $\alpha$ as observed in the lagrangian framework during the time $\delta t$, and in the limit it represents the time derivative of $\alpha$ in the lagrangian system, which will be denoted by $D\alpha/Dt$. It may be also noted that in the limit as $\delta t \to 0$ the ratio $\delta x/\delta t$ becomes the velocity component in the $x$ direction, namely, $u$. Similarly $\delta y/\delta t \to v$ and $\delta z/\delta t \to w$ as $\delta t \to 0$. That is, as $\delta t \to 0$, the expression for the change in $\alpha$ becomes

$$\frac{D\alpha}{Dt} = \frac{\partial \alpha}{\partial t} + u\frac{\partial \alpha}{\partial x} + v\frac{\partial \alpha}{\partial y} + w\frac{\partial \alpha}{\partial z}$$

In vector form this equation may be written as follows:

$$\frac{D\alpha}{Dt} = \frac{\partial \alpha}{\partial t} + (\mathbf{u} \cdot \nabla)\alpha$$

Alternatively, using the Einstein summation convention where repeated subscripts are summed, the tensor form may be written as

$$\frac{D\alpha}{Dt} = \frac{\partial \alpha}{\partial t} + u_k\frac{\partial \alpha}{\partial x_k} \tag{1.1}$$

The term $D\alpha/Dt$ in Eq. (1.1) is the so-called "material derivative." It represents the total change in the quantity $\alpha$ as seen by an observer who is following the fluid and is watching a particular mass of the fluid. The entire right-hand side of Eq. (1.1) represents the total change in $\alpha$ expressed in eulerian coordinates. The term $u_k(\partial\alpha/\partial x_k)$ expresses the fact that in a time-independent flow field in which the fluid properties depend upon the spatial coordinates only, there is a change in $\alpha$ due to the fact that a given fluid element changes its position with time and therefore assumes different values of $\alpha$ as it flows. The term $\partial\alpha/\partial t$ is the familiar eulerian time derivative and expresses the fact that at any point in space the fluid properties may change with time. Then Eq. (1.1) expresses the lagrangian rate of change $D\alpha/Dt$ of $\alpha$ for a given fluid element in terms of the eulerian derivatives $\partial\alpha/\partial t$ and $\partial\alpha/\partial x_k$.

## 1.4 CONTROL VOLUMES

The concept of a control volume, as required to derive the basic conservation equations, has been mentioned in connection with both the lagrangian and the eulerian approaches. Irrespective of which coordinate system is used, there are two principal control volumes from which to choose. One of these is a parallelepiped of sides $\delta x$, $\delta y$, and $\delta z$. Each fluid property, such as the velocity or pressure, is expanded in a Taylor series about the center of the control volume to give expressions for that property at each face of the control volume. The conservation principle is then invoked, and when $\delta x$, $\delta y$, and $\delta z$ are permitted to become vanishingly small, the differential equation for that conservation principle is obtained. Frequently shortcuts are taken and the control volume is taken to have sides of length $dx$, $dy$, and $dz$ with only the first term of the Taylor series being carried out.

The second type of control volume is arbitrary in shape, and each conservation principle is applied to an integral over the control volume. For example, the mass within the control volume is $\int_V \rho \, dV$, where $\rho$ is the fluid density and the integration is carried out over the entire volume $V$ of the fluid contained within the control volume. The result of applying each conservation principle will be an integro-differential equation of the type

$$\int_V L\alpha \, dV = 0$$

where $L$ is some differential operator and $\alpha$ is some property of the fluid. But since the control volume $V$ was arbitrarily chosen, the only way this equation can be satisfied is by setting $L\alpha = 0$, which gives the differential equation of the conservation law. If the integrand in the above equation was not equal to zero, it would be possible to redefine the control volume $V$ in such a way that the integral of $L\alpha$ was not equal to zero.

Each of these two types of control volumes has some merit, and in this book each will be used at some point, depending upon which gives the better insight to the physics of the situation under discussion. The arbitrary control volume will be used in the derivation of the basic conservation laws, since it seems to detract less from the principles being imposed. Needless to say the results obtained by the two methods are identical.

## 1.5 REYNOLDS' TRANSPORT THEOREM

The method which has been selected to derive the basic equations from the conservation laws is to use the continuum concept and to follow an arbitrarily shaped control volume in a lagrangian frame of reference. The combination of the arbitrary control volume and the lagrangian coordinate system means that material derivatives of volume integrals will be encountered. As was mentioned in the previous section, it is necessary to transform such terms into equivalent expressions involving volume integrals of eulerian derivatives. The theorem which permits such a transformation is called Reynolds' transport theorem.

Consider a specific mass of fluid and follow it for a short period of time $\delta t$ as it flows. Let $\alpha$ be any property of the fluid such as its mass, momentum in some direction, or its energy. Since a specific mass of fluid is being considered and since $x_0$, $y_0$, $z_0$, and $t$ are the independent variables in the lagrangian framework, the quantity $\alpha$ will be a function of $t$ only as the control volume moves with the fluid. That is, $\alpha = \alpha(t)$ only and the rate of change of the integral of $\alpha$ will be defined by the following limit:

$$\frac{D}{Dt} \int_{V(t)} \alpha(t) \, dV = \lim_{\delta t \to 0} \left\{ \frac{1}{\delta t} \left[ \int_{V(t+\delta t)} \alpha(t+\delta t) \, dV - \int_{V(t)} \alpha(t) \, dV \right] \right\}$$

where $V(t)$ is the control volume containing the specified mass of fluid and which may change its size and shape as it flows. The quantity $\alpha(t + \delta t)$ integrated over $V(t)$ will now be subtracted, then added again inside the above limit.

$$\frac{D}{Dt} \int_{V(t)} \alpha(t) \, dV = \lim_{\delta t \to 0} \left\{ \frac{1}{\delta t} \left[ \int_{V(t+\delta t)} \alpha(t+\delta t) \, dV - \int_{V(t)} \alpha(t+\delta t) \, dV \right] \right. \\ \left. + \frac{1}{\delta t} \left[ \int_{V(t)} \alpha(t+\delta t) \, dV - \int_{V(t)} \alpha(t) \, dV \right] \right\}$$

The first two integrals inside this limit correspond to holding the integrand fixed and permitting the control volume $V$ to vary while the second two integrals correspond to holding $V$ fixed and permitting the integrand $\alpha$ to vary. The latter component of the change is, by definition, the integral of the familiar eulerian derivative with respect to time. Then the expression for the lagrangian derivative of the integral of $\alpha$ may be written in the following form:

$$\frac{D}{Dt} \int_{V(t)} \alpha(t) \, dV = \lim_{\delta t \to 0} \left\{ \frac{1}{\delta t} \left[ \int_{V(t+\delta t) - V(t)} \alpha(t+\delta t) \, dV \right] \right\} + \int_{V(t)} \frac{\partial \alpha}{\partial t} \, dV$$

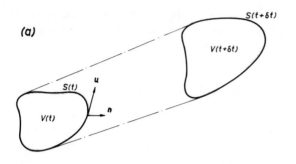

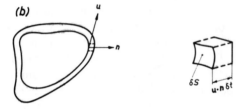

**FIGURE 1.2**
(*a*) Arbitrarily shaped control volume at times $t$ and $t + \delta t$ and (*b*) superposition of the control volume at these times showing an element $\delta V$ of the volume change.

The remaining limit, corresponding to the volume $V$ changing while $\alpha$ remains fixed, may be evaluated from geometric considerations.

Figure 1.2*a* shows the control volume $V(t)$ which encloses the mass of fluid being considered both at time $t$ and at time $t + \delta t$. During this time interval the control volume has moved downstream and has changed its size and shape. The surface which encloses $V(t)$ is denoted by $S(t)$, and at any point on this surface the velocity may be denoted by **u** and the unit outward normal by **n**.

Figure 1.2*b* shows the control volume $V(t + \delta t)$ superimposed on $V(t)$, and an element of the difference in volumes is detailed. The perpendicular distance from any point on the inner surface to the outer surface is $\mathbf{u} \cdot \mathbf{n} \, \delta t$, so that an element of surface area $\delta S$ will correspond to an element of volume change $\delta V$ in which $\delta V = \mathbf{u} \cdot \mathbf{n} \, \delta t \, \delta S$. Then the volume integral inside the limit in the foregoing equation may be transformed into a surface integral in which $dV$ is replaced by $\mathbf{u} \cdot \mathbf{n} \, \delta t \, dS$.

$$\frac{D}{Dt} \int_{V(t)} \alpha(t) \, dV = \lim_{\delta t \to 0} \left\{ \left[ \int_{S(t)} \alpha(t + \delta t) \mathbf{u} \cdot \mathbf{n} \, dS \right] \right\} + \int_{V(t)} \frac{\partial \alpha}{\partial t} \, dV$$

$$= \int_{S(t)} \alpha(t) \mathbf{u} \cdot \mathbf{n} \, dS + \int_{V(t)} \frac{\partial \alpha}{\partial t} \, dV$$

Having completed the limiting process, the lagrangian derivative of a volume integral has been converted into a surface integral and a volume integral in which the integrands contain only eulerian derivatives. As was mentioned in the previous section, it is necessary to obtain each term in the conservation equations as the volume integral of something. The foregoing form of Reynolds' transport theorem may be put in this desired form by converting the surface integral to a volume integral by use of Gauss' theorem, which is formulated in Appendix A. In this way the surface-integral term becomes

$$\int_{S(t)} \alpha(t) \mathbf{u} \cdot \mathbf{n} \, dS = \int_{V(t)} \mathbf{\nabla} \cdot (\alpha \mathbf{u}) \, dV$$

Substituting this result into the foregoing expression and combining the two volume integrals gives the preferred form of Reynolds' transport theorem.

$$\frac{D}{Dt} \int_V \alpha \, dV = \int_V \left[ \frac{\partial \alpha}{\partial t} + \mathbf{\nabla} \cdot (\alpha \mathbf{u}) \right] dV$$

Or, in tensor notation,

$$\frac{D}{Dt} \int_V \alpha \, dV = \int_V \left[ \frac{\partial \alpha}{\partial t} + \frac{\partial}{\partial x_k} (\alpha u_k) \right] dV \tag{1.2}$$

Equation (1.2) relates the lagrangian derivative of a volume integral of a given mass to a volume integral in which the integrand has eulerian derivatives only.

Having established the method to be used to derive the basic conservation equations and having established the necessary background material, it remains to invoke the various conservation principles. The first such principle to be treated will be the conservation of mass.

## 1.6  CONSERVATION OF MASS

Consider a specific mass of fluid whose volume $V$ is arbitrarily chosen. If this given fluid mass is followed as it flows, its size and shape will be observed to change but its mass will remain unchanged. This is the principle of mass conservation which applies to fluids in which no nuclear reactions are taking place. The mathematical equivalence of the statement of mass conservation is to set the lagrangian derivative $D/Dt$ of the mass of fluid contained in $V$, which is $\int_V \rho \, dV$, equal to zero. That is, the equation which expresses conservation of mass is

$$\frac{D}{Dt} \int_V \rho \, dV = 0$$

This equation may be converted to a volume integral in which the integrand contains only eulerian derivatives by use of Reynolds' transport theorem [Eq. (1.2)], in which the fluid property $\alpha$ is, in this case, the mass density $\rho$.

$$\int_V \left[ \frac{\partial \rho}{\partial t} + \frac{\partial}{\partial x_k} (\rho u_k) \right] dV = 0$$

Since the volume $V$ was arbitrarily chosen, the only way in which the above equation can be satisfied for all possible choices of $V$ is for the integrand to be zero. Then the equation expressing conservation of mass becomes

$$\frac{\partial \rho}{\partial t} + \frac{\partial}{\partial x_k}(\rho u_k) = 0 \tag{1.3a}$$

Equation (1.3a) expresses more than the fact that mass is conserved. Since it is a partial differential equation, the implication is that the velocity is continuous. For this reason Eq. (1.3a) is usually called the *continuity equation*. The derivation which has been given here is for a single-phase fluid in which no change of phase is taking place. If two phases were present, such as water and steam, the starting statement would be that the rate at which the mass of fluid 1 is increasing is equal to the rate at which the mass of fluid 2 is decreasing. The generalization to cases of multiphase fluids and to cases of nuclear reactions is obvious. Since such cases cause no changes in the basic ideas or principles, they will not be included in this treatment of the fundamentals.

In many practical cases of fluid flow the variation of density of the fluid may be ignored, as for most cases of the flow of liquids. In such cases the fluid is said to be *incompressible*, which means that as a given mass of fluid is followed, not only will its mass be observed to remain constant but its volume, and hence its density, will be observed to remain constant. Mathematically, this statement may be written as

$$\frac{D\rho}{Dt} = 0$$

In order to use this special simplification, the continuity equation is first expanded by use of a vector identity which is given in Appendix A.

$$\frac{\partial \rho}{\partial t} + u_k \frac{\partial \rho}{\partial x_k} + \rho \frac{\partial u_k}{\partial x_k} = 0$$

The first and second terms in this form of the continuity equation will be recognized as being the eulerian form of the material derivative as given by Eq. (1.1). That is, an alternative form of Eq. (1.3a) is

$$\frac{D\rho}{Dt} + \rho \frac{\partial u_k}{\partial x_k} = 0 \tag{1.3b}$$

This mixed form of the continuity equation in which one term is given as a lagrangian derivative and the other as an eulerian derivative is not useful for actually solving fluid-flow problems. However, it is frequently used in the manipulations which reduce the governing equations to alternative forms, and for this reason it has been identified for future reference. An immediate example of such a case is the incompressible fluid under discussion. Since $D\rho/Dt = 0$ for such a fluid, Eq. (1.3b) shows that the continuity equation assumes the simpler form $\rho(\partial u_k/\partial x_k) = 0$. Since $\rho$ cannot be zero in general,

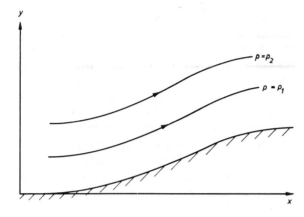

**FIGURE 1.3**
Flow of a density stratified fluid in which $D\rho/Dt = 0$ but $\partial\rho/\partial x$, $\partial\rho/\partial y \neq 0$.

the continuity equation for an incompressible fluid becomes

$$\frac{\partial u_k}{\partial x_k} = 0 \qquad \text{(incompressible)} \qquad\qquad (1.3c)$$

It should be noted that Eq. (1.3c) is valid not only for the special case of $D\rho/Dt = 0$ in which $\rho = $ constant everywhere, but also for stratified-fluid flows of the type depicted in Fig. 1.3. A fluid particle which follows the lines $\rho = \rho_1$ or $\rho = \rho_2$ will have its density remain fixed at $\rho = \rho_1$ or $\rho = \rho_2$ so that $D\rho/Dt = 0$. However, $\rho$ is not constant everywhere, so that $\partial\rho/\partial x \neq 0$ and $\partial\rho/\partial y \neq 0$. Such density stratifications may occur in the ocean (owing to salinity variations) or in the atmosphere (owing to temperature variations). However, in the majority of cases in which the fluid may be considered to be incompressible, the density is constant everywhere.

Equation (1.3), in either the general form (1.3a) or the incompressible form (1.3c), is the first condition which has to be satisfied by the velocity and the density. No dynamical relations have been used to this point, but the conservation-of-momentum principle will utilize dynamics.

## 1.7   CONSERVATION OF MOMENTUM

The principle of conservation of momentum is, in effect, an application of Newton's second law of motion to an element of the fluid. That is, when considering a given mass of fluid in a lagrangian frame of reference, it is stated that the rate at which the momentum of the fluid mass is changing is equal to the net external force acting on the mass. Some individuals prefer to think of forces only and restate this law in the form that the inertia force (due to acceleration of the element) is equal to the net external force acting on the element.

The external forces which may act on a mass of the fluid may be classed as either body forces, such as gravitational or electromagnetic forces, or surface forces, such as pressure forces or viscous stresses. Then if **f** is a vector which represents the resultant of the body forces per unit mass, the net external body force acting on a mass of volume $V$ will be $\int_V \rho \mathbf{f}\, dV$. Also, if **P** is a surface vector which represents the resultant surface force per unit area, the net external surface force acting on the surface $S$ containing $V$ will be $\int_s \mathbf{P}\, dS$.

According to the statement of the physical law which is being imposed in this section, the sum of the resultant forces evaluated above is equal to the rate of change of momentum (or inertia force). The mass per unit volume is $\rho$ and its momentum is $\rho \mathbf{u}$, so that the momentum contained in the volume $V$ is $\int_V \rho \mathbf{u}\, dV$. Then, if the mass of the arbitrarily chosen volume $V$ is observed in the lagrangian frame of reference, the rate of change of momentum of the mass contained within $V$ will be $(D/Dt)\int_V \rho \mathbf{u}\, dV$. Thus the mathematical equation which results from imposing the physical law of conservation of momentum is

$$\frac{D}{Dt}\int_V \rho \mathbf{u}\, dV = \int_s \mathbf{P}\, ds + \int_V \rho \mathbf{f}\, dV$$

In general, there are nine components of stress at any given point, one normal component and two shear components on each coordinate plane. These nine components of stress are most easily illustrated by use of a cubical element in which the faces of the cube are orthogonal to the cartesian coordinates, as shown in Fig. 1.4, and in which the stress components will act at a point as the length of the cube tends to zero. In Fig. 1.4 the cartesian coordinates $x$, $y$, and $z$ have been denoted by $x_1$, $x_2$, and $x_3$, respectively. This permits the components of stress to be identified by a double-subscript notation. In this notation, a particular component of the stress may be represented by the quantity $\sigma_{ij}$, in which the first subscript indicates that this stress component acts on the plane $x_i = $ constant and the second subscript indicates that it acts in the $x_j$ direction.

The fact that the stress may be represented by the quantity $\sigma_{ij}$, in which $i$ and $j$ may be 1, 2, or 3, means that the stress at a point may be represented by a

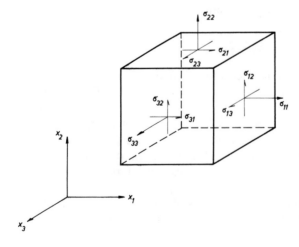

**FIGURE 1.4**
Representation of the nine components of stress which may act at a point in a fluid.

tensor of rank 2. However, on the surface of our control volume it was observed that there would be a vector force at each point, and this force was represented by **P**. The surface force vector **P** may be related to the stress tensor $\sigma_{ij}$ as follows: The three stress components acting on the plane $x_1$ = constant are $\sigma_{11}$, $\sigma_{12}$, and $\sigma_{13}$. Since the unit normal vector acting on this surface is $n_1$, the resulting force acting in the $x_1$ direction is $P_1 = \sigma_{11}n_1$. Likewise, the forces acting in the $x_2$ direction and the $x_3$ direction are, respectively, $P_2 = \sigma_{12}n_1$ and $P_3 = \sigma_{13}n_1$. Then, for an arbitrarily oriented surface whose unit normal has components $n_1$, $n_2$, and $n_3$, the surface force will be given by $P_j = \sigma_{ij}n_i$ in which $i$ is summed from 1 to 3. That is, in tensor notation the equation expressing conservation of momentum becomes

$$\frac{D}{Dt}\int_V \rho u_j \, dV = \int_s \sigma_{ij} n_i \, dS + \int_V \rho f_j \, dV$$

The left-hand side of this equation may be converted to a volume integral in which the integrand contains only eulerian derivatives by use of Reynolds' transport theorem, Eq. (1.2), in which the fluid property $\alpha$ here is the momentum per unit volume $\rho u_j$ in the $x_j$ direction. At the same time the surface integral on the right-hand side may be converted into a volume integral by use of Gauss' theorem as given in Appendix B. In this way the equation which evolved from Newton's second law becomes

$$\int_V \left[ \frac{\partial}{\partial t}(\rho u_j) + \frac{\partial}{\partial x_k}(\rho u_j u_k) \right] dV = \int_V \frac{\partial \sigma_{ij}}{\partial x_i} \, dV + \int_V \rho f_j \, dV$$

All these volume integrals may be collected to express this equation in the form $\int_V \{\ \} \, dV = 0$, where the integrand is a differential equation in eulerian coordinates. As before, the arbitrariness of the choice of the control volume $V$ is now used to show that the integrand of the above integro-differential equation must be zero. This gives the following differential equation to be satisfied by the field variables in order that the basic law of dynamics may be satisfied:

$$\frac{\partial}{\partial t}(\rho u_j) + \frac{\partial}{\partial x_k}(\rho u_j u_k) = \frac{\partial \sigma_{ij}}{\partial x_i} + \rho f_j$$

The left-hand side of this equation may be further simplified if the two terms involved are expanded in which the quantity $\rho u_j u_k$ is considered to be the product of $\rho u_k$ and $u_j$.

$$\rho \frac{\partial u_j}{\partial t} + u_j \frac{\partial \rho}{\partial t} + u_j \frac{\partial}{\partial x_k}(\rho u_k) + \rho u_k \frac{\partial u_j}{\partial x_k} = \frac{\partial \sigma_{ij}}{\partial x_i} + \rho f_j$$

The second and third terms on the left-hand side of this equation are now seen to sum to zero, since they amount to the continuity equation (1.3a) multiplied by the velocity $u_j$. With this simplification, the equation which expresses conservation of momentum becomes

$$\rho \frac{\partial u_j}{\partial t} + \rho u_k \frac{\partial u_j}{\partial x_k} = \frac{\partial \sigma_{ij}}{\partial x_i} + \rho f_j \tag{1.4}$$

It is useful to recall that this equation came from an application of Newton's second law to an element of the fluid. The left-hand side of Eq. (1.4) represents the rate of change of momentum of a unit volume of the fluid (or the inertia force per unit volume). The first term is the familiar temporal acceleration term, while the second term is a convective acceleration and accounts for local accelerations (around obstacles, etc.) even when the flow is steady. Note also that this second term is nonlinear, since the velocity appears quadratically. On the right-hand side of Eq. (1.4) are the forces which are causing the acceleration. The first of these is due to the gradient of surface shear stresses while the second is due to body forces, such as gravity, which act on the mass of the fluid. A clear understanding of the physical significance of each of the terms in Eq. (1.4) is essential when approximations to the full governing equations must be made. The surface-stress tensor $\sigma_{ij}$ has not been fully explained up to this point, but it will be investigated in detail in a later section.

## 1.8   CONSERVATION OF ENERGY

The principle of conservation of energy amounts to an application of the first law of thermodynamics to a fluid element as it flows. The first law of thermodynamics applies to a thermodynamic system which is originally at rest and, after some event, is finally at rest again. Under these conditions it is stated that the change in internal energy, due to the event, is equal to the sum of the total work done on the system during the course of the event and any heat which was added. Although a specified mass of fluid in a lagrangian frame of reference may be considered to be a thermodynamic system, it is, in general, never at rest and therefore never in equilibrium. However, in the thermodynamic sense a flowing fluid is seldom far from a state of equilibrium, and the apparent difficulty may be overcome by considering the instantaneous energy of the fluid to consist of two parts, intrinsic or internal energy and kinetic energy. That is, when applying the first law of thermodynamics, the energy referred to is considered to be the sum of the internal energy per unit mass $e$ and the kinetic energy per unit mass $\frac{1}{2}\mathbf{u} \cdot \mathbf{u}$. In this way the modified form of the first law of thermodynamics which will be applied to an element of the fluid states that the rate of change of the total energy (intrinsic plus kinetic) of the fluid as it flows is equal to the sum of the rate at which work is being done on the fluid by external forces and the rate at which heat is being added by conduction.

With this basic law in mind, we again consider any arbitrary mass of fluid of volume $V$ and follow it in a lagrangian frame of reference as it flows. The total energy of this mass per unit volume is $\rho e + \frac{1}{2}\rho \mathbf{u} \cdot \mathbf{u}$, so that the total energy contained in $V$ will be $\int_V (\rho e + \frac{1}{2}\rho \mathbf{u} \cdot \mathbf{u}) \, dV$. As was established in the previous section, there are two types of external forces which may act on the fluid mass under consideration. The work done on the fluid by these forces is given by the product of the velocity and the component of each force which is collinear with the velocity. That is, the work done is the scalar product of the velocity vector and the force vector. One type of force which may act on the fluid is a surface stress whose magnitude per unit area is represented by the vector $\mathbf{P}$. Then the total work done owing to such forces will be $\int_s \mathbf{u} \cdot \mathbf{P} \, dS$, where $S$ is the surface area enclosing $V$. The other type of force which may act on the fluid is a body force whose magnitude per unit mass is denoted by the vector $\mathbf{f}$. Then the total work done on the fluid due to such forces will be $\int_V \mathbf{u} \cdot \rho \mathbf{f} \, dV$. Finally, an expression for the heat added to the fluid is required. Let the vector $\mathbf{q}$ denote the conductive heat flux *leaving* the control volume. Then the quantity of heat leaving the fluid mass per unit time per unit surface area will be $\mathbf{q} \cdot \mathbf{n}$, where $\mathbf{n}$ is the unit outward normal, so that the net amount of heat leaving the fluid per unit time will be $\int_s \mathbf{q} \cdot \mathbf{n} \, dS$.

Having evaluated each of the terms which appear in the physical law which is to be imposed, the statement may now be written down in analytic form. In doing so, it must be borne in mind that the physical law is being applied to a specific, though arbitrarily chosen, mass of fluid so that lagrangian derivatives must be employed. In this way the expression of the statement that the rate of change of total energy is equal to the rate at which work is being done plus the rate at which heat is being *added* becomes

$$\frac{D}{Dt} \int_V \left( \rho e + \tfrac{1}{2}\rho \mathbf{u} \cdot \mathbf{u} \right) dV = \int_s \mathbf{u} \cdot \mathbf{P} \, dS + \int_V \mathbf{u} \cdot \rho \mathbf{f} \, dV - \int_s \mathbf{q} \cdot \mathbf{n} \, dS$$

This equation may be converted to one involving eulerian derivatives only by use of Reynolds' transport theorem, Eq. (1.2), in which the fluid property $\alpha$ is here the total energy per unit volume ($\rho e + \tfrac{1}{2}\rho \mathbf{u} \cdot \mathbf{u}$). The resulting integro-differential equation is

$$\int_V \left\{ \frac{\partial}{\partial t}\left( \rho e + \tfrac{1}{2}\rho \mathbf{u} \cdot \mathbf{u} \right) + \frac{\partial}{\partial x_k}\left[ \left( \rho e + \tfrac{1}{2}\rho \mathbf{u} \cdot \mathbf{u} \right)u_k \right] \right\} dV$$

$$= \int_s \mathbf{u} \cdot \mathbf{P} \, dS + \int_V \mathbf{u} \cdot \rho \mathbf{f} \, dV - \int_s \mathbf{q} \cdot \mathbf{n} \, dS$$

The next step is to convert the two surface integrals into volume integrals so that the arbitrariness of $V$ may be exploited to obtain a differential equation only. Using the fact that the force vector $\mathbf{P}$ is related to the stress tensor $\sigma_{ij}$ by the equation $P_j = \sigma_{ij}n_i$, as was shown in the previous section, the first surface integral may be converted to a volume integral as follows:

$$\int_s \mathbf{u} \cdot \mathbf{P} \, dS = \int_s u_j \sigma_{ij} n_i \, dS = \int_V \frac{\partial}{\partial x_i}(u_j \sigma_{ij}) \, dV$$

Here use has been made of Gauss' theorem as documented in Appendix A.

Gauss' theorem may be applied directly to the heat-flux term to give

$$\int_s \mathbf{q} \cdot \mathbf{n}\, dS = \int_s q_j n_j\, dS = \int_V \frac{\partial q_j}{\partial x_j}\, dV$$

Since the stress tensor $\sigma_{ij}$ has been brought into the energy equation, it is necessary to use the tensor notation from this point on. Then the expression for conservation of energy becomes

$$\int_V \left\{ \frac{\partial}{\partial t}\left(\rho e + \tfrac{1}{2}\rho u_j u_j\right) + \frac{\partial}{\partial x_k}\left[\left(\rho e + \tfrac{1}{2}\rho u_j u_j\right)u_k\right]\right\} dV$$

$$= \int_V \frac{\partial}{\partial x_i}(u_j \sigma_{ij})\, dV + \int_V u_j \rho f_j\, dV - \int_V \frac{\partial q_j}{\partial x_j}\, dV$$

Having converted each term to volume integrals, the conservation equation may be considered to be of the form $\int_V \{\ \} dV = 0$, where the choice of $V$ is arbitrary. Then the quantity inside the brackets in the integrand must be zero, which results in the following differential equation:

$$\frac{\partial}{\partial t}\left(\rho e + \tfrac{1}{2}\rho u_j u_j\right) + \frac{\partial}{\partial x_k}\left[\left(\rho e + \tfrac{1}{2}\rho u_j u_j\right)u_k\right] = \frac{\partial}{\partial x_i}(u_j \sigma_{ij}) + u_j \rho f_j - \frac{\partial q_j}{\partial x_j}$$

This equation may be made considerably simpler by using the equations which have been already derived, as will now be demonstrated. The first term on the left-hand side may be expanded by considering $\rho e$ and $\tfrac{1}{2}\rho u_j u_j$ to be the products $(\rho)(e)$ and $(\rho)(\tfrac{1}{2}u_j u_j)$, respectively. Then

$$\frac{\partial}{\partial t}\left(\rho e + \tfrac{1}{2}\rho u_j u_j\right) = \rho \frac{\partial e}{\partial t} + e \frac{\partial \rho}{\partial t} + \rho \frac{\partial}{\partial t}\left(\tfrac{1}{2}u_j u_j\right) + \tfrac{1}{2}u_j u_j \frac{\partial \rho}{\partial t}$$

Similarly, the second term on the left-hand side of the basic equation may be expanded by considering $\rho e u_k$ to be the product $(e)(\rho u_k)$ and $\tfrac{1}{2}\rho u_j u_j u_k$ to be the product $(\tfrac{1}{2}u_j u_j)(\rho u_k)$. Thus

$$\frac{\partial}{\partial x_k}\left[\left(\rho e + \tfrac{1}{2}\rho u_j u_j\right)u_k\right] = e \frac{\partial}{\partial x_k}(\rho u_k) + \rho u_k \frac{\partial e}{\partial x_k}$$

$$+ \tfrac{1}{2}u_j u_j \frac{\partial}{\partial x_k}(\rho u_k) + \rho u_k \frac{\partial}{\partial x_k}\left(\tfrac{1}{2}u_j u_j\right)$$

In this last equation, the quantity $(\partial/\partial x_k)(\rho u_k)$, which appears in the first and third terms on the right-hand side, may be replaced by $-\partial \rho/\partial t$ in view of the continuity equation (1.3a). Hence it follows that

$$\frac{\partial}{\partial x_k}\left[\left(\rho e + \tfrac{1}{2}\rho u_j u_j\right)u_k\right] = -e \frac{\partial \rho}{\partial t} + \rho u_k \frac{\partial e}{\partial x_k} - \tfrac{1}{2}u_j u_j \frac{\partial \rho}{\partial t} + \rho u_k \frac{\partial}{\partial x_k}\left(\tfrac{1}{2}u_j u_j\right)$$

Now when the two components constituting the left-hand side of the basic conservation equation are added, the two terms with minus signs above are canceled by corresponding terms with plus signs to give

$$\frac{\partial}{\partial t}\left(\rho e + \tfrac{1}{2}\rho u_j u_j\right) + \frac{\partial}{\partial x_k}\left[\left(\rho e + \tfrac{1}{2}\rho u_j u_j\right)u_k\right]$$

$$= \rho\frac{\partial e}{\partial t} + \rho u_k\frac{\partial e}{\partial x_k} + \rho\frac{\partial}{\partial t}\left(\tfrac{1}{2}u_j u_j\right) + \rho u_k\frac{\partial}{\partial x_k}\left(\tfrac{1}{2}u_j u_j\right)$$

$$= \rho\frac{\partial e}{\partial t} + \rho u_k\frac{\partial e}{\partial x_k} + \rho u_j\frac{\partial u_j}{\partial t} + \rho u_j u_k\frac{\partial u_j}{\partial x_k}$$

Then, noting that

$$\frac{\partial}{\partial x_i}(u_j\sigma_{ij}) = u_j\frac{\partial\sigma_{ij}}{\partial x_i} + \sigma_{ij}\frac{\partial u_j}{\partial x_i}$$

the equation which expresses the conservation of energy becomes

$$\rho\frac{\partial e}{\partial t} + \rho u_k\frac{\partial e}{\partial x_k} + \rho u_j\frac{\partial u_j}{\partial t} + \rho u_j u_k\frac{\partial u_j}{\partial x_k} = u_j\frac{\partial\sigma_{ij}}{\partial x_i} + \sigma_{ij}\frac{\partial u_j}{\partial x_i} + u_j\rho f_i - \frac{\partial q_j}{\partial x_j}$$

Now it can be seen that the third and fourth terms on the left-hand side are canceled by the first and third terms on the right-hand side, since these terms collectively amount to the product of $u_j$ with the momentum equation (1.4). Thus the equation which expresses conservation of thermal energy becomes

$$\rho\frac{\partial e}{\partial t} + \rho u_k\frac{\partial e}{\partial x_k} = \sigma_{ij}\frac{\partial u_j}{\partial x_i} - \frac{\partial q_j}{\partial x_j} \tag{1.5}$$

The terms which were dropped in the last simplification were the mechanical-energy terms. The equation of conservation of momentum, Eq. (1.4), may be regarded as an equation of balancing forces with $j$ as the free subscript. Therefore, the scalar product of each force with the velocity vector, or the multiplication by $u_j$, gives the rate of doing work by the mechanical forces, which is the mechanical energy. On the other hand, Eq. (1.5) is a balance of thermal energy, which is what is left when the mechanical energy is subtracted from the balance of total energy, and is usually referred to as simply the *energy equation*.

As was the case with the equation of momentum conservation, it is instructive to interpret each of the terms appearing in Eq. (1.5) physically. The entire left-hand side represents the rate of change of internal energy, the first term being the temporal change while the second is due to local convective changes caused by the fluid flowing from one area to another. The entire right-hand side represents the cause of the change in internal energy. The first of these terms represents the conversion of mechanical energy into thermal energy due to the action of the surface stresses. As will be seen later, part of this conversion is reversible and part is irreversible. The final term in the equation represents the rate at which heat is being added by conduction from outside.

## 1.9 DISCUSSION OF CONSERVATION EQUATIONS

The basic conservation laws, Eqs. (1.3a), (1.4), and (1.5), represent five scalar equations which the fluid properties must satisfy as the fluid flows. The continuity and the energy equations are scalar equations, while the momentum equation is a vector equation which represents three scalar equations. Two equations of state may be added to bring the number of equations up to seven, but our basic conservation laws have introduced seventeen unknowns. These unknowns are the scalars $\rho$ and $e$, the density and the internal energy, respectively; the vectors $u_j$ and $q_j$, the velocity and heat flux, respectively, each vector having three components; and the stress tensor $\sigma_{ij}$, which has, in general, nine independent components.

In order to obtain a complete set of equations, the stress tensor $\sigma_{ij}$ and the heat-flux vector $q_j$ must be further specified. This leads to the so-called "constitutive equations" in which the stress tensor is related to the deformation tensor and the heat-flux vector is related to temperature gradients. Although the latter relation is very simple, the former is quite complicated and requires either an intimate knowledge of tensor analysis or a clear understanding of the physical interpretation of certain tensor quantities. For this reason, prior to establishing the constitutive relations the tensor equivalents of rotation and rate of shear will be established.

## 1.10 ROTATION AND RATE OF SHEAR

It is the purpose of this section to consider the rotation of a fluid element about its own axis and the shearing of a fluid element and to identify the tensor quantities which represent these physical quantities. This is most easily done by considering an infinitesimal fluid element of rectangular cross section and observing its change in shape and orientation as it flows.

Figure 1.5 shows a two-dimensional element of fluid (or the projection of a three-dimensional element) whose dimensions at time $t = 0$ are $\delta x$ and $\delta y$. The fluid element is rectangular at time $t = 0$, and its centroid coincides with the origin of a fixed-coordinate system. For purposes of identification, the corners of the fluid element have been labeled $A$, $B$, $C$, and $D$.

After a short time interval $\delta t$, the centroid of the fluid element will have moved downstream to some new location as shown in Fig. 1.5. The distance the centroid will have moved in the $x$ direction will be given by

$$\Delta x = \int_0^{\delta t} u[x(t), y(t)]\, dt$$

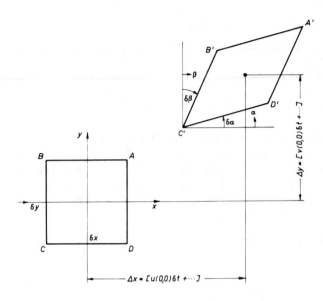

**FIGURE 1.5**
An infinitesimal element of fluid at time $t = 0$ (indicated by $ABCD$) and at time $t = \delta t$ (indicated by $A'B'C'D'$).

Since the values of $x$ and $y$ must be close to zero for short times such as $\delta t$, the velocity component $u$ may be expanded in a Taylor series about the point $(0, 0)$ to give

$$\Delta x = \int_0^{\delta t} \left[ u(0,0) + x(t)\frac{\partial u}{\partial x}(0,0) + y(t)\frac{\partial u}{\partial y}(0,0) + \cdots \right] dt$$

where the dots represents terms which are smaller than those presented and which will eventually vanish as the limit of $\delta t \to 0$ is taken. Integrating the leading term explicitly gives

$$\Delta x = u(0,0)\,\delta t + \int_0^{\delta t} \left[ x(t)\frac{\partial u}{\partial x}(0,0) + y(t)\frac{\partial u}{\partial y}(0,0) + \cdots \right] dt$$

$$= u(0,0)\,\delta t + \cdots$$

similarly

$$\Delta y = v(0,0)\,\delta t + \cdots$$

As well as moving bodily, the fluid element will rotate and will be distorted as indicated by the corners, which are labeled $A'$, $B'$, $C'$, and $D'$ to represent the element at time $t = \delta t$. The rotation of the side $CD$ to its new position $C'D'$ is indicated by the angle $\delta\alpha$, where $\alpha$ is positive when measured counterclockwise. Similarly, the rotation of the side $BC$ to its new position $B'C'$ is indicated by the angle $\delta\beta$, where $\beta$ is positive when measured clockwise. Expressions for $\delta\alpha$ and $\delta\beta$ in terms of the velocity components may be obtained as follows:

From the geometry of the element as it appears at time $t = \delta t$,

$$\delta \alpha = \tan^{-1}\left(\frac{y \text{ component of } D'C'}{x \text{ component of } D'C'}\right)$$

$$= \tan^{-1}\left\{\frac{\left[v\left(\tfrac{1}{2}\delta x, -\tfrac{1}{2}\delta y\right)\delta t + \cdots\right] - \left[v\left(-\tfrac{1}{2}\delta x, -\tfrac{1}{2}\delta y\right)\delta t + \cdots\right]}{\delta x + \cdots}\right\}$$

where $v$ is evaluated first at the point $D$, whose coordinates are $(\tfrac{1}{2}\delta x, -\tfrac{1}{2}\delta y)$, and secondly at the point $C$, whose cordinates are $(-\tfrac{1}{2}\delta x, -\tfrac{1}{2}\delta y)$. The $x$ component of the side $D'C'$ will be only slightly different from $\delta x$, and it turns out that the precise departure from this value need not be evaluated explicitly.

Expanding the velocity component $v$ in a Taylor series about the point $(0,0)$ results in the following expression for $\delta \alpha$:

$$\delta \alpha = \tan^{-1}\left\{\frac{\left[v(0,0) + \tfrac{1}{2}\delta x(\partial v/\partial x)(0,0) - \tfrac{1}{2}\delta y(\partial v/\partial y)(0,0) + \cdots\right]\delta t}{\delta x(1 + \cdots)}\right.$$

$$\left. - \frac{\left[v(0,0) - \tfrac{1}{2}\delta x(\partial v/\partial x)(0,0) - \tfrac{1}{2}\delta y(\partial v/\partial y)(0,0) + \cdots\right]\delta t}{\delta x(1 + \cdots)}\right\}$$

$$= \tan^{-1}\left\{\frac{\left[\delta x(\partial v/\partial x)(0,0) + \cdots\right]\delta t}{\delta x(1 + \cdots)}\right\}$$

$$= \tan^{-1}\left\{\frac{\left[(\partial v/\partial x)(0,0) + \cdots\right]\delta t}{(1 + \cdots)}\right\}$$

$$= \tan^{-1}\left\{\left[\frac{\partial v}{\partial x}(0,0) + \cdots\right]\delta t\right\}$$

Since the argument of the arctangent is small, the entire right-hand side may be expanded to give

$$\delta \alpha = \left[\frac{\partial v}{\partial x}(0,0) + \cdots\right]\delta t + \cdots$$

$$\frac{\delta \alpha}{\delta t} = \frac{\partial v}{\partial x}(0,0) + \cdots$$

This expression represents the change in the angle $\alpha$ per unit time so that in the limit as $\delta x$, $\delta y$, and $\delta t$ all tend to zero, this expression becomes

$$\dot{\alpha} = \frac{\partial v}{\partial x}(0,0)$$

where $\dot{\alpha}$ is the time derivative of the angle $\alpha$. By an identical procedure it follows that the time derivative of the angle $\beta$ is given by

$$\dot{\beta} = \frac{\partial u}{\partial y}(0,0)$$

Recall that $\alpha$ is measured counterclockwise and $\beta$ is measured clockwise. Thus the rate of clockwise rotation of the fluid element about its centroid is given by

$$\tfrac{1}{2}(\dot{\beta} - \dot{\alpha}) = \frac{1}{2}\left(\frac{\partial u}{\partial y} - \frac{\partial v}{\partial x}\right)$$

Likewise the shearing action is measured by the rate at which the sides $B'C'$ and $D'C'$ are approaching each other and is therefore given by the quantity

$$\tfrac{1}{2}(\dot{\beta} + \dot{\alpha}) = \frac{1}{2}\left(\frac{\partial u}{\partial y} + \frac{\partial v}{\partial x}\right)$$

The foregoing analysis was carried out in two dimensions which may be considered as the projection of a three-dimensional element on the $xy$ plane. If the analysis is carried out in the other planes, it may be verified that the rate of rotation of the element about its own axes and the rate of shearing are given by

$$\text{Rate of rotation} = \frac{1}{2}\left(\frac{\partial u_i}{\partial x_j} - \frac{\partial u_j}{\partial x_i}\right) \qquad (1.6a)$$

$$\text{Rate of shearing} = \frac{1}{2}\left(\frac{\partial u_i}{\partial x_j} + \frac{\partial u_j}{\partial x_i}\right) \qquad (1.6b)$$

That is, both the rate of rotation and the rate of shearing may be represented by tensors of rank 2. It will be noted that the rate-of-rotation tensor is antisymmetric and therefore has only three independent components while the rate-of-shearing tensor is symmetric and therefore has six independent components. These two quantities are actually the antisymmetric part and the symmetric part of another tensor called the deformation-rate tensor, as may be shown as follows: Define the *deformation-rate tensor* $e_{ij}$ as

$$e_{ij} = \frac{\partial u_i}{\partial x_j}$$

$$= \frac{1}{2}\left(\frac{\partial u_i}{\partial x_j} - \frac{\partial u_j}{\partial x_i}\right) + \frac{1}{2}\left(\frac{\partial u_i}{\partial x_j} + \frac{\partial u_j}{\partial x_i}\right)$$

That is, the antisymmetric part of the deformation-rate tensor represents the rate of rotation of a fluid element in that flow field about its own axes while the symmetric part of the deformation-rate tensor represents the rate of shearing of the fluid element.

## 1.11  CONSTITUTIVE EQUATIONS

In this section the nine elements of the stress tensor $\sigma_{ij}$ will be related to the nine elements of the deformation-rate tensor $e_{kl}$ by a set of parameters. All these parameters except two will be evaluated analytically and the remaining two, which are the viscosity coefficients, must be determined empirically. In order to achieve this end, the postulates for a newtonian fluid will be introduced directly. Water and air are by far the most abundant fluids on earth, and they behave like newtonian fluids, as do many other common fluids. It should be pointed out, however, that some fluids do not behave in a newtonian manner, and their special characteristics are among the topics of current research. One example is the class of fluids called *viscoelastic fluids*, whose properties may be used to reduce the drag of a body. Since this book is concerned with the classical fundamentals only, the newtonian fluid will be treated directly. If the various steps are clearly understood, there should be no conceptual difficulty in following the details of some of the more complex fluids such as viscoelastic fluids.

Certain observations and postulates will now be made concerning the stress tensor. The precise manner in which the postulates are made is largely a matter of taste, but when the newtonian fluid is being treated, the resulting equations are always the same. The following are the four conditions which the stress tensor is supposed to satisfy:

1. When the fluid is at rest, the stress is hydrostatic and the pressure exerted by the fluid is the thermodynamic pressure.
2. The stress tensor $\sigma_{ij}$ is linearly related to the deformation-rate tensor $e_{kl}$ and depends only on that tensor.
3. Since there is no shearing action in a solid-body rotation of the fluid, no shear stresses will act during such a motion.
4. There are no preferred directions in the fluid, so that the fluid properties are point functions.

Condition 1 requires that the stress tensor $\sigma_{ij}$ be of the form

$$\sigma_{ij} = -p\delta_{ij} + \tau_{ij}$$

where $\tau_{ij}$ depends upon the motion of the fluid only and is called the *shear-stress tensor*. The quantity $p$ is the thermodynamic pressure and $\delta_{ij}$ is the Kronecker delta. The pressure term is negative, since the sign convention being used here is that normal stresses are positive when they are tensile in nature.

The remaining unknown in the constitutive equation for stress is the shear-stress tensor $\tau_{ij}$. Condition 2 postulates that the stress tensor, and hence the shear-stress tensor, is linearly related to the deformation-rate tensor. This is the distinguishing feature of *newtonian fluids*. In general, the shear-stress tensor could depend upon some power of the velocity gradients other than unity, and it could depend upon the velocity itself as well as the velocity gradient. The condition postulated here can be verified experimentally in simple flow fields in most common fluids, and the results predicted for more complex flow fields yield results which agree with physical observations. This is the sole justification for condition 2.

There are nine elements in the shear-stress tensor $\tau_{ij}$, and each of these elements may be expressed as a linear combination of the nine elements in the deformation-rate tensor $e_{kl}$ (just as a vector may be represented as a linear combination of components of the base vectors). That is, each of the nine elements of $\tau_{ij}$ will in general be a linear combination of the nine elements of $e_{kl}$ so that 81 parameters are needed to relate $\tau_{ij}$ to $e_{kl}$. This means that a tensor of rank 4 is required so that the general form of $\tau_{ij}$ will be, according to condition 2,

$$\tau_{ij} = \alpha_{ijkl}\frac{\partial u_k}{\partial x_l}$$

It was shown in the previous section that the tensor $\partial u_k/\partial x_l$, like any other tensor of rank 2, could be broken down into an antisymmetric part and a symmetric part. Here the antisymmetric part corresponds to the rate of rotation of a fluid element and the symmetric part corresponds to the shearing rate. According to condition 3, if the flow field is executing a simple solid-body rotation, there should be no shear stresses in the fluid. But for a solid-body rotation the antisymmetric part of $\partial u_k/\partial x_l$, namely, $\frac{1}{2}(\partial u_k/\partial x_l - \partial u_l/\partial x_k)$, will not be zero. Hence, in order that condition 3 may be satisfied, the coefficients of this part of the deformation-rate tensor must be zero. That is, the constitutive relation for stress must be of the form

$$\tau_{ij} = \tfrac{1}{2}\beta_{ijkl}\left(\frac{\partial u_k}{\partial x_l} + \frac{\partial u_l}{\partial x_k}\right)$$

The 81 elements of the fourth-rank tensor $\beta_{ijkl}$ are still undetermined, but condition 4 has yet to be imposed. This condition is the so-called condition of *isotropy*, which guarantees that the results obtained should be independent of the orientation of the coordinate system chosen. In Appendix B, the summary of some useful tensor relations, it is pointed out that the most general isotropic tensor of rank 4 is of the form

$$\lambda\delta_{ij}\delta_{kl} + \mu\left(\delta_{ik}\delta_{jl} + \delta_{il}\delta_{jk}\right) + \gamma\left(\delta_{ik}\delta_{jl} - \delta_{il}\delta_{jk}\right)$$

where $\lambda$, $\mu$, and $\gamma$ are scalars. The proof of this is straightforward but tedious. The general tensor is subjected to a series of coordinate rotations and inflections, and the condition of invariance is applied. In this way the 81 quantities contained in the general tensor are reduced to three independent quantities in the isotropic case. In the case of the fourth-rank tensor which relates the shear-stress tensor to the deformation-rate tensor, namely, $\beta_{ijkl}$, not only must it be isotropic but it must be symmetric in view of condition 3. That is, the coefficient $\gamma$ must be zero in this case so that the expression for the shear stress becomes

$$\tau_{ij} = \tfrac{1}{2}\left[\lambda\delta_{ij}\delta_{kl} + \mu\left(\delta_{ik}\delta_{jl} + \delta_{il}\delta_{jk}\right)\right]\left(\frac{\partial u_k}{\partial x_l} + \frac{\partial u_l}{\delta x_k}\right)$$

Using the fact that $\delta_{kl} = 0$ unless $l = k$ shows that

$$\tfrac{1}{2}\lambda\delta_{ij}\delta_{kl}\left(\frac{\partial u_k}{\partial x_l} + \frac{\partial u_l}{\partial x_k}\right) = \lambda\delta_{ij}\frac{\partial u_k}{\partial x_k}$$

in which $l$ has been replaced by $k$. Likewise, replacing $k$ by $i$ and $l$ by $j$ shows that

$$\tfrac{1}{2}\mu\delta_{ik}\delta_{jl}\left(\frac{\partial u_k}{\partial x_l} + \frac{\partial u_l}{\partial x_k}\right) = \tfrac{1}{2}\mu\left(\frac{\partial u_i}{\partial x_j} + \frac{\partial u_j}{\partial x_i}\right)$$

and replacing $l$ by $i$ and $k$ by $j$ shows that

$$\tfrac{1}{2}\mu\delta_{il}\delta_{jk}\left(\frac{\partial u_k}{\partial x_l} + \frac{\partial u_l}{\partial x_k}\right) = \tfrac{1}{2}\mu\left(\frac{\partial u_j}{\partial x_i} + \frac{\partial u_i}{\partial x_j}\right)$$

Hence the expression for the shear-stress tensor becomes

$$\tau_{ij} = \lambda\delta_{ij}\frac{\partial u_k}{\partial x_k} + \mu\left(\frac{\partial u_i}{\partial x_j} + \frac{\partial u_j}{\partial x_i}\right)$$

Thus the constitutive relation for stress in a newtonian fluid becomes

$$\sigma_{ij} = -p\delta_{ij} + \lambda\delta_{ij}\frac{\partial u_k}{\partial x_k} + \mu\left(\frac{\partial u_i}{\partial x_j} + \frac{\partial u_j}{\partial x_i}\right) \tag{1.7}$$

which shows that the stress is represented by a second-order symmetric tensor.

The nine elements of the stress tensor $\sigma_{ij}$ have now been expressed in terms of the pressure and the velocity gradients, which have all been previously introduced, and two coefficients $\lambda$ and $\mu$. These coefficients cannot be determined analytically and must be determined empirically. Up to this point both $\lambda$ and $\mu$ are just coefficients, but their nature and physical significance will be discussed in the next section.

The second constitutive relation involves the heat-flux vector $q_j$, which is due to conduction alone. Fourier's law of heat conduction states that the heat flux by conduction is proportional to the negative temperature gradient so that

$$q_j = -k\frac{\partial T}{\partial x_j} \tag{1.8}$$

This is the constitutive equation for the heat flux, where the proportionality factor $k$ in Fourier's law is the thermal conductivity of the fluid. In using Eq. (1.8), it is implicitly assumed that the concept of temperature, as employed in equilibrium thermodynamics, also applies to a moving fluid.

## 1.12 VISCOSITY COEFFICIENTS

It was pointed out in the previous section that the parameters $\lambda$ and $\mu$, which appear in the constitutive equations for stress, must be determined experimentally. It is the purpose of this section to establish a physical interpretation of these two parameters and thus show the manner in which they may be evaluated.

Consider a simple shear flow of an incompressible fluid in which the velocity components are defined by

$$u = u(y)$$
$$v = w = 0$$

That is, only the $x$ component of velocity is nonzero, and that component is a function of $y$ only. From the definition of this flow field the components of the stress tensor may be evaluated from Eq. (1.7) to give

$$\sigma_{12} = \sigma_{21} = \mu \frac{du}{dy}$$

$$\sigma_{11} = \sigma_{22} = \sigma_{33} = -p$$

$$\sigma_{13} = \sigma_{31} = \sigma_{23} = \sigma_{32} = 0$$

That is, the normal components of the stress are defined by the thermodynamic pressure, and the nonzero shear components of the stress are proportional to the velocity gradient with the parameter $\mu$ as the proportionality factor. But, from Newton's law of viscosity, the proportionality factor between the shear stress and the velocity gradient in a simple shear flow is the dynamic viscosity. Hence the quantity $\mu$ which appears in the constitutive equation for stress is the *dynamic viscosity* of the fluid. Frequently the *kinematic viscosity*, defined by $\nu = \mu/\rho$, is used instead of the dynamic viscosity.

The parameter $\lambda$ which appears in Eq. (1.7) is usually referred to as the *second viscosity coefficient*. In order to establish its significance, the average normal stress component $\bar{p}$ will be calculated.

$$-\bar{p} = \tfrac{1}{3}(\sigma_{11} + \sigma_{22} + \sigma_{33})$$

This average normal stress is the mechanical pressure in the fluid and it is equal to one-third of the trace of the stress tensor. Since the mechanical pressure is either purely hydrostatic or hydrostatic plus a component which is induced by the stresses which result from the motion of the fluid, it will, in general, be different from the thermodynamic pressure $p$. Using Eq. (1.7), the mechanical pressure $\bar{p}$ may be evaluated as follows:

$$-\bar{p} = \frac{1}{3}\left[\left(-p + \lambda\frac{\partial u_k}{\partial x_k} + 2\mu\frac{\partial u}{\partial x}\right) + \left(-p + \lambda\frac{\partial u_k}{\partial x_k} + 2\mu\frac{\partial v}{\partial y}\right)\right.$$
$$\left. + \left(-p + \lambda\frac{\partial u_k}{\partial x_k} + 2\mu\frac{\partial w}{\partial z}\right)\right]$$

$$= -p + \lambda\frac{\partial u_k}{\partial x_k} + \tfrac{2}{3}\mu\frac{\partial u_k}{\partial x_k}$$

$$= -p + \left(\lambda + \tfrac{2}{3}\mu\right)\frac{\partial u_k}{\partial x_k}$$

That is, the difference between the thermodynamic pressure and the mechanical pressure is proportional to the divergence of the velocity vector. The proportionality factor is usually referred to as the *bulk viscosity* and is denoted by $K$. That is,

$$p - \bar{p} = K\frac{\partial u_k}{\partial x_k}$$

where $K = \lambda + \frac{2}{3}\mu$. Of the three viscosity coefficients $\mu$, $\lambda$, and $K$, only two are independent and the third is defined by the above equation. For purposes of physical interpretation of these viscosity coefficients it is preferred to discuss $\mu$ (which has already been done) and $K$, leaving $\lambda$ to be defined by $\lambda = K - \frac{2}{3}\mu$.

In order to identify the physical significance of the bulk viscosity, some of the results of the kinetic theory of gases will be used. The mechanical pressure is a measure of the translational energy of the molecules only, whereas the thermodynamic pressure is a measure of the total energy, which includes vibrational and rotational modes of energy as well as the translational mode. For liquids, other forms of energy are also included such as intermolecular attraction. These different modes of molecular energy have different relaxation times, so that in a flow field it is possible to have energy transferred from one mode to another. The bulk viscosity is a measure of this transfer of energy from the translational mode to the other modes, as may be seen from the relation $p - \bar{p} = K(\partial u_k/\partial x_k)$. For example, during the passage through a shock wave the vibrational modes of energy are excited at the expense of the translational modes, so that the bulk viscosity will be nonzero in this case.

The above discussion has been for a polyatomic molecule of a liquid or a gas. If the fluid is a monatomic gas, the only mode of molecular energy is the translational mode. Then, for such a gas the mechanical pressure and the thermodynamic pressure are the same, so that the bulk viscosity is zero. That is,

$$\lambda = -\frac{2}{3}\mu$$

which is called *Stokes' relation*, so that there is only one independent viscosity coefficient in the case of monatomic gases. For polyatomic gases and for liquids the departure from $K = 0$ is frequently small, and many authors incorporate Stokes' relation in the constitutive relation (1.7) for stress. In any case, for incompressible fluids Eq. (1.7) shows that it is immaterial whether $\lambda = -\frac{2}{3}\mu$ or not, for then the term involving $\lambda$ is zero by virtue of the continuity equation.

## 1.13 NAVIER-STOKES EQUATIONS

The equation of momentum conservation (1.4) together with the constitutive relation for a newtonian fluid [Eqs. (1.7)] yield the famous Navier-Stokes equations, which are the principal conditions to be satisfied by a fluid as it flows. Having obtained an expression for the stress tensor, the term $\partial\sigma_{ij}/\partial x_i$ which appears in Eq. (1.4) may be evaluated explicitly as follows:

$$\frac{\partial \sigma_{ij}}{\partial x_i} = \frac{\partial}{\partial x_i}\left[-p\delta_{ij} + \lambda\delta_{ij}\frac{\partial u_k}{\partial x_k} + \mu\left(\frac{\partial u_i}{\partial x_j} + \frac{\partial u_j}{\partial x_i}\right)\right]$$

$$= -\frac{\partial p}{\partial x_j} + \frac{\partial}{\partial x_j}\left(\lambda\frac{\partial u_k}{\partial x_k}\right) + \frac{\partial}{\partial x_i}\left[\mu\left(\frac{\partial u_i}{\partial x_j} + \frac{\partial u_j}{\partial x_i}\right)\right]$$

where, in the first two terms, $i$ has been replaced by $j$, since it is only when $i = j$ that these terms are nonzero. Substituting this result into Eq. (1.4) gives

$$\rho\frac{\partial u_j}{\partial t} + \rho u_k\frac{\partial u_j}{\partial x_k} = -\frac{\partial p}{\partial x_j} + \frac{\partial}{\partial x_j}\left(\lambda\frac{\partial u_k}{\partial x_k}\right) + \frac{\partial}{\partial x_i}\left[\mu\left(\frac{\partial u_i}{\partial x_j} + \frac{\partial u_j}{\partial x_i}\right)\right] + \rho f_j$$

$$(1.9a)$$

Equations (1.9$a$) are known as the *Navier-Stokes equations*, and they represent three scalar equations corresponding to the three possible values of the free subscript $j$. In the most frequently encountered situations the fluid may be assumed to be incompressible and the dynamic viscosity may be assumed to be constant. Under these conditions the second term on the right-hand side of Eqs. (1.9$a$) is identically zero and the viscous-shear term becomes

$$\frac{\partial}{\partial x_i}\left[\mu\left(\frac{\partial u_i}{\partial x_j} + \frac{\partial u_j}{\partial x_i}\right)\right] = \mu\left[\frac{\partial}{\partial x_j}\left(\frac{\partial u_i}{\partial x_i}\right) + \frac{\partial^2 u_j}{\partial x_i\,\partial x_i}\right] = \mu\frac{\partial^2 u_j}{\partial x_i\,\partial x_i}$$

That is, the viscous-shear term is proportional to the laplacian of the velocity vector, and the constant of proportionality is the dynamic viscosity. Then the Navier-Stokes equations for an incompressible fluid of constant density become

$$\rho\frac{\partial u_j}{\partial t} + \rho u_k\frac{\partial u_j}{\partial x_k} = -\frac{\partial p}{\partial x_j} + \mu\frac{\partial^2 u_j}{\partial x_i\,\partial x_i} + \rho f_j \qquad (1.9b)$$

In the special case of negligible viscous effects, Eqs. (1.9$a$) become

$$\rho\frac{\partial u_j}{\partial t} + \rho u_k\frac{\partial u_j}{\partial x_k} = -\frac{\partial p}{\partial x_j} + \rho f_j \qquad (1.9c)$$

Equations (1.9$c$) are known as *Euler equations*.

## 1.14 ENERGY EQUATION

The term $\sigma_{ij}(\partial u_j/\partial x_i)$ which appears in the equation of energy conservation (1.5) may now be evaluated explicitly by use of Eq. (1.7).

$$\sigma_{ij}\frac{\partial u_j}{\partial x_i} = \left[-p\delta_{ij} + \lambda\delta_{ij}\frac{\partial u_k}{\partial x_k} + \mu\left(\frac{\partial u_i}{\partial x_j} + \frac{\partial u_j}{\partial x_i}\right)\right]\frac{\partial u_j}{\partial x_i}$$

Using the fact that in the first two terms of the stress tensor $i = j$ for the nonzero elements, this expression becomes

$$\sigma_{ij}\frac{\partial u_j}{\partial x_i} = -p\frac{\partial u_k}{\partial x_k} + \lambda\left(\frac{\partial u_k}{\partial x_k}\right)^2 + \mu\left(\frac{\partial u_i}{\partial x_j} + \frac{\partial u_j}{\partial x_i}\right)\frac{\partial u_j}{\partial x_i}$$

It will be recalled that the term $\sigma_{ij}(\partial u_j/\partial x_i)$ represents the work done by the surface forces. The first term in the expression for this work done, namely, $-p(\partial u_k/\partial x_k)$, represents the reversible transfer of energy due to compression. The remaining two terms are collectively called the *dissipation function* and are denoted by $\Phi$. That is,

$$\Phi = \lambda\left(\frac{\partial u_k}{\partial x_k}\right)^2 + \mu\left(\frac{\partial u_i}{\partial x_j} + \frac{\partial u_j}{\partial x_i}\right)\frac{\partial u_j}{\partial x_i} \tag{1.10}$$

The reason $\Phi$ is called the dissipation function is that it is a measure of the rate at which mechanical energy is being converted into thermal energy. This may be readily verified by considering an incompressible fluid in a cartesian-coordinate system. Then

$$\Phi = \mu\left(\frac{\partial u_i}{\partial x_j} + \frac{\partial u_j}{\partial x_i}\right)\frac{\partial u_j}{\partial x_i}$$

$$= \mu\left(\frac{\partial u_i}{\partial x_j} + \frac{\partial u_j}{\partial x_i}\right)\left[\frac{1}{2}\left(\frac{\partial u_j}{\partial x_i} - \frac{\partial u_i}{\partial x_j}\right) + \frac{1}{2}\left(\frac{\partial u_j}{\partial x_i} + \frac{\partial u_i}{\partial x_j}\right)\right]$$

$$= \tfrac{1}{2}\mu\left(\frac{\partial u_i}{\partial x_j} + \frac{\partial u_j}{\partial x_i}\right)^2$$

which is a positive definite quantity. This shows that the dissipation function always works to increase irreversibly the internal energy of an incompressible fluid.

In terms of the dissipation function, the total work done by the surface stresses is given by

$$\sigma_{ij}\frac{\partial u_j}{\partial x_i} = -p\frac{\partial u_k}{\partial x_k} + \Phi$$

Using this result and the constitutive relation for the heat flux [Eq. (1.8)] in the equation of conservation of energy, Eq. (1.5), yields the *energy equation* for a newtonian fluid.

$$\rho\frac{\partial e}{\partial t} + \rho u_k\frac{\partial e}{\partial x_k} = -p\frac{\partial u_k}{\partial x_k} + \frac{\partial}{\partial x_j}\left(k\frac{\partial T}{\partial x_j}\right) + \Phi \tag{1.11}$$

where $\Phi$ is defined by Eq. (1.10).

# 1.15 GOVERNING EQUATIONS FOR NEWTONIAN FLUIDS

The equations which govern the motion of a newtonian fluid are the continuity equation (1.3a), the Navier-Stokes equations (1.9a), the energy equation (1.11), and equations of state. For purposes of summary and discussion these equations will be repeated here.

$$\frac{\partial \rho}{\partial t} + \frac{\partial}{\partial x_k}(\rho u_k) = 0 \tag{1.3a}$$

$$\rho \frac{\partial u_j}{\partial t} + \rho u_k \frac{\partial u_j}{\partial x_k} = -\frac{\partial p}{\partial x_j} + \frac{\partial}{\partial x_j}\left(\lambda \frac{\partial u_k}{\partial x_k}\right) + \frac{\partial}{\partial x_i}\left[\mu\left(\frac{\partial u_i}{\partial x_j} + \frac{\partial u_j}{\partial x_i}\right)\right] + \rho f_j \tag{1.9a}$$

$$\rho \frac{\partial e}{\partial t} + \rho u_k \frac{\partial e}{\partial x_k} = -p \frac{\partial u_k}{\partial x_k} + \frac{\partial}{\partial x_j}\left(k \frac{\partial T}{\partial x_j}\right) + \lambda \left(\frac{\partial u_k}{\partial x_k}\right)^2 + \mu\left(\frac{\partial u_i}{\partial x_j} + \frac{\partial u_j}{\partial x_i}\right)\frac{\partial u_j}{\partial x_i} \tag{1.11}$$

$$p = p(\rho, T) \tag{1.12}$$

$$e = e(\rho, T) \tag{1.13}$$

The last two equations are general representations of the thermal and caloric equations of state, respectively. The most frequently encountered form of the thermal equation of state is the ideal-gas law $p = \rho RT$, while the most frequently encountered form of the caloric equation of state is $e = C_v T$, where $C_v$ is the specific heat at constant volume.

The above set of equations represents seven equations which are to be satisfied by seven unknowns. Each of the continuity, energy, and state equations supplies one scalar equation, while the Navier-Stokes equations supply three scalar equations. The seven unknowns are the pressure, density, internal energy, temperature, and velocity components, that is, $p$, $\rho$, $e$, $T$, and $u_j$. The parameters $\lambda$, $\mu$, and $k$ are assumed to be known from experimental data, and they may be constants or specified functions of the temperature and pressure.

It is not always necessary to solve the complete set of equations in order to define the flow field analytically. For example, if compressible effects are thought to be unimportant in the flow field being considered, the incompressible form of the governing equations may be used. The continuity equation and the Navier-Stokes equations are then simpler, as indicated by Eqs. (1.3c) and (1.9b), respectively, but the greatest simplification comes from the fact that the energy equation is mathematically uncoupled from these two equations. The continuity and Navier-Stokes equations offer four scalar equations involving only $p$ and $u_j$. That is, the pressure and velocity fields may be established without reference to the energy equation. Having done this, the temperature field may be established, which may have the trivial solution $T = $ constant. In cases of forced-convection heat transfer in which the flow is turbulent, the continuity and Navier-Stokes equations are frequently replaced by an empirical velocity distribution and the energy equation is solved to yield the temperature distribution. More frequently, however, thermal effects are unimportant and the continuity and Navier-Stokes equations alone must be solved.

The most common type of body force which acts on a fluid is due to gravity, so that the body force $f_j$ which appears in the Navier-Stokes equations is defined in magnitude and direction by the acceleration due to gravity. Sometimes, however, electromagnetic effects are important, and in such cases $\mathbf{f} = (\rho_c \mathbf{E} + \mathbf{J} \times \mathbf{B})$, which is the Lorentz force. Here $\rho_c$ is the charge density, $\mathbf{E}$ is the electric field vector, $\mathbf{J}$ is the electric current density, and $\mathbf{B}$ is the magnetic field vector. The electric and magnetic fields themselves must obey a set of physical laws which are expressed by Maxwell's equations. The solution to such problems requires the simultaneous solution of the equations of fluid mechanics and of electromagnetism. One special case of this type of coupling is the field known as *magnetohydrodynamics*.

It may be also pointed out that the governing equations summarized here contain the equations of hydrostatics and heat conduction as special cases. If the fluid is at rest, the velocity components will all be zero, so that the Navier-Stokes equations (1.9a) become

$$0 = - \frac{\partial p}{\partial x_j} + pf_j$$

If the body force $f_j$ is now set equal to the gravitational force, the equation of hydrostatics is obtained. For example, if gravity acts in the negative $z$ direction, $f_j = -ge_z$, where $e_z$ is the unit vector in the $z$ direction. Then

$$\frac{\partial p}{\partial x_j} = -\rho g e_z$$

which shows that $\partial p/\partial x = \partial p/\partial y = 0$ and $\partial p/\partial z = -\rho g$. In the case of zero velocity the energy equation becomes

$$\rho \frac{\partial e}{\partial t} = \frac{\partial}{\partial x_j} \left( k \frac{\partial T}{\partial x_j} \right)$$

Introducing the enthalpy $h = e + p/\rho$ and using the fact that $p$ and $\rho$ are constant in the stationary fluid gives

$$\rho \frac{\partial h}{\partial t} = \frac{\partial}{\partial x_j} \left( k \frac{\partial T}{\partial x_j} \right)$$

If the fluid is thermally perfect, $h$ will be a function of $T$ only, so that

$$\frac{\partial h}{\partial t} = \frac{\partial h}{\partial T} \frac{\partial T}{\partial t} = C_p \frac{\partial T}{\partial t}$$

where $C_p$ is the specific heat at constant pressure, which is the appropriate process for this case. Then the energy equation becomes

$$\rho C_p \frac{\partial T}{\partial t} = \frac{\partial}{\partial x_j}\left(k\frac{\partial T}{\partial x_j}\right)$$

which is the equation of heat conduction.

## 1.16  BOUNDARY CONDITIONS

The Navier-Stokes equations are, mathematically, a set of three elliptic, second-order partial differential equations. The appropriate type of boundary conditions are therefore Dirichlet or Neumann conditions on a closed boundary. Physically, this usually amounts to specifying the velocity on all solid boundaries. Within the continuum approximation the experimentally determined boundary condition is that there is no slip between the fluid and a solid boundary at the interface. On the molecular scale, slippage is possible, but it is confined within a layer whose dimensions are of the same order as the mean free path between the molecules. Then if **U** represents the velocity of a solid boundary, the boundary condition which should be imposed on our continuum velocity is

$$\mathbf{u} = \mathbf{U} \qquad \text{on solid boundaries} \qquad (1.14)$$

In the case of an infinite expanse of fluid, one common form of Eq. (1.14) is that $\mathbf{u} \to 0$ as $\mathbf{x} \to \infty$.

  If thermal effects are included, a boundary condition on the temperature is also required. As in the case of heat-conduction problems, this may take the form of specifying the temperature or the heat flux on some boundary.

## PROBLEMS

**1.1.** Derive the continuity equation from first principles using an *infinitesimal control volume* of rectangular shape and having dimensions $(\delta x, \delta y, \delta z)$. Identify the net mass flow rate through each surface of this element as well as the rate at which the mass of the element is increasing. The resulting equation should be expressed in terms of the cartesian coordinates $(x, y, z, t)$, the cartesian velocity components $(u, v, w)$ and the fluid density $\rho$.

**1.2.** Derive the continuity equation from first principles using an *infinitesimal control volume* of cylindrical shape and having dimensions $(\delta R, R\,\delta\theta, \delta z)$. Identify the net mass flow rate through each surface of this element as well as the rate at which the mass of the element is increasing. The resulting equation should be expressed in terms of the cylindrical coordinates $(R, \theta, z, t)$, the cylindrical velocity components $(u_R, u_\theta, u_z)$ and the fluid density $\rho$.

**1.3.** Derive the continuity equation from first principles using an *infinitesimal control volume* of spherical shape and having dimensions $(\delta r, r, \delta\theta, r \sin\theta\,\delta\omega)$. Identify the net mass flow rate through each surface of this element as well as the rate at which the mass of the element is increasing. The resulting equation should be expressed in terms of the spherical coordinates $(r, \theta, \omega, t)$, the spherical velocity components $(u_r, u_\theta, u_\omega)$ and the fluid density $\rho$.

**1.4.** Obtain the continuity equation in cylindrical coordinates by expanding Eq. (1.3a) in cylindrical coordinates. To do this, make use of the following relationships connecting the coordinates and the velocity components in cartesian and cylindrical coordinates:

$$x = R \cos \theta$$
$$y = R \sin \theta$$
$$z = z$$
$$u = u_R \cos \theta - u_\theta \sin \theta$$
$$v = u_R \sin \theta + u_\theta \cos \theta$$
$$w = u_z$$

**1.5.** Obtain the continuity equation in spherical coordinates by expanding Eq. (1.3a) in spherical coordinates. Make use of the vector relationships outlined in Appendix A and follow the procedure used in Prob. 1.4.

**1.6.** Evaluate the radial component of the inertia term $(\mathbf{u} \cdot \nabla)\mathbf{u}$ in cylindrical coordinates by using the identities

$$x = R \cos \theta$$
$$y = R \sin \theta$$
$$u\mathbf{e}_x + v\mathbf{e}_y = u_R\mathbf{e}_R + u_\theta\mathbf{e}_\theta$$

and any other vector identities from Appendix A as required. Here $R$ and $\theta$ are cylindrical coordinates, $u_R$ and $u_\theta$ are the corresponding velocity components, and $\mathbf{e}_R, \mathbf{e}_\theta$ are the unit base vectors.

**1.7.** Evaluate the radial component of the inertia term $(\mathbf{u} \cdot \nabla)\mathbf{u}$ in spherical coordinates by use of the vector identities given in Appendix A.

**1.8.** Start with the shear stress tensor $\tau_{ij}$. Write out the independent components of this tensor in cartesian coordinates $(x, y, z)$ using the cartesian representation $(u, v, w)$ for the velocity vector. Specialize these expressions for the case of a monotonic gas for which the Stokes relation applies.

**1.9.** Write out the expression for the dissipation function, $\Phi$, for the same conditions and using the same notation as defined in Prob. 1.8 above.

**1.10.** Write down the equations governing the velocity and pressure in steady, two-dimensional flow of an inviscid, incompressible fluid in which gravity may be neglected. If the fluid is stratified, the density $\rho$ will depend, in general, on both $x$ and $y$. Show that the transformation:

$$u^* = \sqrt{\frac{\rho}{\rho_0}}\, u$$

$$v^* = \sqrt{\frac{\rho}{\rho_0}}\, v$$

in which $\rho_0$ is a constant reference density, transforms these governing equations into those of a constant-density fluid whose velocity components are $u^*$ and $v^*$.

# CHAPTER
# 2

## FLOW
## KINEMATICS

This chapter explores some of the results which may be deduced about the nature of a flowing continuum without reference to the dynamics of the continuum.

The first topic, flow lines, introduces the notions of streamlines, pathlines, and streaklines. These concepts are not only useful for flow-visualization experiments, but they supply the means by which solutions to the governing equations may be interpreted physically.

The concepts of circulation and vorticity are then introduced. Although these quantities are treated only in a kinematic sense at this stage, their full usefulness will become apparent in the later chapters when they are used in the dynamic equations of motion.

The concept of the streamline leads to the concept of a stream tube or a stream filament. Likewise, the introduction of the vorticity vector permits the topic of vortex tubes and vortex filaments to be discussed. Finally, this chapter ends with a discussion of the kinematics of vortex filaments or vortex lines. In this treatment, a useful analogy with the flow of an incompressible fluid is used. The results of this study form part of the so-called "Helmholtz equations," the remaining parts being taken up in the next chapter, which deals with, among other things, the dynamics of vorticity.

## 2.1 FLOW LINES

There are three types of flow lines which are used frequently for flow-visualization purposes. These flow lines are respectively called *streamlines*, *pathlines*, and *streaklines*, and in a general flow field they are all different. The definitions and equations of these various flow lines will be obtained separately below.

## Streamlines

Streamlines are lines whose tangents are everywhere parallel to the velocity vector. Since, in unsteady flow, the velocity vector at a given point will change both its magnitude and its direction with time, it is meaningful to consider only the instantaneous streamlines in the case of unsteady flows.

In order to establish the equations of the streamlines in a given flow field, consider first a two-dimensional flow field in which the velocity vector **u** has components $u$ and $v$ in the $x$ and $y$ directions, respectively. Then, by virtue of the definition of a streamline, its slope in the $xy$ plane, namely, $dy/dx$, must be equal to that of the velocity vector, namely, $v/u$. That is, the equation of the streamline in the $xy$ plane is

$$\frac{dy}{dx} = \frac{v}{u}$$

where, in general, both $u$ and $v$ will be functions of $x$ and $y$. Integration of this equation with respect to $x$ and $y$, holding $t$ fixed, will then yield the equation of the streamline in the $xy$ plane at that instant in time.

In the case of a three-dimensional flow field, the foregoing analysis is valid for the projection of the velocity vector on the $xy$ plane. By similarly treating the projections on the $xz$ plane and on the $yz$ plane, the slopes of the streamlines are found to be

$$\frac{dz}{dx} = \frac{w}{u}$$

$$\frac{dz}{dy} = \frac{w}{v}$$

on the $xz$ and $yz$ planes, respectively. These three equations which define the streamline may be written in the form

$$\frac{dy}{v} = \frac{dx}{u} \qquad \frac{dz}{w} = \frac{dx}{u} \qquad \frac{dz}{w} = \frac{dy}{v}$$

Written in this form, it is clear that these three equations may be expressed in the following, more compact form:

$$\frac{dx}{u} = \frac{dy}{v} = \frac{dz}{w}$$

Integration of these equations for fixed $t$ will yield, for that instant in time, an equation of the form $z = z(x, y)$, which is the required streamline. The easiest way of carrying out the required integration is to try to obtain the parametric equations of the curve $z = z(x, y)$ in the form $x = x(s)$, $y = y(s)$, and $z = z(s)$.

Elimination of the parameter $s$ among these equations will then yield the equation of the streamline in the form $z = z(x, y)$.

Thus a parameter $s$ is introduced whose value is zero at some reference point in space and whose value increases along the streamline. In terms of this parameter, the equations of the streamline become

$$\frac{dx}{u} = \frac{dy}{v} = \frac{dz}{w} = ds$$

These three equations may be combined in tensor notation to give

$$\frac{dx_i}{ds} = u_i(x_i, t) \qquad t \text{ fixed} \tag{2.1}$$

in which it is noted that if the velocity components depend upon time, the instantaneous streamline for any fixed value of $t$ is considered. If the streamline which passes through the point $(x_0, y_0, z_0)$ is required, Eqs. (2.1) are integrated and the initial conditions that when $s = 0$, $x = x_0$, $y = y_0$, and $z = z_0$ are applied. This will result in a set of equations of the form

$$x_i = x_i(x_0, y_0, z_0, t, s)$$

which, as $s$ takes on all real values, traces out the required streamline.

As an illustration of the determination of streamline patterns for a given flow field, consider the two-dimensional flow field defined by

$$u = x(1 + 2t)$$
$$v = y$$
$$w = 0$$

From Eqs. (2.1), the equations to be satisfied by the streamlines in the $xy$ plane are

$$\frac{dx}{ds} = x(1 + 2t)$$

$$\frac{dy}{ds} = y$$

Integration of these equations yields

$$x = C_1 e^{(1 + 2t)s}$$
$$y = C_2 e^{s}$$

which are the parametric equations of the streamlines in the $xy$ plane. In particular, suppose the streamlines which pass through the point $(1, 1)$ are required. Using the initial conditions that when $s = 0$, $x = 1$ and $y = 1$ shows that $C_1 = C_2 = 1$. Then the parametric equations of the streamlines which pass through the point $(1, 1)$ are

$$x = e^{(1 + 2t)s}$$
$$y = e^{s}$$

The fact that the streamlines change with time is evident from the above equations. Suppose the streamline which passes through the point $(1, 1)$ at time $t = 0$ is required, then

$$x = e^s$$
$$y = e^s$$

Hence the equation of the streamline is

$$x = y$$

This streamline is shown in Fig. 2.1 together with other flow lines which are discussed below.

### Pathlines

A pathline is a line which is traced out in time by a given fluid particle as it flows. Since the particle under consideration is moving with the fluid at its local velocity, pathlines must satisfy the equations

$$\frac{dx_i}{dt} = u_i(x_i, t) \tag{2.2}$$

The equation of the pathline which passes through the point $(x_0, y_0, z_0)$ at time $t = 0$ will then be the solution to Eq. (2.2), which satisfies the initial condition that when $t = 0$, $x = x_0$, $y = y_0$, and $z = z_0$. The solution will therefore yield a set of equations of the form

$$x_i = x_i(x_0, y_0, z_0, t)$$

which, as $t$ takes on all values greater than zero, will trace out the required pathline.

As an illustration of the manner in which the equation of a pathline is obtained, consider again the flow field defined by

$$u = x(1 + 2t)$$
$$v = y$$
$$w = 0$$

From Eqs. (2.2), the differential equations to be satisfied by the pathlines are

$$\frac{dx}{dt} = x(1 + 2t)$$

$$\frac{dy}{dt} = y$$

Integration of these equations gives

$$x = C_1 e^{t(1+t)}$$

$$y = C_2 e^t$$

These are the parametric equations of all the pathlines in the $xy$ plane for this particular flow field. In particular, if the pathline of the particle which passed through the point $(1, 1)$ at $t = 0$ is required, these parametric equations become

$$x = e^{t(1+t)}$$

$$y = e^t$$

Eliminating $t$ from these equations shows that the equation of the required pathline is

$$x = y^{1 + \ln y}$$

This pathline is shown in Fig. 2.1, from which it will be seen that the streamline which passes through $(1, 1)$ at $t = 0$ does not coincide with the pathline for the particle which passed through $(1, 1)$ at $t = 0$.

## Streaklines

A streakline is a line which is traced out by a neutrally buoyant marker fluid which is continuously injected into a flow field at a fixed point in space. The marker fluid may be smoke (if the main flow involves air or some other gas) or a dye (if the main flow involves water or some other liquid).

A particle of the marker fluid which is at the location $(x, y, z)$ at time $t$ must have passed through the injection point $(x_0, y_0, z_0)$ at some earlier time $t = \tau$. Then the time history of this particle may be obtained by solving the equations for the pathline [Eqs. (2.2)] subject to the initial conditions that $x = x_0$, $y = y_0$, and $z = z_0$ when $t = \tau$. Then as $\tau$ takes on all possible values in the range $-\infty \leq \tau \leq t$, all fluid particles on the streakline will be obtained. That is, the equation of the streakline through the point $(x_0, y_0, z_0)$ is obtained by solving Eqs. (2.2) subject to the initial conditions that when $t = \tau$, $x = x_0$, $y = y_0$, and $z = z_0$. This will yield an expression of the form

$$x_i = x_i(x_0, y_0, z_0, t, \tau)$$

Then as $\tau$ takes on the values $\tau \leq t$, these equations will define the instantaneous location of that streakline.

As an illustrative example, consider the flow field which was used to illustrate the streamline and the pathline. Then the equations to be solved for the streakline are

$$\frac{dx}{dt} = x(1 + 2t)$$

$$\frac{dy}{dt} = y$$

which integrate to give

$$x = C_1 e^{t(1+t)}$$

$$y = C_2 e^t$$

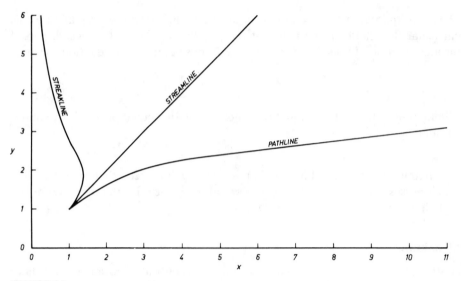

**FIGURE 2.1**

Comparison of the streamline through the point $(1, 1)$ at $t = 0$ with the pathline of the particle which passed through the point $(1, 1)$ at $t = 0$ and the streakline through the point $(1, 1)$ at $t = 0$ for the flow field $u = x(1 + 2t)$, $v = y$, $w = 0$.

Using the initial conditions that $x = y = 1$ when $t = \tau$, these equations become

$$x = e^{t(1+t)-\tau(1+\tau)}$$

$$y = e^{t-\tau}$$

These are the parametric equations of the streakline which passes through the point $(1, 1)$, and they are valid for all times $t$. In particular, at $t = 0$ these equations become

$$x = e^{-\tau(1+\tau)}$$

$$y = e^{-\tau}$$

Eliminating $\tau$ from these parametric equations shows that the equation of the streakline which passes through the point $(1, 1)$ is, at time $t = 0$,

$$x = y^{1-\ln y}$$

This streakline is shown in Fig. 2.1 along with the streamline and the pathline which were obtained for the same flow field. It will be noticed that none of the three flow lines coincide.

## 2.2 CIRCULATION AND VORTICITY

The *circulation* contained within a closed contour in a body of fluid is defined as the integral around the contour of the component of the velocity vector which is locally tangent to the contour. That is, the circulation $\Gamma$ is defined as

$$\Gamma = \oint \mathbf{u} \cdot d\mathbf{l} \tag{2.3}$$

where $d\mathbf{l}$ represents an element of the contour. The integration is taken counterclockwise around the contour, and the circulation is positive if this integral is positive.

The *vorticity* of an element of fluid is defined as the curl of its velocity vector. That is, the vorticity $\boldsymbol{\omega}$ is defined by

$$\boldsymbol{\omega} = \nabla \times \mathbf{u} \tag{2.4}$$

In tensor notation, Eqs. (2.4) may be written in the form

$$\omega_i = -\varepsilon_{ijk}\frac{\partial u_j}{\partial x_k} = \left(\frac{\partial u_k}{\partial x_j} - \frac{\partial u_j}{\partial x_k}\right)e_i$$

From this definition it is evident, by comparison with Eq. (1.6a), that the vorticity vector is numerically twice the angular speed of rotation of the fluid element about its own axes. That is, the vorticity is equal to twice the antisymmetric part of the deformation-rate tensor $e_{jk}$. It should be noted that a fluid element may travel on a circular streamline while having zero vorticity. Vorticity is proportional to the angular velocity of a fluid element about its principal axes, not that of the center of gravity of the element about some reference point. Thus a particle which is traveling on a circular streamline will have no vorticity, provided that it does not revolve about its center of gravity as it moves.

The vorticity contained in a fluid element is related to the circulation around the element. This relationship may be obtained from an application of Stokes' theorem to the definition of circulation as follows: From Eq. (2.1),

$$\Gamma = \oint \mathbf{u} \cdot d\mathbf{l}$$

$$= \int_A (\nabla \times \mathbf{u}) \cdot \mathbf{n}\, dA$$

where the contour integral has been converted to a surface integral by use of Stokes' theorem, in which $A$ is the area defined by the closed contour around which the circulation is calculated and $\mathbf{n}$ is the unit normal to the surface. Finally, invoking the definition of the vorticity vector, this relationship becomes

$$\Gamma = \int_A \boldsymbol{\omega} \cdot \mathbf{n}\, dA \tag{2.5}$$

Equation (2.5) shows that, for arbitrary choices of contours and enclosing areas $A$, if $\boldsymbol{\omega} = 0$ then $\Gamma = 0$ and vice versa. Flows for which $\boldsymbol{\omega} = 0$ are called *irrotational*, and flows for which this is not so are called *rotational*. The distinction between rotational and irrotational flow fields is an important one from the analytic point of view, as will be seen in later chapters.

## 2.3 STREAM TUBES AND VORTEX TUBES

The concept of a streamline, which was introduced in an earlier section, may be used to define a stream tube which is a region whose sidewalls are made up of streamlines. For any closed contour in a flow field, each point on the contour will have a streamline passing through it. Then, by considering all points on the contour, a series of streamlines are obtained which form a surface, and this surface is called a *stream tube*. Figure 2.2*a* shows a length of stream tube defined by a contour whose area is $A_1$. The corresponding area at some other section is shown as $A_2$, and in general $A_2$ will be different from $A_1$ and the shapes of the two cross sections of the stream tube will be different. If the cross section of a stream tube is infinitesimally small, the stream tube is usually referred to as a *stream filament*.

By analogy with streamlines and stream tubes, the useful concepts of vortex lines and vortex tubes may be introduced. A *vortex line* is a line whose tangents are everywhere parallel to the vorticity vector. Then for any closed contour in a flow field, each point on the contour will have a vortex line passing through it, and the series of vortex lines defined by the closed contour form a *vortex tube*. Figure 2.2*b* shows a length of vortex tube defined by a contour

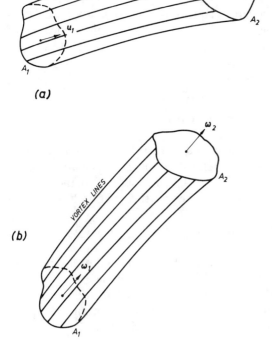

(a)

(b)

**FIGURE 2.2**
(*a*) Stream tube and (*b*) vortex tube subtended by a contour of area $A_1$ in a flow field.

whose area is $A_1$. The crosssectional area and shape at any other section of the vortex tube will, in general, be different. A vortex tube whose area is infinitesimally small is usually referred to as a *vortex filament*.

## 2.4   KINEMATICS OF VORTEX LINES

Certain properties of flow lines may be established by studying the kinematics of vortex lines. The results so obtained form part of what is sometimes referred to as the *Helmholtz theorems of vorticity*. The other parts of the Helmholtz theorems involve the dynamics of vorticity, which will be taken up in the next chapter.

Equation (2.4) defines the vorticity vector as the curl of the velocity vector. Since the divergence of the curl of any vector is identically zero, it follows that

$$\boldsymbol{\nabla} \cdot \boldsymbol{\omega} = 0$$

Since the vorticity vector is divergence-free, it follows that there can be no sources or sinks of vorticity in the fluid itself. That is, vortex lines either must form closed loops or must terminate on the boundaries of the fluid. The boundaries of the fluid may be either a solid surface or a free surface.

The fact that the vorticity vector is divergence-free leads to an analogy with the flow of an incompressible fluid. In this analogy, the counterpart of the velocity vector is the vorticity vector, and the counterpart of the volume flow rate is the circulation. To establish this analogy, a sequence of operations will be performed first on the velocity vector for an incompressible flow field and then on the vorticity vector.

The continuity equation for an incompressible fluid is

$$\boldsymbol{\nabla} \cdot \mathbf{u} = 0$$

Integrating this expression over some volume $V$ gives

$$\int_V \boldsymbol{\nabla} \cdot \mathbf{u} \, dV = 0$$

By use of Gauss' theorem this volume integral may be converted to the equivalent surface integral

$$\int_s \mathbf{u} \cdot \mathbf{n} \, ds = 0$$

where the surface $s$ encloses the volume $V$. Now consider the surface $s$ to be the entire outer surface of an element of a stream tube or stream filament, as shown in Fig. 2.2$a$, including the ends. Then, since $\mathbf{u} \cdot \mathbf{n} = 0$ on the walls of the stream tube by definition, it follows that

$$\int_{A_1} \mathbf{u} \cdot \mathbf{n} \, ds + \int_{A_2} \mathbf{u} \cdot \mathbf{n} \, ds = 0$$

Since **n** is defined as the outward unit normal,

$$\int_{A_1} \mathbf{u} \cdot \mathbf{n}\, ds = -Q_1$$

and

$$\int_{A_2} \mathbf{u} \cdot \mathbf{n}\, ds = Q_2$$

where $Q_1$ is volume flow rate crossing the area $A_1$ and $Q_2$ is the volume flow rate crossing the area $A_2$. That is, the fact that the vector **u** is divergence-free leads to the result

$$Q_1 = Q_2$$

which states that the volume of fluid crossing the area $A_1$ per unit time is equal to that crossing the area $A_2$ per unit time. Since the fluid was assumed to be incompressible, this result appears intuitively obvious.

Turning now to the vorticity vector, it was shown that

$$\nabla \cdot \boldsymbol{\omega} = 0$$

so that

$$\int_V \nabla \cdot \boldsymbol{\omega}\, dV = 0$$

and

$$\int_s \boldsymbol{\omega} \cdot \mathbf{n}\, ds = 0$$

where the surface $s$ encloses the volume $V$. Now consider the surface $s$ to be the entire outer surface of an element of a vortex tube or a vortex filament as shown in Fig. 2.2$b$, including the ends. By definition of the vortex lines which make up the surface of the vortex tube, $\boldsymbol{\omega} \cdot \mathbf{n} = 0$ on the walls of the vortex tube. Then

$$\int_{A_1} \boldsymbol{\omega} \cdot \mathbf{n}\, ds + \int_{A_2} \boldsymbol{\omega} \cdot \mathbf{n}\, ds = 0$$

But from Eq. (2.5),

$$\int_{A_1} \boldsymbol{\omega} \cdot \mathbf{n}\, ds = -\Gamma_1$$

and

$$\int_{A_2} \boldsymbol{\omega} \cdot \mathbf{n}\, ds = \Gamma_2$$

Hence the fact that $\boldsymbol{\omega}$ is divergence-free results in the condition

$$\Gamma_1 = \Gamma_2$$

That is, the circulation around the limiting contour of the area $A_1$ is equal to that around $A_2$. Alternatively, this result may be stated in the form that the circulation at each cross section of a vortex tube is the same. This means that if the crosssectional area of the vortex tube increases, the average value of the vorticity across that section must decrease, just as the average velocity would decrease to satisfy continuity. In fact, the result $\Gamma_1 = \Gamma_2$ may be put in the form of the simple, one-dimensional continuity equation. If $\omega_1$ denotes the average vorticity across the area $A_1$ and $\omega_2$ denotes that across $A_2$, the result

$$\int_{A_1} \boldsymbol{\omega} \cdot \mathbf{n} \, ds = \int_{A_2} \boldsymbol{\omega} \cdot \mathbf{n} \, ds$$

becomes

$$\omega_1 A_1 = \omega_2 A_2 \qquad (2.6)$$

The fact that the vorticity vector $\boldsymbol{\omega}$ is divergence-free means that vortex tubes must terminate on themselves, at a solid boundary or at a free surface. Smoke rings terminate on themselves, while a vortex tube in a free surface flow may have one end at the solid boundary forming the bottom and the other end at the free surface.

## PROBLEMS

**2.1.** Consider the two-dimensional flow field defined by the following velocity components:

$$u = \frac{v}{1 + t} \qquad v = 1 \qquad w = 0$$

For this flow field find the equation of:
(a) The streamline through the point $(1, 1)$ at $t = 0$
(b) The pathline for a particle released at the point $(1, 1)$ at $t = 0$
(c) The streakline at $t = 0$ which passes through the point $(1, 1)$

**2.2.** Show that the streamlines and particle paths coincide for the flow $u_i = x_i/(1 + t)$. (From Aris, "Vectors, Tensors, and the Basic Equations of Fluid Mechanics," Prentice-Hall, Inc., Englewood Cliffs, N.J., 1962)

**2.3.** The velocity components for a particular flow field are as follows:

$$u = 16x^2 + y \qquad v = 10 \qquad w = yz^2$$

(a) Determine the circulation, $\Gamma$, for this flow field around the following contour by integrating the velocity around it:

$$0 \le x \le 10 \qquad y = 0$$
$$0 \le y \le 5 \qquad x = 10$$
$$0 \le x \le 10 \qquad y = 5$$
$$0 \le y \le 5 \qquad x = 0$$

(b) Calculate the vorticity vector, $\boldsymbol{\omega}$, for the given flow field and hence evaluate:

$$\int_A \boldsymbol{\omega} \cdot \mathbf{n} \, dA$$

where $A$ is the area of the rectangle defined in (a), and $\mathbf{n}$ is the unit normal to that area. Compare the result obtained in (b) with that obtained in (a).

**2.4.** Consider the two-dimensional velocity distribution defined by:

$$u = -\frac{x}{x^2 + y^2} \qquad v = \frac{y}{x^2 + y^2}$$

Determine the circulation for this flow field around the following contour by integrating the velocity around it:

$$-1 \le x \le +1 \qquad y = -1$$
$$-1 \le y \le +1 \qquad x = +1$$
$$-1 \le x \le +1 \qquad y = +1$$
$$-1 \le y \le +1 \qquad x = -1$$

**2.5.** The velocity components for a particular two-dimensional flow field are defined as follows:

$$u = -\frac{y}{x^2 + y^2} \qquad v = \frac{x}{x^2 + y^2}$$

(a) Using the same contour as defined in Prob. 2.4, determine the circulation for the given flow field.

(b) Calculate the vorticity vector for the given flow field.

(c) Calculate the divergence of the velocity vector for the given flow field.

**2.6.** Calculate the vorticity at any point $(R, \theta)$ for each of the following two-dimensional flow fields:

(a) $u_R = 0$, $u_\theta = \omega R$

(b) $u_R = 0$, $u_\theta = \Gamma / 2\pi R$

In the above, $R$ and $\theta$ are cylindrical coordinates while $\omega$ and $\Gamma$ are constants.

# CHAPTER
# 3

## SPECIAL FORMS OF THE GOVERNING EQUATIONS

Some alternative forms of the governing equations, as derived in Chap. 1, will be discussed here. The results are all obtained from the governing equations under various degrees of approximation such as negligible viscous effects. Some of the results are frequently referred to as *theorems*. They are used either as alternatives to the general equations derived in Chap. 1, under the specified restrictions, or as supplementary information to these equations.

The first result established is Kelvin's theorem. This theorem establishes the conditions under which irrotational motion remains irrotational and so justifies the simplifying methods of analysis which are utilized for irrotational flows. Then the Bernoulli equations are derived. These equations are integrals of the Euler equations under certain conditions. They are used to relate the pressure and velocity fields when the velocity is established separately from, for example, the condition of irrotationality. Crocco's equation is derived next. This equation relates the entropy of the fluid to the vorticity and shows that under certain conditions isentropic flows are irrotational, and vice versa. Finally, the vorticity equation is derived for a fluid of constant density and viscosity. This equation is useful in the study of rotational flows.

## 3.1  KELVIN'S THEOREM

This theorem states that for an inviscid fluid in which the density is constant, or in which the pressure depends on the density alone, and for which any body forces which exist are conservative, the vorticity of each fluid particle will be preserved. Kelvin's theorem covers the remainder of the Helmholtz theorems of vorticity which were not treated in Sec. 2.4 during the discussion of the kinematics of vortex lines. Although Kelvin's theorem appears to be kinematic in nature, the dynamic equations of motion are required in the proof.

Suppose that any body force $f_j$ per unit mass which may act on the fluid is conservative, such as gravity. Then $f_i$ may be written as the gradient of some scalar function $G$, giving

$$f_j = \frac{\partial G}{\partial x_j}$$

Then, from Eqs. (1.9c), the equations of motion for an inviscid fluid subjected to only conservative body forces are

$$\rho \frac{\partial u_j}{\partial t} + \rho u_k \frac{\partial u_j}{\partial x_k} = -\frac{\partial p}{\partial x_j} + \rho \frac{\partial G}{\partial x_j}$$

Or, in terms of the material derivative,

$$\frac{Du_j}{Dt} = -\frac{1}{\rho}\frac{\partial p}{\partial x_j} + \frac{\partial G}{\partial x_j}$$

It is this form of the momentum equations, which are valid for an inviscid fluid which is subjected to only conservative body forces, which will be used to prove Kelvin's theorem.

In order to determine the rate of change of vorticity associated with a given fluid element, the material derivative of the circulation $\Gamma$ will be calculated. From Eq. (2.3),

$$\frac{D\Gamma}{Dt} = \frac{D}{Dt}\oint u_j \, dx_j$$

$$= \oint \left[ \frac{Du_j}{Dt} dx_j + u_j \frac{D(dx_j)}{Dt} \right]$$

The quantity $D(dx_j)/Dt$ is the material derivative of an element $dx_j$ of the contour around which the circulation is to be calculated. Its value may be established as follows:

$$\frac{D(dx_j)}{Dt} = d\left(\frac{Dx_j}{Dt}\right) = d\left(\frac{\partial x_j}{\partial t} + u_k \frac{\partial x_j}{\partial x_k}\right) = du_j$$

Here the material derivative has been converted into its eulerian equivalent, using Eq. (1.1) in which $t$ and the spatial coordinates are independent. Thus

$\partial x_j/\partial t = 0$ and $\partial x_j/\partial x_k = \delta_{jk}$, which is zero unless $k = j$, at which time its value is unity. This shows that the value of $Dx_j/Dt$ is $u_j$ and hence the value of $D(dx_j)Dt$ is $du_j$. In this way the expression for the rate of change of circulation becomes

$$\frac{D\Gamma}{Dt} = \oint \left( \frac{Du_j}{Dt}\, dx_j + u_j\, du_j \right)$$

The quantity $Du_j/Dt$ will now be eliminated from this expression by using the momentum equations which were derived above for an inviscid fluid in which any body forces were conservative. Thus the rate of change of circulation becomes

$$\frac{D\Gamma}{Dt} = \oint \left( -\frac{1}{\rho}\frac{\partial p}{\partial x_j}\, dx_j + \frac{\partial G}{\partial x_j}\, dx_j + u_j\, du_j \right)$$

$$= \oint \left[ -\frac{dp}{\rho} + dG + \tfrac{1}{2}d(u_j u_j) \right]$$

where it has been observed that $(\partial p/\partial x_j)\, dx_j = dp$, which is the total spatial variation of $p$ and likewise $(\partial G/\partial x_j)\, dx_j = dG$. It is now observed that, since the integration is to be carried out around a closed contour, the integral of $dG$ and that of $d(u_j u_j)$ are both zero, since the body force and the velocity are both assumed to be single-valued. Then

$$\frac{D\Gamma}{Dt} = -\oint \frac{dp}{\rho}$$

Now if $\rho =$ constant, the remaining integral is zero for the same reason that the other integrals were zero. However, this integral is zero under less restrictive conditions also. Suppose the pressure $p$ may be considered to be a function of the density $\rho$ only as, for example, in isentropic flows. Then for some function $g$,

$$p = g(\rho)$$

so that $\qquad\qquad\qquad dp = g'(\rho)\, d\rho$

The expression for $D\Gamma/Dt$ now becomes

$$\frac{D\Gamma}{Dt} = -\oint \frac{g'(\rho)}{\rho}\, d\rho$$

That is, this integral falls into the same category as the two previous integrals, and its value around any closed contour is zero. This gives the result which is known as *Kelvin's theorem*:

$$\frac{D\Gamma}{Dt} = 0 \tag{3.1}$$

Equation (3.1) says that, if we follow a given contour as it flows, the vorticity inside that contour will not change. Recall that the right-hand side of Eq. (3.1) could be proved to be zero by considering the fluid to be inviscid, the body forces to be conservative, and either the density to be constant or the pressure to be a function of the density only. Relaxing any of these conditions leads to, in general, a nonzero term on the right-hand side of Eq. (3.1). Thus it may be deduced that vorticity may be changed by the action of viscosity, the application of nonconservative body forces, or density variations which are not simply related to the pressure variation.

It should be noted that Eq. (3.1) applies to a simply connected region. That is, for any closed contour in the fluid which contains only fluid, there will be some definite value of the circulation $\Gamma$. Equation (3.1) asserts that, under the conditions specified in the derivation, the value of $\Gamma$ will not change around that contour even though the contour itself may be deformed by the flow. A closed contour which originally does not include a body cannot at any subsequent time contain a body such as a two-dimensional airfoil. There is therefore no conflict in the fact that such an airfoil may have a circulation around it while immersed in an irrotational flow.

From Kelvin's theorem and the results established in Sec. 2.4, it is evident that the total vorticity associated with a vortex filament is fixed and will not change as the vortex filament flows with the fluid. Distortion of the vortex filament may take place, but the total vorticity associated with it will remain the same. The vortex filament will always consist of the same fluid particles as it flows, and if the vortex filament is elongated, the vorticity at any section of the filament will decrease so that the total vorticity associated with the filament remains fixed.

The principal use of Kelvin's theorem is in the study of incompressible, inviscid fluid flows. If a body is moving through such a fluid, or if a uniform flow of such a fluid passes around a body, then the vorticity far from the body will be zero. Then, according to Kelvin's theorem, the vorticity in the fluid will everywhere be zero, even adjacent to the body. Then the condition $\nabla \times \mathbf{u} = 0$ may be used to replace the Euler equations so that the condition of irrotationality becomes the alternative form of the equations of motion for the fluid. Again it is emphasized that this kinematic equivalent is valid only because of Kelvin's theorem, and in turn, the Euler equations were used to prove Kelvin's theorem.

## 3.2 BERNOULLI EQUATION

For an inviscid fluid in which any body forces are conservative and either the flow is steady or it is irrotational, the equations of momentum conservation may be integrated to yield a single scalar equation which is called the Bernoulli equation.

In the previous section it was pointed out that the equations of motion for an inviscid fluid in which any body forces were conservative could be written in the form

$$\rho \frac{\partial u_j}{\partial t} + \rho u_k \frac{\partial u_j}{\partial x_k} = -\frac{\partial p}{\partial x_j} + \rho \frac{\partial G}{\partial x_j}$$

Using a vector identity given in Appendix A, the second term on the left-hand side of these equations may be rewritten as follows:

$$u_k \frac{\partial u_j}{\partial x_k} = (\mathbf{u} \cdot \nabla)\mathbf{u} = \nabla(\tfrac{1}{2}\mathbf{u} \cdot \mathbf{u}) - \mathbf{u} \times (\nabla \times \mathbf{u})$$

$$= \nabla(\tfrac{1}{2}\mathbf{u} \cdot \mathbf{u}) - \mathbf{u} \times \boldsymbol{\omega}$$

In this way the Euler equations may be written in the following vector form:

$$\frac{\partial \mathbf{u}}{\partial t} + \nabla(\tfrac{1}{2}\mathbf{u} \cdot \mathbf{u}) - \mathbf{u} \times \boldsymbol{\omega} = -\frac{1}{\rho}\nabla p + \nabla G$$

It is now proposed to show that the term $(1/\rho)\nabla p$, which appears on the right-hand side of this equation, may be written as $\nabla(\int dp/\rho)$. To do this, we form the scalar product of an element of a space curve $d\mathbf{l}$, such as an element of a streamline, with the vector quantity $(1/\rho)\nabla p$.

$$d\mathbf{l} \cdot \left(\frac{1}{\rho}\nabla p\right) = \frac{1}{\rho}d\mathbf{l} \cdot \nabla p = \frac{1}{\rho}dp$$

Here the result $d\mathbf{l} \cdot \nabla = dx(\partial/\partial x) + dy(\partial/\partial y) + dz(\partial/\partial z) = d$ has been used, where the scalar operator $d$ is the total spatial derivative. Then, using $d$ and its inverse integral operation, it follows that

$$d\mathbf{l} \cdot \left(\frac{1}{\rho}\nabla p\right) = d\int \frac{dp}{\rho} = d\mathbf{l} \cdot \nabla\left(\int \frac{dp}{\rho}\right)$$

where, again, the equivalence of $d$ and $d\mathbf{l} \cdot \nabla$ has been used. The vectors which form the scalar product with $d\mathbf{l}$ in this last equation must be equal since $d\mathbf{l}$ was arbitrarily chosen; hence it follows that

$$\frac{1}{\rho}\nabla p = \nabla\left(\int \frac{dp}{\rho}\right)$$

Using this result, the Euler equations become

$$\frac{\partial \mathbf{u}}{\partial t} + \nabla\left(\int \frac{dp}{\rho} + \tfrac{1}{2}\mathbf{u} \cdot \mathbf{u} - G\right) = \mathbf{u} \times \boldsymbol{\omega} \tag{3.2a}$$

The vector equation $(3.2a)$ may be integrated for steady flow and for unsteady or steady irrotational flow.

Considering first steady flow, Eqs. $(3.2a)$ become

$$\nabla\left(\int \frac{dp}{\rho} + \tfrac{1}{2}\mathbf{u} \cdot \mathbf{u} - G\right) = \mathbf{u} \times \boldsymbol{\omega}$$

Forming the scalar product of the velocity vector $\mathbf{u}$ with this equation gives

$$\mathbf{u} \cdot \nabla\left(\int \frac{dp}{\rho} + \tfrac{1}{2}\mathbf{u} \cdot \mathbf{u} - G\right) = \mathbf{u} \cdot (\mathbf{u} \times \boldsymbol{\omega})$$

But the vector product of $\mathbf{u}$ with $\boldsymbol{\omega}$ will yield a vector which is perpendicular to $\mathbf{u}$; hence the quantity $\mathbf{u} \cdot (\mathbf{u} \times \boldsymbol{\omega})$ is zero. Furthermore, the operator $\mathbf{u} \cdot \nabla$ is the steady-state form of the material derivative. Thus the above equation states that, as we flow along a streamline in steady flow, the quantity $\int dp/\rho + \frac{1}{2}\mathbf{u} \cdot \mathbf{u} - G$ remains constant. That is,

$$\int \frac{dp}{\rho} + \tfrac{1}{2}\mathbf{u} \cdot \mathbf{u} - G = \text{constant along each streamline} \qquad (3.2b)$$

This result is referred to as the *Bernoulli integral* or the *Bernoulli equation*. It should be recalled that it is valid for the steady flow of a fluid in which viscous effects are negligible and in which any body forces are conservative. In many cases the flow around some body originates in a uniform flow, and in such cases, and in some other cases, the constant on the right-hand side of Eq. (3.2b) is the same for each streamline. Then the quantity $\int dp/\rho + \frac{1}{2}\mathbf{u} \cdot \mathbf{u} - G$ is constant everywhere. The constant is usually referred to as the *Bernoulli constant*.

Equation (3.2a) may also be integrated under slightly different circumstances from those which led to Eq. (3.2b). Rather than considering steady flows, consider irrotational flows. Then the vorticity $\boldsymbol{\omega}$ will be zero so that (3.2a) becomes

$$\frac{\partial \mathbf{u}}{\partial t} + \nabla\left(\int \frac{dp}{\rho} + \tfrac{1}{2}\mathbf{u} \cdot \mathbf{u} - G\right) = 0$$

Now since $\boldsymbol{\omega} = \nabla \times \mathbf{u} = 0$, it follows that the velocity vector $\mathbf{u}$ may be written as the gradient of some scalar, say $\phi$, since $\nabla \times \nabla\phi = 0$ for any function $\phi$. The quantity $\phi$ is known as the *velocity potential*, and it will be used extensively in Chap. 4. Then, replacing $\mathbf{u}$ by $\nabla\phi$ in the above equation gives

$$\nabla\left(\frac{\partial \phi}{\partial t} + \int \frac{dp}{\rho} + \tfrac{1}{2}\nabla\phi \cdot \nabla\phi - G\right) = 0$$

Forming the scalar product of this vector equation with an element of space curve $d\mathbf{l}$ gives

$$d\left(\frac{\partial \phi}{\partial t} + \int \frac{dp}{\rho} + \tfrac{1}{2}\nabla\phi \cdot \nabla\phi - G\right) = 0$$

where again the fact that $d\mathbf{l} \cdot \nabla = d$ has been used, where $d$ is the total spatial derivative. Thus integration yields

$$\frac{\partial \phi}{\partial t} + \int \frac{dp}{\rho} + \tfrac{1}{2}\nabla\phi \cdot \nabla\phi - G = F(t) \qquad (3.2c)$$

where $F(t)$ is some function of time which may be added after integrating over the space coordinates. $F(t)$ is usually referred to as the *unsteady Bernoulli constant*, even though it is not strictly a constant. Recall that Eq. (3.2c) is valid for irrotational motion of a fluid in which viscous effects are negligible and in which any body forces are conservative. Kelvin's theorem usually helps to establish the condition of irrotationality by relating the flow under consideration to a simpler form of the flow far upstream.

## 3.3  CROCCO'S EQUATION

This equation relates the vorticity of a flow field to the entropy of the fluid. Under certain conditions it will be shown that isentropic flows are irrotational, and vice versa. Then, if it is known that a flow field is essentially isentropic, the mathematical simplifications associated with irrotational motion may be employed. This simplification will be employed in the chapters which deal with compressible fluid flow, and it is justified by Crocco's equation.

In order to establish Crocco's equation, consider the flow of an inviscid fluid in which there are no body forces. Then, from Eqs. (1.9c), the Euler equations which guarantee dynamic equilibrium become

$$\frac{\partial \mathbf{u}}{\partial t} + (\mathbf{u} \cdot \nabla)\mathbf{u} = -\frac{1}{\rho} \nabla p$$

The nonlinear term may be expanded as follows sing a vector identity given in Appendix A:

$$(\mathbf{u} \cdot \nabla)\mathbf{u} = \nabla(\tfrac{1}{2}\mathbf{u} \cdot \mathbf{u}) - \mathbf{u} \times (\nabla \times \mathbf{u})$$

Hence the Euler equation becomes

$$\frac{\partial \mathbf{u}}{\partial t} + \nabla(\tfrac{1}{2}\mathbf{u} \cdot \mathbf{u}) - \mathbf{u} \times \boldsymbol{\omega} = -\frac{1}{\rho} \nabla p$$

It is this form of the Euler equation which is the starting point for the derivation of Crocco's equation. In order to relate the dynamics of the flow to its thermodynamics, it is proposed to eliminate the pressure $p$ and the density $\rho$, which appear in the term on the right-hand side of the above equation, in favor of the enthalpy $h$ and the entropy $s$. To do this, we use the first law of thermodynamics and the definition of the entropy. From Appendix E, a change in internal energy $de$ is caused by work done on the fluid $-pd(1/\rho)$ and by any heat which is added to the fluid $dq$. That is,

$$de = -pd\left(\frac{1}{\rho}\right) + dq$$

$$= -pd\left(\frac{1}{\rho}\right) + T\,ds$$

where the last relation follows from the definition of the entropy. Now $p$ and $\rho$ have been related to $e$ and $s$. In order to eliminate $e$ in favor of the enthalpy $h$, we use the equation which defines the enthalpy, namely, $e = h - p/\rho$. Then the foregoing thermodynamic relation becomes

$$dh - d\left(\frac{p}{\rho}\right) = -pd\left(\frac{1}{\rho}\right) + T\,ds$$

Since $d(p/\rho) = pd(1/\rho) + dp/\rho$, this equation simplifies to

$$-\frac{1}{\rho} dp = T ds - dh$$

Using again the result established in the previous section that $d\mathbf{l} \cdot \nabla = d$, it follows that

$$-\frac{1}{\rho} \nabla p = T \nabla s - \nabla h$$

This result will now be used to eliminate the pressure and the density which appear on the right-hand side of the Euler equations.

$$\frac{\partial \mathbf{u}}{\partial t} + \nabla(\tfrac{1}{2}\mathbf{u} \cdot \mathbf{u}) - \mathbf{u} \times \boldsymbol{\omega} = T \nabla s - \nabla h$$

Rearranging this vector equation slightly yields the result known as *Crocco's equation*:

$$\mathbf{u} \times \boldsymbol{\omega} + T \nabla s = \nabla(h + \tfrac{1}{2}\mathbf{u} \cdot \mathbf{u}) + \frac{\partial \mathbf{u}}{\partial t} \tag{3.3a}$$

Equations (3.3a) are valid for flows in which viscous effects are negligible and in which there are no body forces.

Under conditions of steady, adiabatic flow, Eq. (3.3a) may be reduced to a scalar equation. To show this, it will first be shown that for adiabatic flow of an inviscid fluid in which there are no body forces, the quantity $h_0 = h + \tfrac{1}{2}\mathbf{u} \cdot \mathbf{u}$ is constant along each streamline. The quantity $h_0$ is called the *stagnation enthalpy*.

From Prob. 3.1, the energy equation for adiabatic flow of an inviscid fluid is

$$\rho \frac{Dh}{Dt} = \frac{Dp}{Dt}$$

The Euler equations for a flow without body forces are

$$\rho \frac{D\mathbf{u}}{Dt} = -\nabla p$$

Forming the scalar product of this equation with the velocity vector $\mathbf{u}$ gives

$$\rho \frac{D}{Dt}\left(\frac{1}{2}\mathbf{u} \cdot \mathbf{u}\right) = -\mathbf{u} \cdot \nabla p$$

Adding this equation to the energy equation derived above yields

$$\rho \frac{D}{Dt}(h + \tfrac{1}{2}\mathbf{u} \cdot \mathbf{u}) = \frac{Dp}{Dt} - \mathbf{u} \cdot \nabla p$$

or

$$\rho \frac{Dh_0}{Dt} = \frac{\partial p}{\partial t}$$

Then, for steady flow, the right-hand side of this equation will be zero. That is, for steady, adiabatic flow of an inviscid fluid in which there are no body forces, the quantity $Dh_0/Dt$ will be zero. Hence the stagnation enthalpy $h_0$ will be constant along each streamline.

Equation (3.3$a$) was derived for an inviscid fluid which is not subjected to any body forces. Then, if, in addition, the flow is steady and adiabatic, Eq. (3.3$a$) becomes

$$\mathbf{u} \times \boldsymbol{\omega} + T\, \nabla s = \nabla h_0$$

where the quantity $h_0$ is constant along each streamline. Hence $\nabla h_0$ will be a vector perpendicular to the streamlines. But $\mathbf{u} \times \boldsymbol{\omega}$ is also perpendicular to the streamlines, so that the remaining vector, namely, $T\, \nabla s$, must also be perpendicular to the streamlines. Then the above vector equation may be written in the following scalar form:

$$U\Omega + T\,\frac{ds}{dn} = \frac{dh_0}{dn} \qquad (3.3b)$$

Here $U$ and $\Omega$ are, respectively, the magnitudes of the velocity vector $\mathbf{u}$ and the vorticity vector $\boldsymbol{\omega}$. The coordinate $n$ is perpendicular to the streamlines locally. Equation (3.3$b$) is valid for steady, adiabatic flow of an inviscid fluid in which there are no body forces.

Usually when the stagnation enthalpy is constant along each streamline, it is constant everywhere. That is, the value of $h_0$ along each streamline is the same. Under these conditions $dh_0/dn = 0$, so that Eq. (3.3$b$) becomes

$$U\Omega + T\,\frac{ds}{dn} = 0 \qquad (3.3c)$$

In this form Crocco's equation clearly shows that if $s$ is constant, $\Omega$ must be zero. Likewise if $\Omega$ is zero, $ds/dn$ must be zero so that $s$ must be constant. That is, isentropic flows are irrotational and irrotational flows are isentropic. This result is true, in general, only for steady flows of inviscid fluids in which there are no body forces and in which the stagnation enthalpy is constant.

## 3.4  VORTICITY EQUATION

The equation to be satisfied by the vorticity vector $\boldsymbol{\omega}$ for a fluid of constant density and constant viscosity will be derived in this section. Such an equation is useful in the study of viscous flows in incompressible fluids, which is the topic of Part III of this book. One reason that the vorticity equation is of interest is that it enables us to learn more about the physics of given flow fields. Also, in the analysis of some flow fields it is frequently possible to make some statement about the vorticity distribution which facilitates the analysis if the problem is posed in terms of the vorticity.

From Eqs. (1.9b), the Navier-Stokes equations for a fluid of constant density and viscosity are

$$\frac{\partial \mathbf{u}}{\partial t} + (\mathbf{u} \cdot \nabla)\mathbf{u} = -\nabla\left(\frac{p}{\rho}\right) + \nu \nabla^2 \mathbf{u}$$

Replacing the nonlinear term by its equivalent form given by the vector identities in Appendix A, this vector equation becomes

$$\frac{\partial \mathbf{u}}{\partial t} + \nabla(\tfrac{1}{2}\mathbf{u} \cdot \mathbf{u}) - \mathbf{u} \times (\nabla \times \mathbf{u}) = -\nabla\left(\frac{p}{\rho}\right) + \nu \nabla^2 \mathbf{u}$$

The vorticity equation is obtained by taking the curl of this equation and noting that the curl of the gradient of any scalar is zero. Hence

$$\frac{\partial \boldsymbol{\omega}}{\partial t} - \nabla \times (\mathbf{u} \times \boldsymbol{\omega}) = \nu \nabla^2 \boldsymbol{\omega}$$

Using a vector identity given in Appendix A, the second term on the left-hand side may be expanded to give

$$\nabla \times (\mathbf{u} \times \boldsymbol{\omega}) = \mathbf{u}(\nabla \cdot \boldsymbol{\omega}) - \boldsymbol{\omega}(\nabla \cdot \mathbf{u}) - (\mathbf{u} \cdot \nabla)\boldsymbol{\omega} + (\boldsymbol{\omega} \cdot \nabla)\mathbf{u}$$

But $\nabla \cdot \boldsymbol{\omega} = 0$, since the divergence of the curl of any vector is zero and $\nabla \cdot \mathbf{u} = 0$ from the continuity equation. Hence the *vorticity equation* becomes

$$\frac{\partial \boldsymbol{\omega}}{\partial t} + (\mathbf{u} \cdot \nabla)\boldsymbol{\omega} = (\boldsymbol{\omega} \cdot \nabla)\mathbf{u} + \nu \nabla^2 \boldsymbol{\omega} \tag{3.4a}$$

For two-dimensional flows the vorticity vector $\boldsymbol{\omega}$ will be perpendicular to the plane of the flow, so that $(\boldsymbol{\omega} \cdot \nabla)\mathbf{u}$ will be zero. Then

$$\frac{\partial \boldsymbol{\omega}}{\partial t} + (\mathbf{u} \cdot \nabla)\boldsymbol{\omega} = \nu \nabla^2 \boldsymbol{\omega} \tag{3.4b}$$

The vorticity equation, in either the general form (3.4a) or the two-dimensional form (3.4b), has another advantage over and above those mentioned in the preliminary remarks. It will be noted from these equations that the pressure $p$ does not appear explicitly. Thus the vorticity and velocity fields may be obtained without any knowledge of the pressure field.

In order to determine the pressure distribution in terms of the vorticity, the Navier-Stokes equations are again used in the form

$$\frac{\partial \mathbf{u}}{\partial t} + (\mathbf{u} \cdot \nabla)\mathbf{u} = -\nabla\left(\frac{p}{\rho}\right) + \nu \nabla^2 \mathbf{u}$$

Taking the divergence of this equation and using the result of Prob. 3.2 together with the continuity equation $\nabla \cdot \mathbf{u} = 0$, it follows that the equation to be satisfied by the pressure $p$ is

$$\nabla^2\left(\frac{p}{\rho}\right) = \boldsymbol{\omega} \cdot \boldsymbol{\omega} + \mathbf{u} \cdot (\nabla^2 \mathbf{u}) - \tfrac{1}{2}\nabla^2(\mathbf{u} \cdot \mathbf{u}) \tag{3.5}$$

From the foregoing results we see that the vorticity satisfies a diffusion equation while the pressure satisfies a Poisson equation.

## PROBLEMS

**3.1.** In vector form, the thermal energy equation (1.11) is

$$\rho \frac{De}{Dt} = -p \, \nabla \cdot \mathbf{u} + \nabla \cdot (k \, \nabla T) + \Phi$$

By using the definition of the enthalpy $h$, show that an equivalent form of this equation is

$$\rho \frac{Dh}{Dt} = \frac{Dp}{Dt} + \nabla \cdot (k \, \nabla T) + \Phi \tag{3.6}$$

**3.2.** Show that, for an incompressible fluid, the following identity holds between the velocity vector $\mathbf{u}$ and the vorticity vector $\boldsymbol{\omega}$:

$$\nabla \cdot [(\mathbf{u} \cdot \nabla)\mathbf{u}] = \tfrac{1}{2} \nabla^2 (\mathbf{u} \cdot \mathbf{u}) - \mathbf{u} \cdot (\nabla^2 \mathbf{u}) - \boldsymbol{\omega} \cdot \boldsymbol{\omega}$$

**3.3.** In cylindrical coordinates, the velocity components for uniform flow around a circular cylinder are

$$u_R = U\left(1 - \frac{a^2}{R^2}\right) \cos \theta$$

$$u_\theta = U\left(1 + \frac{a^2}{R^2}\right) \sin \theta$$

Here $U$ is the constant magnitude of the velocity approaching the cylinder and $a$ is the radius of the cylinder. If compressible and viscous effects are negligible, determine the pressure $p(R, \theta)$ at any point in the fluid in the absence of any body forces. Take the pressure far from the cylinder to be constant and equal to $p_0$.

Specialize the result obtained above to obtain an expression for the pressure $p(a, \theta)$ on the surface of the cylinder.

## FURTHER READING—PART I

Part I of this book has been concerned with the derivation of the equations governing the motion of a fluid. The number of books dealing with fluid mechanics in which these equations are derived is large. The following represents a sample of some of these books.

Aris, Rutherford: "Vectors, Tensors, and the Basic Equations of Fluid Mechanics," Prentice-Hall, Inc., Englewood, N.J., 1964.

Batchelor, G. K.: "An Introduction to Fluid Dynamics," Cambridge University Press, London, 1967.

Lagerstrom, P. A.: Laminar Flow Theory, in F. K. Moore (ed.): "Theory of Laminar Flows," Princeton University Press, Princeton, N.J., 1964.

Panton, Ronald L.: "Incompressible Flow," John Wiley & Sons, New York, N.Y., 1984.

Serrin, James: Mathematical Principles of Classical Fluid Mechanics, in S. Flügge (ed.): "Handbuch der Physik," vol. VIII/1, Springer-Verlag OHG, Berlin, 1959.

Yih, Chia-Shun: "Fluid Mechanics," McGraw-Hill Book Company, New York, N.Y., 1969.

# PART
# II

# IDEAL-FLUID FLOW

This part of the book deals with the flow of ideal fluids, that is, fluids which are inviscid and incompressible. The results are therefore limited to flow fields in which viscous effects of the fluid are negligible and compressibility of the fluid is unimportant. Then, any phenomena which are predicted by the governing equations will be due to the inertia of the fluid. The mathematical simplification which results from neglecting viscous and compressible effects is great, and consequently the topic of ideal-fluid flow is, mathematically, the best understood.

Part II contains Chaps. 4, 5, and 6. Chapter 4 deals with two-dimensional potential flows. Apart from some fundamental flows, the flow around some two-dimensional bodies such as cylinders, ellipses, and airfoils is covered. Chapter 5 treats three-dimensional potential flows including the flow around submerged bodies such as spheres. Finally Chap. 6 deals with surface waves on liquids. This chapter includes traveling waves, standing waves, and waves at the interface of two fluids.

## Governing Equations and Boundary Conditions

Since the fluid is assumed to be incompressible, the equation of mass conservation is Eq. (1.3c). The equations of momentum conservation for an inviscid fluid are the Euler equations, which are expressed by Eqs. (1.9c). That is, the equations governing the velocity and pressure fields for an ideal fluid are

$$\mathbf{\nabla} \cdot \mathbf{u} = 0 \tag{II.1}$$

$$\frac{\partial \mathbf{u}}{\partial t} + (\mathbf{u} \cdot \mathbf{\nabla})\mathbf{u} = -\frac{1}{\rho}\mathbf{\nabla}p + \mathbf{f} \tag{II.2}$$

Equations (II.1) and (II.2) are sufficient to establish the velocity and the pressure in the flow independent of any temperature distribution which may exist. It was pointed out in Chap. 1 that compressibility is the fluid property which couples the equations of thermodynamics to those of dynamics so that in the study of ideal fluids the equations of thermodynamics need not be solved concurrently with the equations of motion. The study of ideal-fluid flows is frequently referred to as *hydrodynamics*, and Eqs. (II.1) and (II.2) are frequently called the *equations of hydrodynamics*.

Within macroscopic length scales, the proper boundary condition to be satisfied by the velocity is the no-slip boundary condition which is expressed by Eq. (1.14). It is not possible to satisfy this boundary condition with the Euler equations. The reason lies in the fact that the Euler equations are one order lower than the Navier-Stokes equations because the viscous terms are absent in the former equations. Thus the true boundary condition must be relaxed somehow under the approximation of negligible viscous effects. Since it is primarily viscous effects which prohibit a fluid from slipping along a solid boundary, the condition of no tangential slip at boundaries is relaxed. That is, the condition of no normal velocity at a solid boundary is retained but the condition of no tangential velocity is dropped. Thus the boundary condition which should be used with the Euler equations is

$$\mathbf{u} \cdot \mathbf{n} = \mathbf{U} \cdot \mathbf{n} \quad \text{on solid boundaries} \tag{II.3}$$

where $\mathbf{n}$ is the unit normal to the surface of the body and $\mathbf{U}$ is the velocity vector of the body. Comparison of Eq. (II.3) with Eq. (1.14) shows that the former constitutes one component of the true boundary condition and the two tangential components are unspecified. Physically, this means that the condition of no slip on a solid boundary has become the condition that the surface of the body must be a streamline. Any boundary condition which is to be satisfied far from the body, such as the flow becoming uniform, is unaffected by the inviscid approximation.

## Potential Flows

If the flow of an ideal fluid about a body originates in an irrotational flow, such as a uniform flow, for example, then Kelvin's theorem [Eq. (3.1)] guarantees that the flow will remain irrotational even near the body. That is, the vorticity vector **ω** will be zero everywhere in the fluid. Then, since $\nabla \times \nabla\phi = 0$ for any scalar function $\phi$, the condition of irrotationality will be satisfied identically by choosing

$$\mathbf{u} = \nabla\phi \tag{II.4}$$

The function $\phi$ is called the *velocity potential*, and flow fields which are irrotational, and so can be represented in the form of Eq. (II.4), are frequently referred to as *potential flows*. In order to find the equation which the velocity potential $\phi$ satisfies, the expression for **u** given by Eq. (II.4) is substituted into the continuity equation (II.1) to give

$$\nabla^2\phi = 0 \tag{II.5}$$

Thus by solving Eq. (II.5) and utilizing Eq. (II.4), the velocity field may be established without directly using the equations of motion [Eqs. (II.2)]. This is so because the condition of irrotationality has been used, and this condition is justified by Kelvin's theorem, which uses Eqs. (II.2) in its proof. However, the equations of motion must be used directly to obtain the pressure distribution. Solving Eq. (II.5) for the velocity potential $\phi$ determines the velocity distribution only, and in order to determine the pressure, use must be made of the equations of dynamics. Rather than use Eqs. (II.2), their integrated form, that is the Bernoulli equation, will be used. Using Eq. (3.2c), the pressure may be determined from the following relation:

$$\frac{\partial\phi}{\partial t} + \frac{p}{\rho} + \tfrac{1}{2}\nabla\phi \cdot \nabla\phi - G = F(t) \tag{II.6}$$

Having determined the velocity potential $\phi$, this becomes a simple algebraic equation for the pressure.

From the foregoing, it is evident that a simpler form of the governing equations exists for potential flows. Rather than solving Eqs. (II.1) and (II.2) directly, Eq. (II.5), together with the appropriate boundary conditions, may be solved to yield the velocity potential and hence the velocity field. Having done this, Eq. (II.6) may be used to establish the pressure field. This formulation has certain simplifying features. First, it will be noticed that the differential equation to be solved, given by Eq. (II.5), is linear, whereas Eq. (II.2) is nonlinear. Of course, the nonlinearity cannot be completely circumvented, and indeed it appears in the term $\nabla\phi \cdot \nabla\phi$ in the Bernoulli equation. However, in this equation it poses no difficulty in the analysis. One of the most useful properties of linear differential equations is that different solutions may be superimposed to yield other solutions. This property will be used extensively in the following chapters.

# CHAPTER

# 4

# TWO-DIMENSIONAL POTENTIAL FLOWS

It was pointed out in the introduction to Part II that potential flows may be analyzed in a much simpler way than general fluid flows. Within the category of potential flows, the two-dimensional subset lends itself to even greater simplification. It will be shown in this chapter that the simplification is so great that solutions to Eqs. (II.5) and (II.6) may be obtained without actually solving any differential equations. This is achieved through use of the powerful tool of complex variable theory.

The chapter begins by introducing the stream function, which together with the velocity potential, leads to the definition of a complex potential. Through this complex potential, some elementary solutions corresponding to sources, sinks, and vortices are examined. The superposition of such elementary solutions then leads to the solution for the flow around a circular cylinder. The method of conformal transformations is then introduced, and the Joukowski transformation is used to establish the solutions for the flow around ellipses and airfoils. The Schwarz-Christoffel transformation is then introduced and used to study the flow in regions involving sharp corners. Included in this chapter are examples of free-surface configurations.

## 4.1 STREAM FUNCTION

The velocity potential $\phi$ was defined in such a way that it automatically satisfied the condition of irrotationality. The continuity equation then showed that $\phi$ had to be a solution of Laplace's equation. A second function may be defined by a complementary procedure for two-dimensional incompressible fluid flows. That is, a function may be defined in such a way that it automatically satisfies the continuity equation, and the equation which it must satisfy will be determined by the condition of irrotationality.

The continuity equation, in cartesian coordinates, for the flow field under consideration is

$$\frac{\partial u}{\partial x} + \frac{\partial v}{\partial y} = 0$$

Now introduce a function $\psi$ which is defined as follows:

$$u = \frac{\partial \psi}{\partial y} \tag{4.1a}$$

$$v = -\frac{\partial \psi}{\partial x} \tag{4.1b}$$

With this definition, the continuity equation is satisfied identically for all functions $\psi$. The function $\psi$ is called the *stream function*, and by virtue of its definition it is valid for all two-dimensional flows, both rotational and irrotational.

The equation which the stream function $\psi$ must satisfy is obtained from the condition of irrotationality. Denoting the components of the vorticity vector $\boldsymbol{\omega}$ by $(\xi, \eta, \zeta)$, it is first observed that, in two dimensions, the only nonzero component of the vorticity vector is $\zeta$, the component which is perpendicular to the plane of the flow. Secondly, it is noted that $\zeta = \partial v/\partial x - \partial u/\partial y$. Thus, the condition of irrotationality is

$$\frac{\partial v}{\partial x} - \frac{\partial u}{\partial y} = 0$$

Substituting for $u$ and $v$ from Eqs. (4.1) shows that $\psi$ must satisfy the following equation:

$$\frac{\partial^2 \psi}{\partial x^2} + \frac{\partial^2 \psi}{\partial y^2} = 0 \tag{4.2}$$

That is, the stream function $\psi$, like the velocity potential $\phi$, must satisfy Laplace's equation. The stream function $\psi$ has some useful properties which will now be derived.

The flow lines which correspond to $\psi = $ constant are the streamlines of the flow field. To show this, it is noted that $\psi$ is a function of both $x$ and $y$ in general so that the total variation in $\psi$ associated with a change in $x$ and a change in $y$ may be calculated from the expression

$$d\psi = \frac{\partial \psi}{\partial x}\, dx + \frac{\partial \psi}{\partial y}\, dy$$

$$= -v\, dx + u\, dy$$

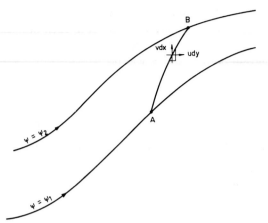

**FIGURE 4.1**
Two streamlines showing the components of the volumetric flow rate across
an element of control surface joining the
streamlines.

where Eqs. (4.1) have been used. Then the equation of the line $\psi = $ constant
will be

$$0 = -v\,dx + u\,dy$$

or
$$\left(\frac{dy}{dx}\right)_\psi = \frac{v}{u}$$

where the subscript denotes that this expression for $dy/dx$ is valid for $\psi$ held
constant. But it was shown in Chap. 2 that this is precisely the equation of the
streamlines in the $xy$ plane. Hence the lines corresponding to $\psi = $ constant are
the streamlines, and each value of the constant defines a different streamline. It
is this property of the function $\psi$ which justifies the name *stream function*.

Another property of the stream function $\psi$ is that the difference of its
values between two streamlines gives the volume of fluid which is flowing
between these two streamlines. To show this, consider two streamlines corresponding to $\psi = \psi_1$ and $\psi = \psi_2$ as shown in Fig. 4.1. A control surface $AB$ of
arbitrary shape but positive slope is shown joining these two streamlines, and an
element of this surface shows the positive volumetric flow rates crossing it in the
$x$ and $y$ directions per unit depth perpendicular to the flow field. Then the total
volume of fluid flowing between the streamlines per unit time per unit depth of
flow field will be

$$Q = \int_A^B u\,dy - \int_A^B v\,dx$$

But it was observed earlier that $d\psi = -v\,dx + u\,dy$, so that, integrating this
expression between the two points $A$ and $B$, it follows that

$$\psi_2 - \psi_1 = -\int_A^B v\,dx + \int_A^B u\,dy$$

Comparing these two expressions confirms that $\psi_2 - \psi_1 = Q$.

Finally, it should be noted that the streamlines $\psi$ = constant and the lines $\phi$ = constant, which are called *equipotential lines*, are orthogonal to each other. This may be shown by noting that if $\phi$ depends upon both $x$ and $y$, the total change in $\phi$ associated with changes in both $x$ and $y$ will be

$$d\phi = \frac{\partial \phi}{\partial x}\, dx + \frac{\partial \phi}{\partial y}\, dy$$

$$= u\, dx + v\, dy$$

where Eq. (II.4) has been used. Then the lines corresponding to $\phi$ = constant will be defined by

$$0 = u\, dx + v\, dy$$

or

$$\left( \frac{dy}{dx} \right)_\phi = -\frac{u}{v}$$

That is,

$$\left( \frac{dy}{dx} \right)_\phi = -\frac{1}{(dy/dx)_\psi}$$

In words, the slope of the lines $\phi$ = constant is the negative reciprocal of the slope of the lines $\psi$ = constant, so that these sets of lines must be orthogonal. This property of the streamlines and the equipotential lines is the basis of a numerical procedure for solving two-dimensional potential-flow problems. The method is referred to as the *flow net*.

## 4.2 COMPLEX POTENTIAL AND COMPLEX VELOCITY

The velocity components $u$ and $v$ may be expressed in terms of either the velocity potential or the stream function. From Eqs. (II.4) and (4.1), these expressions are

$$u = \frac{\partial \phi}{\partial x} = \frac{\partial \psi}{\partial y}$$

$$v = \frac{\partial \phi}{\partial y} = -\frac{\partial \psi}{\partial x}$$

That is, the functions $\phi$ and $\psi$ are related by the expressions

$$\frac{\partial \phi}{\partial x} = \frac{\partial \psi}{\partial y}$$

$$\frac{\partial \phi}{\partial y} = -\frac{\partial \psi}{\partial x}$$

But these will be recognized as the Cauchy-Riemann equations for the functions $\phi(x, y)$ and $\psi(x, y)$. Then consider the *complex potential F(z)*, which is defined as follows:

$$F(z) = \phi(x, y) + i\psi(x, y) \tag{4.3}$$

where $z = x + iy$. Now if $F(z)$ is an analytic function, it follows that $\phi$ and $\psi$ will automatically satisfy the Cauchy-Riemann equations. That is, for every analytic function $F(z)$ the real part is automatically a valid velocity potential and the imaginary part is a valid stream function.

The foregoing result suggests a very simple way of establishing solutions to the equations of two-dimensional potential flows. By equating the real part of a given analytic function to $\phi$ and the imaginary part to $\psi$, the theory of complex variables guarantees that $\nabla^2\phi = 0$ and $\nabla^2\psi = 0$ as required. The flow field corresponding to that analytic function may be determined by studying the streamlines $\psi = $ constant. The corresponding velocity components may be calculated from Eqs. (II.4) or (4.1), and the pressure may be obtained using Eq. (II.6). This approach has the disadvantage of being inverse in the sense that a problem is first solved and then examined to see what the physical problem was in the first place. However, for teaching purposes this is of no consequence. Another disadvantage is that the method cannot be generalized to three-dimensional potential flows. On the other hand, this approach avails itself of the powerful results of complex variable theory and avoids the difficulties of solving partial differential equations. For these reasons the complex-potential approach will be used in this chapter.

Another quantity of prime interest, apart from the complex potential $F(z)$, is the derivative of $F(z)$ with respect to $z$. Since $F(z)$ is supposed to be analytic, $dF/dz$ will be a point function whose value is independent of the direction in which it is calculated. Then, denoting this derivative by $W$, its value will be given by

$$W(z) = \frac{dF}{dz} = \frac{\partial F}{\partial x}$$

$$= \frac{\partial \phi}{\partial x} + i\frac{\partial \psi}{\partial x}$$

that is,

$$W(z) = \frac{dF}{dz} = u - iv \tag{4.4}$$

where use has been made of Eqs. (4.3), (II.4), and (4.1b). In view of this result the quantity $W(z)$ is called the *complex velocity*, although its imaginary part is $-iv$. Equation (4.4) offers a convenient alternative to Eqs. (II.4) and (4.1) for finding the velocity components corresponding to a given complex potential.

A useful property of the complex velocity is that, when multiplied by its own complex conjugate, it gives the scalar product of the velocity vector with itself. To show this, consider $W(z)$ and its complex conjugate $\overline{W}(z)$. Then

$$W\overline{W} = (u - iv)(u + iv)$$

$$W\overline{W} = u^2 + v^2 \tag{4.5}$$

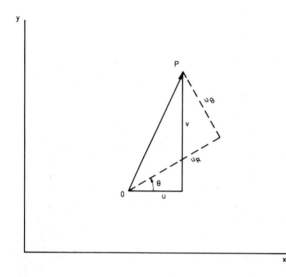

**FIGURE 4.2**
Decomposition of a velocity vector
$OP$ into its cartesian components
$(u, v)$ and its cylindrical components
$(u_R, u_\theta)$.

The significance of this result is that the quantity $\mathbf{u} \cdot \mathbf{u} = \nabla\phi \cdot \nabla\phi = u^2 + v^2$ appears in the Bernoulli equation.

Frequently it is advantageous to work in cylindrical coordinates rather than cartesian coordinates. An expression for the complex velocity may be readily obtained in cylindrical coordinates by converting the cartesian components of the velocity vector $(u, v)$ to cylindrical components $(u_R, u_\theta)$. Figure 4.2 shows a velocity vector $OP$ decomposed into its cartesian components (shown solid) and also its cylindrical components (shown dotted). From this figure each of the cartesian velocity components may be expressed in terms of the two cylindrical components as follows:

$$u = u_R \cos\theta + u_\theta \cos\left(\frac{\pi}{2} - \theta\right) = u_R \cos\theta - u_\theta \sin\theta$$

$$v = u_R \sin\theta + u_\theta \sin\left(\frac{\pi}{2} - \theta\right) = u_R \sin\theta + u_\theta \cos\theta$$

Substituting these expressions into Eq. (4.4) gives the expression for the complex velocity $W$ in terms of $u_R$ and $u_\theta$.

$$W = (u_R \cos\theta - u_\theta \sin\theta) - i(u_R \sin\theta + u_\theta \cos\theta)$$
$$= u_R(\cos\theta - i \sin\theta) - iu_\theta(\cos\theta - i \sin\theta)$$

that is, $\qquad W = (u_R - iu_\theta)e^{-i\theta} \qquad\qquad (4.6)$

The foregoing results [Eqs. (4.3) to (4.6)] are sufficient to establish the flow fields, which are represented by simple analytic functions.

## 4.3 UNIFORM FLOWS

The simplest analytic function of $z$ is proportional to $z$ itself, and the corresponding flow fields are uniform flows.

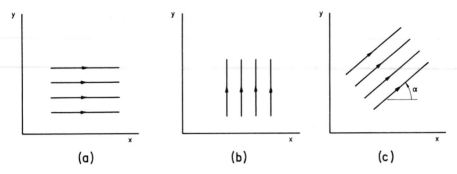

**FIGURE 4.3**
Uniform flow in ($a$) the $x$ direction, ($b$) the $y$ direction, and ($c$) an angle $\alpha$ to the $x$ direction.

First, consider $F(z)$ to be proportional to $z$ where the constant of proportionality is real. That is,

$$F(z) = cz$$

where $c$ is real. Then, from Eq. (4.4),

$$W(z) = u - iv = c$$

Then, by equating real and imaginary parts of this equation, the velocity components corresponding to this complex potential are

$$u = c$$
$$v = 0$$

But this is just the velocity field for a *uniform rectilinear flow* as shown in Fig. 4.3$a$. Thus the complex potential for such a flow whose velocity magnitude is $U$ in the positive $x$ direction will be

$$F(z) = Uz \tag{4.7a}$$

Next consider the complex potential to be proportional to $z$ with an imaginary constant of proportionality. Then

$$F(z) = -icz$$

where $c$ is real. The minus sign has been included to make the velocity component positive when $c$ is positive. For this complex potential

$$W(z) = u - iv = -ic$$

so that the velocity components are

$$u = 0$$
$$v = c$$

This is a uniform vertical flow as shown in Fig. 4.3$b$. Then the complex potential for such a flow whose velocity magnitude is $V$ in the positive $y$ direction will be

$$F(z) = -iVz \tag{4.7b}$$

Finally, consider a complex constant of proportionality so that

$$F(z) = ce^{-i\alpha}z$$

where $c$ and $\alpha$ are real. For this complex potential

$$W(z) = u - iv = c \cos \alpha - ic \sin \alpha$$

Hence the velocity components of the flow field are

$$u = c \cos \alpha$$

$$v = c \sin \alpha$$

This corresponds to a uniform flow inclined at an angle $\alpha$ to the $x$ axis as shown in Fig. 4.3c. Hence the complex potential for such a flow whose velocity magnitude is $V$ will be

$$F(z) = Ve^{-i\alpha}z \tag{4.7c}$$

This last result, of course, contains the two previous results as special cases corresponding to $\alpha = 0$ and $\alpha = \pi/2$.

## 4.4 SOURCE, SINK, AND VORTEX FLOWS

Complex potentials which correspond to the flow fields generated by sources, sinks, and vortices are obtained by considering $F(z)$ to be proportional to log $z$. When considering log $z$, we consider the principal part of this multivalued function corresponding to $0 < \theta < 2\pi$.

Consider, first, the constant of proportionality to be real. Then

$$F(z) = c \log z$$

$$= c \log Re^{i\theta}$$

$$= c \log R + ic\theta$$

Hence, from Eq. (4.3),

$$\phi = c \log R$$

$$\psi = c\theta$$

That is, the equipotential lines are the circles $R =$ constant and the streamlines are the radial lines $\theta =$ constant. This gives a flow field as shown in Fig. 4.4a in which the streamlines are shown solid and the direction of the flow is shown for $c > 0$. The direction of the flow is readily confirmed by evaluating the velocity components. In view of the geometry of the flow, cylindrical coordinates are preferred, so that

$$W(z) = \frac{c}{z} = \frac{c}{R}e^{-i\theta}$$

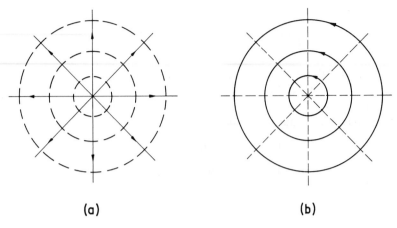

**(a)** **(b)**

**FIGURE 4.4**
Streamlines (shown solid) and equipotential lines (dotted) for (a) source flow and (b) vortex flow in the positive sense.

Comparison with Eq. (4.6) shows that the velocity components are

$$u_R = \frac{c}{R}$$

$$u_\theta = 0$$

which confirms the directions indicated in Fig. 4.4a for $c > 0$.

The flow field indicated in Fig. 4.4a is called a *source*. The velocity is purely radial and its magnitude decreases as the flow leaves the origin. In fact, the origin is a singular point corresponding to infinite velocity, and as the fluid flows radially outwards, its velocity is decreased in such a way that the volume of fluid crossing each circle is constant, as required by the continuity equation.

Sources are characterized by their strength, denoted by $m$, which is defined as the volume of fluid leaving the source per unit time per unit depth of the flow field. From this definition it follows that

$$m = \int_0^{2\pi} u_R R\, d\theta$$

$$= \int_0^{2\pi} c\, d\theta = 2\pi c$$

Here, the result $u_R = c/R$ has been used. Then $c$ may be replaced by $m/2\pi$, giving the following complex potential for a source of strength $m$:

$$F(z) = \frac{m}{2\pi} \log z$$

The source corresponding to this complex potential is located at the origin, the location of the singularity. Then the complex potential for a source of strength

$m$ located at the point $z = z_0$ will be

$$F(z) = \frac{m}{2\pi} \log(z - z_0) \tag{4.8}$$

Clearly, the complex potential for a sink, which is a negative source, is obtained by replacing $m$ by $-m$ in Eq. (4.8).

Now consider the constant of proportionality in the logarithmic complex potential to be imaginary. That is, consider

$$F(z) = -ic \log z$$

where $c$ is real and the minus is included to give a positive vortex. Then, using cylindrical coordinates,

$$F(z) = -ic \log Re^{i\theta}$$
$$= c\theta - ic \log R$$

Then, from Eq. (4.3), the velocity potential and the stream function are

$$\phi = c\theta$$
$$\psi = -c \log R$$

That is, the equipotential lines are the radial lines $\theta = $ constant and the streamlines are the circles $R = $ constant as shown in Fig. 4.4$b$. The velocity components may be evaluated by use of the complex velocity.

$$W(z) = -i\frac{c}{z} = -i\frac{c}{R}e^{-i\theta}$$

Comparison with Eq. (4.6) shows that the velocity components are

$$u_R = 0$$
$$u_\theta = \frac{c}{R}$$

Hence the direction of the flow is positive (counterclockwise) for $c > 0$, and the resulting flow field is called a *vortex*.

A vortex is characterized by its strength, which may be measured by the circulation $\Gamma$ which is associated with it. From Eq. (2.3), the circulation $\Gamma$ which is associated with the singularity at the origin is

$$\Gamma = \oint \mathbf{u} \cdot d\mathbf{l}$$
$$= \int_0^{2\pi} u_\theta \mathbf{R}\, d\theta$$
$$= \int_0^{2\pi} c\, d\theta = 2\pi c$$

Here, the result $u_\theta = c/R$ has been used. Then $c$ may be replaced by $\Gamma/2\pi$, giving the following complex potential for a positive (counterclockwise) vortex of strength $\Gamma$.

$$F(z) = -i\frac{\Gamma}{2\pi} \log z$$

The singularity in this expression is located at $z = 0$. That is, the line vortex is located at $z = 0$. Then the complex potential for a positive vortex located at $z = z_0$ will be

$$F(z) = -i\frac{\Gamma}{2\pi} \log(z - z_0) \tag{4.9}$$

The complex potential for a negative vortex would be obtained by replacing $\Gamma$ by $-\Gamma$ in Eq. (4.9). Note, however, that the negative coefficient is associated with the positive vortex.

The flow field represented by Eq. (4.9), which is shown in Fig. 4.4b for $z_0 = 0$, corresponds to a so-called *free vortex*. That is, for any closed contour which does not include the singularity, the circulation will be zero and the flow will be irrotational. All the circulation and vorticity associated with this type of vortex is concentrated at the singularity. This is in contrast with the solid-body rotation vortex which was mentioned in Chap. 2.

The principal application of the source, the sink, and the vortex is in the superposition with other flows to yield more practical flow fields.

## 4.5 FLOW IN A SECTOR

The flows in sharp bends or sectors are represented by complex potentials which are proportional to $z^n$, where $n \geq 1$. A special case of such complex potentials would be $n = 1$, which represents a uniform rectilinear flow. Then, in order that this special case will reduce to Eq. (4.7a), consider the complex potentials

$$F(z) = Uz^n$$

Substituting $z = Re^{i\theta}$ and separating the real and imaginary parts of this function gives

$$F(z) = UR^n \cos n\theta + iUR^n \sin n\theta$$

Then the velocity potential and the stream function are

$$\phi = UR^n \cos n\theta$$

$$\psi = UR^n \sin n\theta$$

From this it is evident that when $\theta = 0$ and when $\theta = \pi/n$, the stream function $\psi$ is zero. That is, the streamline $\psi = 0$ corresponds to the radial lines $\theta = 0$ and

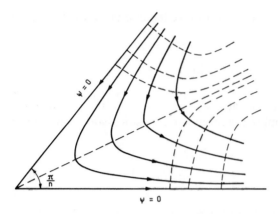

**FIGURE 4.5**
Streamlines (shown solid) and equipotential lines (shown dotted) for flow in a sector.

$\theta = \pi/n$. Between these two lines, the streamlines are defined by $R^n \sin n\theta =$ constant. This gives the flow field shown in Fig. 4.5. The direction of the flow along the streamlines may be determined from the complex velocity as follows:

$$W(z) = nUz^{n-1} = nUR^{n-1}e^{i(n-1)\theta}$$

$$= (nUR^{n-1} \cos n\theta + inUR^{n-1} \sin n\theta)e^{-i\theta}$$

Thus, by comparison with Eq. (4.6), the velocity components are

$$u_R = nUR^{n-1} \cos n\theta$$

$$u_\theta = -nUR^{n-1} \sin n\theta$$

Then, for $0 < \theta < (\pi/2n)$, $u_R$ is positive while $u_\theta$ is negative and for $(\pi/2n) < \theta < (\pi/n)$, $u_R$ is negative and $u_\theta$ remains negative. This establishes the flow directions as indicated in Fig. 4.5.

From the foregoing, the complex potential for the flow in a corner or sector of angle $\pi/n$ is

$$F(z) = Uz^n \qquad (4.10)$$

For $n = 1$, Eq. (4.10) gives the complex potential for a uniform rectilinear flow, and for $n = 2$, it gives the complex potential for the flow in a right-angled corner.

## 4.6  FLOW AROUND A SHARP EDGE

The complex potential for the flow around a sharp edge, such as the edge of a flat plate, is obtained from the function $z^{1/2}$. Then consider the complex

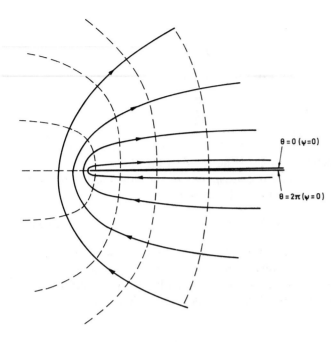

**FIGURE 4.6**
Streamlines (shown solid) and equipotential lines (shown dotted) for flow around a sharp edge.

potential

$$F(z) = cz^{1/2}$$

where $c$ is real and $0 < \theta < 2\pi$. Then, in cylindrical coordinates,

$$F(z) = cR^{1/2}e^{i\theta/2}$$

so that the velocity potential and stream function are

$$\phi = cR^{1/2} \cos \frac{\theta}{2}$$

$$\psi = cR^{1/2} \sin \frac{\theta}{2}$$

Thus the lines $\theta = 0$ and $\theta = 2\pi$ correspond to the streamline $\psi = 0$. The other streamlines are defined by the equation $R^{1/2} \sin \theta/2 = $ constant, which yields the flow pattern shown in Fig. 4.6. The direction of the flow is obtained from the complex velocity as follows:

$$W(z) = \frac{c}{2z^{1/2}} = \frac{c}{2R^{1/2}}e^{-i\theta/2}$$

$$= \frac{c}{2R^{1/2}}\left(\cos \frac{\theta}{2} + i \sin \frac{\theta}{2}\right)e^{-i\theta}$$

Hence the velocity components are

$$u_R = \frac{c}{2R^{1/2}} \cos \frac{\theta}{2}$$

$$u_\theta = -\frac{c}{2R^{1/2}} \sin \frac{\theta}{2}$$

Then, for $0 < \theta < \pi$, $u_R > 0$ and $u_\theta < 0$. Also, for $\pi < \theta < 2\pi$, $u_R < 0$ and $u_\theta < 0$. This gives the direction of flow as indicated in Fig. 4.6.

The flow field shown in Fig. 4.6 corresponds to the flow around a sharp edge, and so the complex potential for such a flow is

$$F(z) = cz^{1/2} \tag{4.11}$$

An important feature of this result is that the corner itself is a singular point at which the velocity components become infinite. Since both $u_R$ and $u_\theta$ vary as the inverse of $R^{1/2}$, it follows that the velocity is singular as the square root of the distance from the edge. This result will be discussed in Sec. 4.15.

## 4.7   FLOW DUE TO A DOUBLET

The function $1/z$ has a singularity at $z = 0$, and in the context of complex potentials, this singularity is called a *doublet*. The quickest way of establishing the flow field which corresponds to the complex potentials which are proportional to $1/z$ would be to follow the methods which were used in the previous sections. However, it turns out that the doublet may be considered to be the coalescing of a source and a sink, and the required complex potential may be obtained through a limiting procedure which uses this fact. This interpretation leads to a better physical understanding of the doublet, and for this reason it will be followed here before studying the flow field.

Referring to the geometry indicated in Fig. 4.7a, consider a source of strength $m$ and a sink of strength $m$, each of which is located on the real axis a small distance $\varepsilon$ from the origin. The complex potential for such a configuration is, from Eq. (4.8),

$$F(z) = \frac{m}{2\pi} \log(z + \varepsilon) - \frac{m}{2\pi} \log(z - \varepsilon)$$

$$= \frac{m}{2\pi} \log\left(\frac{z + \varepsilon}{z - \varepsilon}\right)$$

$$= \frac{m}{2\pi} \log\left(\frac{1 + \varepsilon/z}{1 - \varepsilon/z}\right)$$

If the nondimensional distance $\varepsilon / |z|$ is considered to be small, the argument of

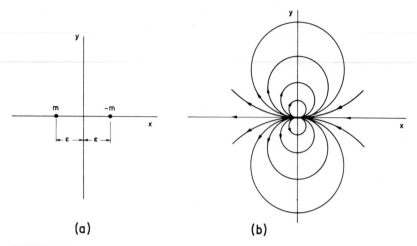

**FIGURE 4.7**
(a) Superposition of source and sink which leads to (b) streamline pattern for limit $\varepsilon \to 0$ with $m\varepsilon$ = constant.

the logarithm may be expanded as follows:

$$F(z) = \frac{m}{2\pi} \log \left\{ \left(1 + \frac{\varepsilon}{z}\right)\left[1 + \frac{\varepsilon}{z} + O\left(\frac{\varepsilon^2}{z^2}\right)\right] \right\}$$

$$= \frac{m}{2\pi} \log\left[1 + 2\frac{\varepsilon}{z} + O\left(\frac{\varepsilon^2}{z^2}\right)\right]$$

where the designation $O(\varepsilon^2/z^2)$ means terms of order $\varepsilon^2/z^2$ or smaller. The logarithm is now in the form $\log(1 + \gamma)$, where $\gamma \ll 1$, so that the equivalent expansion $\gamma + O(\gamma^2)$ may be used. Then

$$F(z) = \frac{m}{2\pi}\left[2\frac{\varepsilon}{z} + O\left(\frac{\varepsilon^2}{z^2}\right)\right]$$

It is now proposed to let $\varepsilon \to 0$ and $m \to \infty$ in such a way that $\lim_{\varepsilon \to 0}(m\varepsilon) = \pi\mu$, where $\mu$ is a constant. Then the complex potential becomes

$$F(z) = \frac{\mu}{z}$$

Thus the complex potential $\mu/z$ may be thought of as being the equivalent of the superposition of a very strong source and a very strong sink which are very close together.

In order to establish the flow field which the above complex potential represents, the stream function will be established as follows:

$$F(z) = \frac{\mu}{x + iy}$$

$$= \mu \frac{x - iy}{x^2 + y^2}$$

$$\therefore \quad \psi = -\mu \frac{y}{x^2 + y^2}$$

Thus the equation of the streamlines $\psi = $ constant is

$$x^2 + y^2 + \frac{\mu}{\psi} y = 0$$

or
$$x^2 + \left( y + \frac{\mu}{2\psi} \right)^2 = \left( \frac{\mu}{2\psi} \right)^2$$

But this is the equation of a circle of radius $\mu/(2\psi)$ whose center is located at $y = -\mu/(2\psi)$. This gives the streamline pattern shown in Fig. 4.7b. Although the direction of the flow along the streamlines may be deduced from the source and sink interpretation, it will be checked by evaluating the velocity components. The complex velocity for this complex potential is

$$W(z) = -\frac{\mu}{z^2} = -\frac{\mu}{R^2} e^{-i2\theta}$$

$$= -\frac{\mu}{R^2} (\cos \theta - i \sin \theta) e^{-i\theta}$$

Hence the velocity components are

$$u_R = -\frac{\mu}{R^2} \cos \theta$$

$$u_\theta = -\frac{\mu}{R^2} \sin \theta$$

These expressions for $u_R$ and $u_\theta$ confirm the flow directions indicated in Fig. 4.7b.

The flow field illustrated in Fig. 4.7b is called a *doublet flow*, and the singularity which is at the heart of the flow field is called a *doublet*. Then the complex potential for a doublet of strength $\mu$ which is located at $z = z_0$ is

$$F(z) = \frac{\mu}{z - z_0} \qquad (4.12)$$

The principal use of the doublet is in the superposition of fundamental flow fields to generate more complex and more practical flow fields. An application of this will be illustrated in the next section.

## 4.8  CIRCULAR CYLINDER WITHOUT CIRCULATION

The fundamental solutions to the foregoing flow situations provide the basis for more general solutions through the principle of superposition. Superposition is valid here, since the governing equation, for either the velocity potential or the stream function, is linear. The first example of superposition of fundamental solutions will be the flow around a circular cylinder.

Consider the superposition of a uniform rectilinear flow and a doublet at the origin. Then, from Eqs. (4.7a) and (4.12), the complex potential for the resulting flow field will be

$$F(z) = Uz + \frac{\mu}{z}$$

It will now be shown that for a certain choice of the doublet strength the circle $R = a$ becomes a streamline. On the circle $R = a$, the value of $z$ is $ae^{i\theta}$, so that the complex potential on this circle is

$$F(z) = Uae^{i\theta} + \frac{\mu}{a}e^{-i\theta}$$

$$= \left( Ua + \frac{\mu}{a} \right) \cos\theta + i\left( Ua - \frac{\mu}{a} \right) \sin\theta$$

Thus the value of the stream function on the circle $R = a$ is

$$\psi = \left( Ua - \frac{\mu}{a} \right) \sin\theta$$

For general values of $\mu$, $\psi$ is clearly variable, but if we choose the strength of the doublet to be $\mu = Ua^2$, then $\psi = 0$ on $R = a$. The flow pattern for this doublet strength is shown in Fig. 4.8a. The flow field due to the doublet encounters that due to the uniform flow and is bent downstream. For clarity, the flow due to the doublet is shown dotted in Fig. 4.8a. It may be seen that the doublet flow is entirely contained within the circle $R = a$, while the uniform flow is deflected by the doublet in such a way that it is entirely outside the circle $R = a$. The circle $R = a$ itself is common to the two flow fields.

Under these conditions, a thin metal cylinder of radius $a$ could be slid into the flow field perpendicular to the uniform flow so that it coincides with the streamline on $R = a$. Clearly the flow due to the doublet and that due to the free stream would be undisturbed by such a cylindrical shell. Having done this, the flow due to the doublet could be removed and the outer flow would remain unchanged. Finally, the inside of the shell could be filled to yield a solid cylinder. That is, for $R \geq a$, the flow field due to the doublet of strength $Ua^2$ and the uniform rectilinear flow of magnitude $U$ give the same flow as that for a uniform flow of magnitude $U$ past a circular cylinder of radius $a$. The latter flow is shown in Fig. 4.8b. Then the complex potential for a uniform flow of

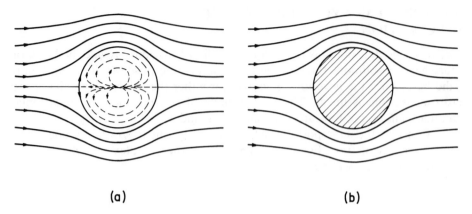

**(a)**                                   **(b)**

**FIGURE 4.8**
(a) Flow field represented by the complex potential $F(z) = U(z + a^2/z)$ and (b) flow around a
circular cylinder of radius $a$.

magnitude $U$ past a circular cylinder of radius $a$ is

$$F(z) = U\left(z + \frac{a^2}{z}\right) \tag{4.13}$$

This result is useful in its own right, but it will also be found useful in later
sections, through the technique of conformal transformations, to obtain additional solutions.

The solution given by Eq. (4.13) for the flow around a circular cylinder
predicts no hydrodynamic force acting on the cylinder. This statement will be
proved quantitatively in a later section, and in the meantime it will be proved
qualitatively. Referring to Fig. 4.8b, it can be seen that the flow is symmetric
about the $x$ axis. That is, for each point on the upper surface there is a
corresponding point on the lower surface, vertically below it, for which the
magnitude of the velocity is the same. Then, from the Bernoulli equation, the
magnitude of the pressure is the same at these two points. Hence, by integrating
$p\,dx$ around the surface of the cylinder, the lift force acting on the cylinder must
be zero. Similarly, owing to the symmetry of the flow about the $y$ axis, the drag
force acting on the cylinder is zero.

Although the foregoing result does not agree with our physical intuition,
the potential-flow solution for the circular cylinder, and indeed for other bodies,
is valuable. The absence of any hydrodynamic force on the cylinder is due to the
neglect of viscosity. It will be seen in Part III that viscous effects create a thin
boundary layer around the cylinder, and this boundary layer separates from the
surface at some point, creating a low-pressure wake. The resulting pressure
distribution creates a drag force. However, it will be pointed out that the viscous
boundary-layer solution is valid only in the thin boundary layer around the
cylinder, and the solution obtained from the boundary-layer equations must be

matched to that given by Eq. (4.13) at the edge of the boundary layer. That is, Eq. (4.13) gives a valid solution outside the thin boundary layer and upstream of the vicinity of the separation point. It also indicates the idealized flow situation which would be approached if viscous effects are minimized. For more stream-lined bodies, such as airfoils, the potential-flow solution is approached over the entire length of the body.

## 4.9 CIRCULAR CYLINDER WITH CIRCULATION

The flow field studied in the previous section not only was irrotational but it produced no circulation around the cylinder itself. It was found that there was no hydrodynamic force acting on the cylinder under these conditions. It will be shown in a later section that it is the circulation around a body which produces any lift force which acts on it. It is therefore of interest to study the flow around a circular cylinder which has a circulation around it.

It was established in a previous section that the streamlines for a vortex flow form concentric circles. Therefore, if a vortex was added at the origin to the flow around a circular cylinder, as described in the previous section, the fact that the circle $R = a$ was a streamline would be unchanged. Thus, from Eqs. (4.13) and (4.9), $z_0$ being zero in the latter, the complex potential for the flow around a circular cylinder with a negative bound vortex around it will be

$$F(z) = U\left(z + \frac{a^2}{z}\right) + \frac{i\Gamma}{2\pi} \log z + c$$

The negative vortex has been used, since it will turn out that this leads to a positive lift. A constant $c$ has been added to the complex potential for the following reason. For no circulation, it was found that not only was $\psi$ constant on $R = a$ but the value of the constant was zero. By adding the vortex, $\psi$ will no longer be zero on $R = a$, although it will have some other constant value. Since it is frequently useful to have the streamline on $R = a$ be $\psi = 0$, it is desirable to adjust things so that this condition is achieved. By adding a constant $c$ to the complex potential, we have the flexibility to choose $c$ in such a way that $\psi = $ constant becomes $\psi = 0$. It should be noted that this adjustment has no effect on the velocity and pressure distributions, since the velocity components are defined by derivatives of $\psi$, so that the absolute value of $\psi$ at any point is of no significance.

In order to evaluate the constant $c$, the value of the stream function on the circle $R = a$ will be computed. Then, putting $z = ae^{i\theta}$, the complex potential becomes

$$F(z) = U(ae^{i\theta} + ae^{-i\theta}) + \frac{i\Gamma}{2\pi} \log ae^{i\theta} + c$$

$$= 2Ua \cos \theta - \frac{\Gamma}{2\pi}\theta + \frac{i\Gamma}{2\pi} \log a + c$$

Hence on the circle $R = a$ the value of $\psi$ is indeed constant, and by choosing $c = -(i\Gamma/2\pi)\log a$, the value of this constant will be zero. With this value of $c$, the complex potential becomes

$$F(z) = U\left(z + \frac{a^2}{z}\right) + \frac{i\Gamma}{2\pi}\log\frac{z}{a} \tag{4.14}$$

which describes a uniform rectilinear flow of magnitude $U$ approaching a circular cylinder of radius $a$ which has a negative vortex of strength $\Gamma$ around it. As required, this result agrees with Eq. (4.13) when $\Gamma = 0$.

In order to visualize the flow field described by Eq. (4.14), the corresponding velocity components will be evaluated from the complex velocity.

$$\begin{aligned}
W(z) &= U\left(1 - \frac{a^2}{z^2}\right) + \frac{i\Gamma}{2\pi}\frac{1}{z} \\
&= U\left(1 - \frac{a^2}{R^2}e^{-i2\theta}\right) + \frac{i\Gamma}{2\pi R}e^{-i\theta} \\
&= \left[U\left(e^{i\theta} - \frac{a^2}{R^2}e^{-i\theta}\right) + \frac{i\Gamma}{2\pi R}\right]e^{-i\theta} \\
&= \left\{U\left(1 - \frac{a^2}{R^2}\right)\cos\theta + i\left[U\left(1 + \frac{a^2}{R^2}\right)\sin\theta + \frac{\Gamma}{2\pi R}\right]\right\}e^{-i\theta}
\end{aligned}$$

Hence, by comparison with Eq. (4.6), the velocity components are

$$u_R = U\left(1 - \frac{a^2}{R^2}\right)\cos\theta \tag{4.15a}$$

$$u_\theta = -U\left(1 + \frac{a^2}{R^2}\right)\sin\theta - \frac{\Gamma}{2\pi R} \tag{4.15b}$$

On the surface of the cylinder, where $R = a$, Eqs. (4.15) become

$$u_R = 0$$

$$u_\theta = -2U\sin\theta - \frac{\Gamma}{2\pi a}$$

The fact that $u_R = 0$ on $R = a$ is to be expected, since this is the boundary condition (II.3). A significant point in the flow field is a point where the velocity components all vanish—that is, a stagnation point. For this flow field the stagnation points are defined by

$$\sin\theta_s = -\frac{\Gamma}{4\pi Ua} \tag{4.16}$$

where $\theta_s$ is the value of $\theta$ corresponding to the stagnation point. For $\Gamma = 0$,

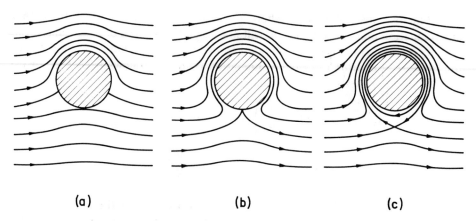

(a)                    (b)                    (c)

**FIGURE 4.9**
Flow of approach velocity $U$ around a circular cylinder of radius $a$ with negative bound circulation of magnitude $\Gamma$ for (a) $0 < \Gamma/(4\pi Ua) < 1$, (b) $\Gamma/(4\pi Ua) = 1$, and (c) $\Gamma/(4\pi Ua) > 1$.

$\sin \theta_s = 0$, so that $\theta_s = 0$ or $\pi$, which agrees with Fig. 4.8b for the circular cylinder without circulation. For nonzero circulation, the value of $\theta_s$ clearly depends upon the magnitude of the parameter $\Gamma/(4\pi Ua)$, and it is convenient to discuss Eq. (4.16) for different ranges of this parameter.

First, consider the range $0 < \Gamma/(4\pi Ua) < 1$. Here $\sin \theta_s < 0$, so that $\theta_s$ must lie in the third and fourth quadrants. There are two stagnation points, and clearly the one which was at $\theta = \pi$ is now located in the third quadrant while the one which was located at $\theta = 0$ is now located in the fourth quadrant. The two stagnation points will be symmetrically located about the $y$ axis in order that $\sin \theta_s = -$ constant may be satisfied. The resulting flow situation is shown in Fig. 4.9a.

Physically, the location of the stagnation points may be explained as follows: The flow due to the vortex and that due to the flow around the cylinder without circulation reinforce each other in the first and second quadrants. On the other hand, these two flow fields oppose each other in the third and fourth quadrants, so that at some point in each of these regions the net velocity is zero. Thus the effect of circulation around the cylinder is to make the front and rear stagnation points approach each other, and for a negative vortex they do so along the lower surface of the cylinder.

Consider next the case when the nondimensional circulation is unity, that is, when $\Gamma/(4\pi Ua) = 1$. Here $\sin \theta_s = -1$, so that $\theta_s = 3\pi/2$. The corresponding flow configuration is shown in Fig. 4.9b. The two stagnation points have been brought together by the action of the bound vortex such that they coincide to form a single stagnation point at the bottom of the cylindrical surface. It is evident that if the circulation is increased above this value, the single stagnation point cannot remain on the surface of the cylinder. It will move off into the fluid as either a single stagnation point or two stagnation points.

Finally, consider the case where $\Gamma/(4\pi U a) > 1$. Since it seems likely that any stagnation points there may be will not lie on the surface of the cylinder, the velocity components must be evaluated from Eqs. (4.15). Then if $R_s$ and $\theta_s$ are the cylindrical coordinates of the stagnation points, it follows from Eqs. (4.15) that $R_s$ and $\theta_s$ must satisfy the equations

$$U\left(1 - \frac{a^2}{R_s^2}\right)\cos\theta_s = 0$$

$$U\left(1 + \frac{a^2}{R_s^2}\right)\sin\theta_s = -\frac{\Gamma}{2\pi R_s}$$

Since it is assumed that the stagnation points are not on the surface of the cylinder, it follows that $R_s \neq a$, so that the first of these equations requires that $\theta_s = \pi/2$ or $3\pi/2$. For these values of $\theta_s$, the second of the above equations becomes

$$U\left(1 + \frac{a^2}{R_s^2}\right) = \pm\frac{\Gamma}{2\pi R_s}$$

where the minus sign corresponds to $\theta_s = \pi/2$ and the plus sign to $\theta_s = 3\pi/2$. Since $U > 0$, the left-hand side of the above equation is positive, and since $\Gamma > 0$, the minus sign must be rejected on the right-hand side. This might have been expected, since for $\Gamma/(4\pi U a) = 1$ the value of $\theta_s$ was $3\pi/2$, whereas the minus sign corresponds to $\theta_s = \pi/2$, which would require a large jump in $\theta_s$ for a small change in $\Gamma$. The equation for $R_s$ now becomes

$$U\left(1 + \frac{a^2}{R_s^2}\right) = \frac{\Gamma}{2\pi R_s}$$

or

$$R_s^2 - \frac{\Gamma}{2\pi U}R_s + a^2 = 0$$

hence

$$R_s = \frac{\Gamma}{4\pi U} \pm \sqrt{\left(\frac{\Gamma}{4\pi U}\right)^2 - a^2}$$

or

$$\frac{R_s}{a} = \frac{\Gamma}{4\pi U a}\left[1 \pm \sqrt{1 - \left(\frac{4\pi U a}{\Gamma}\right)^2}\right]$$

This result shows that as $4\pi U a/\Gamma \to 0$, $R_s \to \infty$ for the plus sign, but the corresponding limit is indeterminate for the minus sign. This difficulty may be overcome by expanding the square root for $4\pi U a/\Gamma \ll 1$ as follows:

$$\frac{R_s}{a} = \frac{\Gamma}{4\pi U a}\left\{1 \pm \left[1 - \frac{1}{2}\left(\frac{4\pi U a}{\Gamma}\right)^2 + \cdots\right]\right\}$$

where the dots indicate terms of order $(4\pi a/\Gamma)^4$ or smaller. In this form it is evident that as $4\pi Ua/\Gamma \to 0$, $R_s \to 0$ for the minus sign. Since this stagnation point would be inside the surface of the cylinder, the minus sign may be rejected, so that the coordinates of the stagnation point in the fluid outside the cylinder are

$$\theta_s = \frac{3\pi}{2}$$

$$\frac{R_s}{a} = \frac{\Gamma}{4\pi Ua}\left[1 + \sqrt{1 - \left(\frac{4\pi Ua}{\Gamma}\right)^2}\right]$$

This gives a single stagnation point below the surface of the cylinder. The corresponding flow configuration is shown in Fig. 4.9c, from which it will be seen that there is a portion of the fluid which perpetually encircles the cylinder.

The flow fields for the circular cylinder with circulation, as shown in Fig. 4.9, exhibit symmetry about the $y$ axis. Then, following the arguments used in the previous section, it may be concluded that there will be no drag force acting on the cylinder. However, the existence of the circulation around the cylinder has destroyed the symmetry about the $x$ axis; so there will be some force acting on the cylinder in the vertical direction. For the negative circulation shown the velocity on the top surface of the cylinder will be higher than that for no circulation, while the velocity on the bottom surface will be lower. Then, from Bernoulli's equation, the pressure on the top surface will be lower than that on the bottom surface, so that the vertical force acting on the cylinder will be upward. That is, a positive lift will exist. In order to determine the magnitude of this lift, a quantitative analysis must be performed, and this will be done in the next section.

The principal interest in the flow around a circular cylinder with circulation is in the study of airfoil theory. By use of conformal transformations the flow around certain airfoil shapes may be transformed into that of the flow around a circular cylinder with circulation.

## 4.10  BLASIUS' INTEGRAL LAWS

In the previous section it was argued that a lift force exists on a circular cylinder which has a circulation around it. However, the magnitude of the force can be established only by quantitative methods. The obvious way to evaluate the magnitude of this force is to establish the velocity components from the complex potential. Knowing the velocity components, the pressure distribution around the surface of the cylinder may be established by use of the Bernoulli equation. Integration of this pressure distribution will then yield the required force acting on the cylinder.

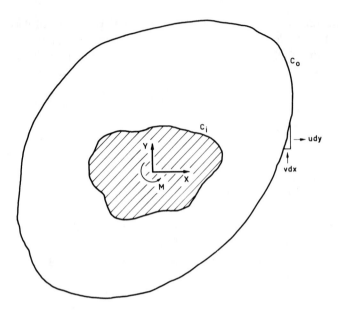

**FIGURE 4.10**
Arbitrarily shaped body enclosed by arbitrary control surface. $X$, $Y$, and $M$ are the drag, lift, and moment acting on the body, respectively.

The difficulty with the foregoing procedure is that it would have to be carried out for each pressure distribution and for each body under consideration. The Blasius laws provide a convenient alternative. It will be shown that if the complex potential for the flow around a body is known, then it is possible to evaluate the forces and the turning moment acting on the body by means of simple contour integrals. These contour integrals, in turn, may be readily evaluated by use of the residue theorem. The Blasius laws are actually two separate laws, one for forces and one for the hydrodynamic moment acting on the body.

In order to establish the forces acting on an arbitrarily shaped body in a flow field, consider such a body as shown in Fig. 4.10. A fixed control contour $C_0$ of arbitrary shape is drawn around the body whose surface is denoted by $C_i$. The forces acting through the center of gravity, as indicated by $X$ and $Y$, are the hydrodynamic forces acting *on the body* in the $x$ and $y$ directions, respectively. Then, for the fluid contained between the surfaces $C_0$ and $C_i$, it may be stated that the net external force acting on the positive $x$ direction must equal the net rate of increase of the $x$ component of the momentum. In Fig. 4.10, an element of positive slope of the surface $C_0$ is shown decomposed in the $x$ and $y$ directions. The components of the volume flow which pass through this element of surface are also indicated. Then the above statement of newtonian mechanics for the $x$ direction may be expressed by the following equation:

$$-X - \int_{C_0} p \, dy = \int_{C_0} \rho(u \, dy - v \, dx)u$$

In writing this equation, it has been noted that there is no transfer of momentum across the surface $C_i$, since it is a streamline, and that the integral of the pressure around $C_i$ yields the force $X$, which acts in the positive direction on the body and hence in the negative direction on the fluid. Also, the mass efflux across the element of the surface $C_0$ is $\rho(u \, dy - v \, dx)$, so that the product of this quantity and the $x$ component of velocity, when integrated around the surfaces $C_0$, gives the net increase in the $x$ component of momentum.

A similar equation may be obtained by applying the same newtonian law to the $y$ direction. Thus, the statement that the net external force acting in the positive $y$ direction must equal the net rate of increase of the $y$ component of the momentum yields the equation

$$-Y + \int_{C_0} p \, dx = \int_{C_0} \rho(u \, dy - v \, dx)v$$

Solving these two equations for the unknown forces $X$ and $Y$ yields the following pair of integrals:

$$X = \int_{C_0} (-p \, dy - \rho u^2 \, dy + \rho uv \, dx)$$

$$Y = \int_{C_0} (p \, dx - \rho uv \, dy + \rho v^2 \, dx)$$

The pressure may be eliminated from these equations by use of the Bernoulli equation, which for the case under consideration, may be written in the form

$$p + \tfrac{1}{2}\rho(u^2 + v^2) = B$$

where $B$ is the Bernoulli constant. Then, by eliminating the pressure $p$, the expressions for $X$ and $Y$ become

$$X = \rho \int_{C_0} \left[ uv \, dx - \tfrac{1}{2}(u^2 - v^2) \, dy \right]$$

$$Y = -\rho \int_{C_0} \left[ uv \, dy + \tfrac{1}{2}(u^2 - v^2) \, dx \right]$$

where the fact that $\int_{C_0} B \, dx = \int_{C_0} B \, dy = 0$ for any constant $B$ around any closed contour $C_0$ has been used.

It will now be shown that the quantity $X - iY$ may be related to a complex integral. Consider the following complex integral involving the complex velocity $W$:

$$i\frac{\rho}{2}\int_{C_0} W^2\,dz = i\frac{\rho}{2}\int_{C_0}(u - iv)^2(dx + i\,dy)$$

$$= i\frac{\rho}{2}\int_{C_0}\{[(u^2 - v^2)\,dx + 2uv\,dy] + i[(u^2 - v^2)\,dy - 2uv\,dx]\}$$

$$= \rho\int_{C_0}\{[uv\,dx - \tfrac{1}{2}(u^2 - v^2)\,dy] + i[uv\,dy + \tfrac{1}{2}(u^2 - v^2)\,dx]\}$$

$$= X - iY$$

The last equality follows by comparison of the expanded form of the complex integral with the expressions derived above for the body forces $X$ and $Y$. That is, the complex force $X - iY$ may be evaluated from

$$X - iY = i\frac{\rho}{2}\int_{C_0} W^2\,dz \qquad (4.17a)$$

where $W(z)$ is the complex velocity for the flow field and $C_0$ is any closed contour which encloses the body under consideration. It should be noted that $X$ and $Y$ were defined as the forces acting on the body through its center of gravity.

Equation (4.17a) constitutes one of the two Blasius laws. Normally, in applying Eq. (4.17a), the contour integral is evaluated with the aid of the residue theorem. An application of this procedure will be covered in the next section.

In order to establish the hydrodynamic moment acting on the body, consider again Fig. 4.10. The quantity $M$ is the moment acting *on the body* about its center of gravity. Then, taking clockwise moments as positive, moment equilibrium of the fluid enclosed between $C_0$ and $C_i$ requires that

$$-M + \int_{C_0}[px\,dx + py\,dy + \rho(u\,dy - v\,dx)uy - \rho(u\,dy - v\,dx)vx] = 0$$

The first two terms under the integral are the components of the pressure force multiplied by their respective perpendicular distances from the center of gravity of the body, which is at the origin of the coordinate system. The remaining two terms under the integral represent the inertia forces, which were evaluated in the discussion of the force equations, multiplied by their respective perpendicular distances from the origin. These inertia forces are equal in magnitude and opposite in direction to the rate of increase of the horizontal and vertical momentum components.

Solving the foregoing equation for the hydrodynamic moment $M$ gives

$$M = \int_{C_0} \left[ px\, dx + py\, dy + \rho\left(u^2 y\, dy + v^2 x\, dx - uvy\, dx - uvx\, dy\right)\right]$$

Substituting $p = B - \rho(u^2 + v^2)/2$ from the Bernoulli equation gives

$$M = \rho \int_{C_0} \left[ -\tfrac{1}{2}(u^2 + v^2)(x\, dx + y\, dy) + (u^2 y\, dy + v^2 x\, dx) - (uvy\, dx + uvx\, dy)\right]$$

where the fact has been used that $\int_{C_0} Bx\, dx = \int_{C_0} By\, dy = 0$ for any constant $B$ and any closed contour $C_0$. Rearranging the above equation shows that the integral for the moment $M$ may be put in the following form:

$$M = -\frac{\rho}{2} \int_{C_0} \left[(u^2 - v^2)(x\, dx - y\, dy) + 2uv(x\, dy + y\, dx)\right]$$

It will now be shown that the quantity $M$ may be related to the real part of a complex integral. Consider the real part, designated by $\mathrm{Re}(\ )$, of the following complex integral:

$$\mathrm{Re}\left(\frac{\rho}{2} \int_{C_0} zW^2\, dz\right) = \mathrm{Re}\left[\frac{\rho}{2} \int_{C_0} (x + iy)(u - iv)^2(dx + i\, dy)\right]$$

$$= \mathrm{Re}\left\{ \frac{\rho}{2} \int_{C_0} \left[(u^2 - v^2)(x\, dx - y\, dy) + 2uv(x\, dy + y\, dx)\right]\right.$$

$$\left. + i\frac{\rho}{2} \int_{C_0} \left[(u^2 - v^2)(x\, dy + y\, dx) - 2uv(x\, dx - y\, dy)\right]\right\}$$

$$= \frac{\rho}{2} \int_{C_0} \left[(u^2 - v^2)(x\, dx - y\, dy) + 2uv(x\, dy + y\, dx)\right]$$

$$= -M$$

The last equality follows from a comparison of the real part of the complex integral with the expression derived for $M$. That is, the hydrodynamic moment acting on a body is given by

$$M = -\frac{\rho}{2}\, \mathrm{Re}\left(\int_{C_0} zW^2\, dz\right) \tag{4.17b}$$

where $W(z)$ is the complex velocity for the flow field and $C_0$ is any closed contour which encloses the body. It should be noted that $M$ is defined as the hydrodynamic moment acting on the body, and it is positive when it acts in the clockwise direction. Equation (4.17b) is the second of the Blasius laws, and the contour integral in this equation is usually evaluated by use of the residue theorem.

## 4.11 FORCE AND MOMENT ON A CIRCULAR CYLINDER

It was observed in an earlier section that a force exists on a circular cylidner which is immersed in a uniform flow and which has a circulation around it. The magnitude of this force may now be evaluated using the results of the previous section.

From Eq. (4.14), the complex potential for a circular cylinder of radius $a$ in a uniform rectilinear flow of magnitude $U$ and having a bound vortex of magnitude $\Gamma$ in the negative direction is

$$F(z) = U\left(z + \frac{a^2}{z}\right) + \frac{i\Gamma}{2\pi} \log \frac{z}{a}$$

Then the complex velocity for this flow field is

$$W(z) = U\left(1 - \frac{a^2}{z^2}\right) + \frac{i\Gamma}{2\pi z}$$

$$\therefore \quad W^2(z) = U^2 - \frac{2U^2 a^2}{z^2} + \frac{U^2 a^4}{z^4} + \frac{iU\Gamma}{\pi z} - \frac{iU\Gamma a^2}{\pi z^3} - \frac{\Gamma^2}{4\pi^2 z^2}$$

But from the Blasius integral law [Eq. (4.17a)]

$$X - iY = i\frac{\rho}{2} \int_{C_0} W^2 \, dz$$

$$= i\frac{\rho}{2}\left[2\pi i \sum (\text{residues of } W^2 \text{ inside } C_0)\right]$$

where the last equality follows from the residue theorem. It is therefore required to evaluate the residue of $W^2(z)$ at each of the singular points which lie inside an arbitrary contour in the fluid which encloses the cylinder. But inspection of the expression derived for $W^2(z)$ above shows that the only singularity is at $z = 0$, corresponding to the doublet and the vortex which are located there. Furthermore, $W^2(z)$ is in the form of its Laurent series about $z = 0$, from which it is seen that the only term of the form $b_1/z$ is the fourth one. Hence, the residue of $W^2(z)$ at $z = 0$ is $iU\Gamma/\pi$. Then the value of the complex force is

$$X - iY = i\frac{\rho}{2}\left[2\pi i\left(\frac{iU\Gamma}{\pi}\right)\right]$$

$$= -i\rho U\Gamma$$

Equating the real and imagainary parts of this equation shows that the drag force $X$ is zero, as was expected, and that the value of the lift force is

$$Y = \rho U\Gamma \tag{4.18}$$

Equation (4.18) is known as the *Kutta-Joukowski law*, and it asserts that, for flow around a circular cylinder, there will be no lift force on the cylinder if there is no circulation around it, and if there is a circulation, the value of the lift force will be given by the product of the magnitude of this circulation with the free-stream velocity and the density of the fluid. It should be noted that the right-hand side of Eq. (4.18) is positive, so that the negative circulation which acted on the cylinder led to a positive, that is, upward, lift force.

In order to evaluate the hydrodynamic moment $M$ acting on the cylinder, the quantity $zW^2$ must be evaluated. From the expression for $W^2(z)$ which was established above,

$$zW^2(z) = U^2z - \frac{2U^2a^2}{z} + \frac{U^2a^4}{z^3} + \frac{iU\Gamma}{\pi} - \frac{iU\Gamma a^2}{\pi z^2} - \frac{\Gamma^2}{4\pi^2z}$$

But from the Blasius integral law [Eq. (4.17$b$)],

$$M = -\frac{\rho}{2} \operatorname{Re} \left( \int_{C_0} zW^2 \, dz \right)$$

$$= -\frac{\rho}{2} \operatorname{Re} \left[ 2\pi i \sum (\text{residues of } zW^2 \text{ inside } C_0) \right]$$

where again the residue theorem has been used. But the quantity $zW^2(z)$, as evaluated above, is already in the form of its Laurent series about $z = 0$. From this, it is evident that the only singularity is at $z = 0$, and the residue there comes from the second and last terms in the expansion. Hence

$$M = -\frac{\rho}{2} \operatorname{Re} \left[ 2\pi i \left( -2U^2a^2 - \frac{\Gamma^2}{4\pi^2} \right) \right]$$

$$= 0$$

That is, as might be expected, there is no hydrodynamic moment acting on the cylinder.

## 4.12  CONFORMAL TRANSFORMATIONS

Many complicated flow boundaries may be transformed into regular flow boundaries, such as the ones already studied, by the technique of conformal transformations. Before using this fact, it is necessary to study the effect of conformal transformations on the complex potential, the complex velocity, sources, sinks, and vortices. In carrying out this study, it will be considered that some geometric shape in the $z$ plane whose coordinates are $x$ and $y$ is mapped into some other shape in the $\zeta$ plane whose coordinates are $\xi$ and $\eta$ by means of the transformation

$$\zeta = f(z)$$

where $f$ is an analytic function. This situation is depicted in Fig. 4.11.

The basis of the complex potential was that both the velocity potential and the stream function had to satisfy Laplace's equation. Hence, in order to

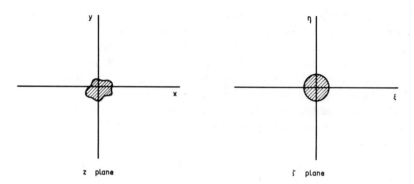

**FIGURE 4.11**
Original and mapped planes for the mapping $\zeta = f(z)$, where $f$ is an analytic function.

establish the effect of a conformal transformation on complex potentials, their effect on Laplace's equation should be studied. This will be done by transforming the second derivatives with respect to $x$ and $y$ into derivatives with respect to the new coordinates, namely, $\xi$ and $\eta$. Then, considering $\phi$ to be a function of $\xi$ and $\eta$

$$\frac{\partial \phi}{\partial x} = \frac{\partial \xi}{\partial x} \frac{\partial \phi}{\partial \xi} + \frac{\partial \eta}{\partial x} \frac{\partial \phi}{\partial \eta}$$

where $\partial \xi / \partial x$ and $\partial \eta / \partial x$ will be known from the equation of the mapping, $\zeta = f(z)$. Now in order to transform $\partial^2 \phi / \partial x^2$, each of the two terms on the right-hand side of the expression for $\partial \phi / \partial x$ must be differentiated with respect to $x$. Then, using the product rule and considering $\phi$ to be a function of $\xi$ and $\eta$,

$$\frac{\partial}{\partial x}\left(\frac{\partial \xi}{\partial x} \frac{\partial \phi}{\partial \xi}\right) = \frac{\partial^2 \xi}{\partial x^2} \frac{\partial \phi}{\partial \xi} + \frac{\partial \xi}{\partial x}\left(\frac{\partial \xi}{\partial x} \frac{\partial^2 \phi}{\partial \xi^2} + \frac{\partial \eta}{\partial x} \frac{\partial^2 \phi}{\partial \xi \partial \eta}\right)$$

and

$$\frac{\partial}{\partial x}\left(\frac{\partial \eta}{\partial x} \frac{\partial \phi}{\partial \eta}\right) = \frac{\partial^2 \eta}{\partial x^2} \frac{\partial \phi}{\partial \eta} + \frac{\partial \eta}{\partial x}\left(\frac{\partial \xi}{\partial x} \frac{\partial^2 \phi}{\partial \xi \partial \eta} + \frac{\partial \eta}{\partial x} \frac{\partial^2 \phi}{\partial \eta^2}\right)$$

Hence the expression for $\partial^2 \phi / \partial x^2$, in terms of derivatives with respect to $\xi$ and $\eta$, becomes

$$\frac{\partial^2 \phi}{\partial x^2} = \left(\frac{\partial \xi}{\partial x}\right)^2 \frac{\partial^2 \phi}{\partial \xi^2} + \left(\frac{\partial \eta}{\partial x}\right)^2 \frac{\partial^2 \phi}{\partial \eta^2} + 2 \frac{\partial \xi}{\partial x} \frac{\partial \eta}{\partial x} \frac{\partial^2 \phi}{\partial \xi \partial \eta} + \frac{\partial^2 \xi}{\partial x^2} \frac{\partial \phi}{\partial \xi} + \frac{\partial^2 \eta}{\partial x^2} \frac{\partial \phi}{\partial \eta}$$

The corresponding expression for $\partial^2 \phi / \partial y^2$ is

$$\frac{\partial^2 \phi}{\partial y^2} = \left(\frac{\partial \xi}{\partial y}\right)^2 \frac{\partial^2 \phi}{\partial \xi^2} + \left(\frac{\partial \eta}{\partial y}\right)^2 \frac{\partial^2 \phi}{\partial \eta^2} + 2 \frac{\partial \xi}{\partial y} \frac{\partial \eta}{\partial y} \frac{\partial^2 \phi}{\partial \xi \partial \eta} + \frac{\partial^2 \xi}{\partial y^2} \frac{\partial \phi}{\partial \xi} + \frac{\partial^2 \eta}{\partial y^2} \frac{\partial \phi}{\partial \eta}$$

Now since $\phi$ must satisfy Laplace's equation in the original plane, that is, the $xy$

plane, the sum of the above two quantities must be zero. Then

$$\left[\left(\frac{\partial \xi}{\partial x}\right)^2 + \left(\frac{\partial \xi}{\partial y}\right)^2\right]\frac{\partial^2 \phi}{\partial \xi^2} + \left[\left(\frac{\partial \eta}{\partial x}\right)^2 + \left(\frac{\partial \eta}{\partial y}\right)^2\right]\frac{\partial^2 \phi}{\partial \eta^2} + 2\left(\frac{\partial \xi}{\partial x}\frac{\partial \eta}{\partial x} + \frac{\partial \xi}{\partial y}\frac{\partial \eta}{\partial y}\right)\frac{\partial^2 \phi}{\partial \xi \partial \eta}$$

$$+ \left(\frac{\partial^2 \xi}{\partial x^2} + \frac{\partial^2 \xi}{\partial y^2}\right)\frac{\partial \phi}{\partial \xi} + \left(\frac{\partial^2 \eta}{\partial x^2} + \frac{\partial^2 \eta}{\partial y^2}\right)\frac{\partial \phi}{\partial \eta} = 0$$

This is the equation which has to be satisfied by $\phi(\xi, \eta)$ in the $\zeta$ plane due to any transformation $\zeta = f(z)$ corresponding to $\partial^2\phi/\partial x^2 + \partial^2\phi/\partial y^2 = 0$ in the $z$ plane. So far, no restrictions have been imposed on the transformation. But if the transformation is conformal, the mapping function $f$ will be analytic and the real and imaginary parts of the new variable $\zeta$ will be harmonic. That is, $\partial^2\xi/\partial x^2 + \partial^2\xi/\partial y^2 = 0$ and $\partial^2\eta/\partial x^2 + \partial^2\eta/\partial y^2 = 0$, so that the terms involving these quantities in the equation for $\phi$ will be zero. Also, $\xi(x, y)$ and $\eta(x, y)$ must satisfy the Cauchy-Riemann equations if the mapping function is analytic. That is,

$$\frac{\partial \xi}{\partial x} = \frac{\partial \eta}{\partial y}$$

and

$$\frac{\partial \xi}{\partial y} = -\frac{\partial \eta}{\partial x}$$

then

$$\left(\frac{\partial \xi}{\partial x}\frac{\partial \eta}{\partial x} + \frac{\partial \xi}{\partial y}\frac{\partial \eta}{\partial y}\right) = \left(\frac{\partial \xi}{\partial x}\frac{\partial \eta}{\partial x} - \frac{\partial \eta}{\partial x}\frac{\partial \xi}{\partial x}\right) = 0$$

Using this result, the equation to be satisfied by $\phi$ becomes

$$\left[\left(\frac{\partial \xi}{\partial x}\right)^2 + \left(\frac{\partial \xi}{\partial y}\right)^2\right]\frac{\partial^2 \phi}{\partial \xi^2} + \left[\left(\frac{\partial \eta}{\partial x}\right)^2 + \left(\frac{\partial \eta}{\partial y}\right)^2\right]\frac{\partial^2 \phi}{\partial \eta^2} = 0$$

Using the Cauchy-Riemann equations to eliminate first $\xi$, then $\eta$, then shows that the following pair of equations must be satisfied:

$$\left[\left(\frac{\partial \eta}{\partial x}\right)^2 + \left(\frac{\partial \eta}{\partial y}\right)^2\right]\left(\frac{\partial^2 \phi}{\partial \xi^2} + \frac{\partial^2 \phi}{\partial \eta^2}\right) = 0$$

$$\left[\left(\frac{\partial \xi}{\partial x}\right)^2 + \left(\frac{\partial \xi}{\partial y}\right)^2\right]\left(\frac{\partial^2 \phi}{\partial \xi^2} + \frac{\partial^2 \phi}{\partial \eta^2}\right) = 0$$

But these equations must be satisfied for all analytic mapping functions; hence it follows that

$$\frac{\partial^2 \phi}{\partial \xi^2} + \frac{\partial^2 \phi}{\partial \eta^2} = 0$$

That is, Laplace's equation in the $z$ plane transforms into Laplace's equation in the $\zeta$ plane, provided these two planes are related by a conformal transformation. Then, since both $\phi$ and $\psi$ must satisfy Laplace's equation, it follows that a complex potential in the $z$ plane is also a valid complex potential in the $\zeta$ plane, and vice versa. This means that if the solution for some simple body is known in one of these planes, say the $\zeta$ plane, then the solution for the more complex body may be obtained by substituting $\zeta = f(z)$ in the complex potential $F(\zeta)$.

Consider now what happens to the complex velocity under a conformal transformation. Starting in the $z$ plane with the definition of complex velocity,

$$
\begin{aligned}
W(z) &= \frac{dF(z)}{dz} \\[2mm]
&= \frac{d\zeta}{dz} \frac{dF(\zeta)}{d\zeta} \\[2mm]
W(z) &= \frac{d\zeta}{dz} W(\zeta)
\end{aligned}
\tag{4.19}
$$

That is, complex velocities are not, in general, mapped one to one, but they are proportional to each other, and the proportionality factor depends on the mapping function.

Finally, the effect of a conformal transformation on the strength of the basic singularities will be investigated. That is, the strength of transformed sources, sinks, and vortices will be established. This is most readily done by first proving the general relation that the integral of the complex velocity around any closed contour in the flow field equals $\Gamma + im$, where $\Gamma$ is the net strength of any vortices inside the contour and $m$ is the net strength of any sources and sinks inside the contour.

To prove this relation, consider any closed contour $C$ such as the one shown in Fig. 4.12. An element $dl$ of this contour is shown resolved into its coordinate components. Then the net strength of all the sources inside $C$ (sinks being considered negative sources) and the net strength of all the vortices inside $C$ will be given by

$$
m = \int_C \mathbf{u} \cdot \mathbf{n}\, dl = \int_C (u\, dy - v\, dx)
$$

$$
\Gamma = \int_C \mathbf{u} \cdot d\mathbf{l} = \int_C (u\, dx + v\, dy)
$$

Now consider the integral around $C$ of the complex velocity $W(z)$.

$$
\begin{aligned}
\int_C W(z)\, dz &= \int_C (u - iv)(dx + i\, dy) \\[2mm]
&= \int_C (u\, dx + v\, dy) + i \int_C (u\, dy - v\, dx) \\[2mm]
&= \Gamma + im
\end{aligned}
$$

where the last equality follows from a comparison with the expressions derived

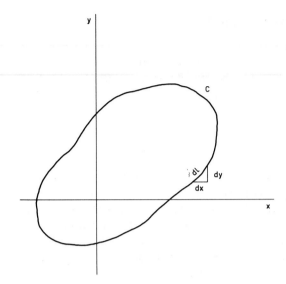

**FIGURE 4.12**
Arbitrary closed contour $C$ with an element $dl$ resolved into its coordinate components.

for $m$ and $\Gamma$. This general result will now be applied to a single vortex $\Gamma_z$ and a single source $m_z$ located in the $z$ plane. Then

$$\Gamma_z + im_z = \int_{C_z} W(z)\, dz$$

$$= \int_{C_z} W(\zeta)\frac{d\zeta}{dz}\, dz$$

$$= \int_{C_\zeta} W(\zeta)\, d\zeta$$

$$= \Gamma_\zeta + im_\zeta$$

where $C_z$ is some closed contour in the $z$ plane and $C_\zeta$ is its counterpart in the mapped plane. $\Gamma_\zeta$ and $m_\zeta$ are the corresponding vortex and source strengths in the $\zeta$ plane, and the above result shows that the vortex and source strengths are the same in the $\zeta$ plane as in the $z$ plane. That is, sources, sinks, and vortices map into sources, sinks, and vortices of the same strength under a conformal transformation.

In summary, if the complex potential for the flow around some body is known in the $\zeta$ plane, then the complex potential for the body corresponding to the conformal mapping $\zeta = f(z)$ may be obtained by substituting this transformation into the complex potential $F(\zeta)$. Complex velocities, on the other hand, do not transform one to one but are related by Eq. (4.19). Sources, sinks, and vortices maintain the same strength under conformal transformations.

## 4.13  JOUKOWSKI TRANSFORMATION

One of the most important transformations in the study of fluid mechanics is the Joukowski transformation. By means of this transformation and the basic flow solutions already studied, it is possible to obtain solutions for the flow around ellipses and a family of airfoils. The Joukowski transformation is of the form

$$z = \zeta + \frac{c^2}{\zeta} \tag{4.20}$$

where the constant $c^2$ is usually taken to be real. A general property of the Joukowski transformation is that for large values of $|\zeta|$, $z \to \zeta$. That is, far from the origin the transformation becomes the identity mapping, so that the complex velocity in the two planes is the same far from the origin. This means that if a uniform flow of a certain magnitude is approaching a body in the $z$ plane at some angle of attack, a uniform flow of the same magnitude and angle of attack will approach the corresponding body in the $\zeta$ plane.

From Eq. (4.20),

$$\frac{dz}{d\zeta} = 1 - \frac{c^2}{\zeta^2}$$

so that there is a singular point in the Joukowski transformation at $\zeta = 0$. Since we are normally dealing with the flow around some body, the point $\zeta = 0$ is normally not in the fluid, and so this singularity is of no consequence. There are also two critical points of the transformation, that is, points at which $dz/d\zeta$ vanishes, at $\zeta = \pm c$. Since smooth curves passing through critical points of a mapping may become corners in the transformed plane, it is of interest to investigate the consequence of a smooth curve passing through the critical points of the Joukowski transformation. To do this, consider an arbitrary point $z$ and its counterpart $\zeta$ as shown in Fig. 4.13a. Let the point $\zeta$ be measured by the radii $\rho_1$ and $\rho_2$ and the angles $\nu_1$ and $\nu_2$ relative to the two critical points $\zeta = c$ and $\zeta = -c$, respectively. But according to the Joukowski transformation the points $\zeta = \pm c$ map into the points $z = \pm 2c$. Then let the mapping of the point $\zeta$ be measured by the radii $R_1$ and $R_2$ and the angles $\theta_1$ and $\theta_2$ relative to the two points $z = 2c$ and $z = -2c$, respectively.

From Eq. (4.20),

$$z + 2c = \frac{(\zeta + c)^2}{\zeta}$$

and

$$z - 2c = \frac{(\zeta - c)^2}{\zeta}$$

$$\therefore \quad \frac{z - 2c}{z + 2c} = \left(\frac{\zeta - c}{\zeta + c}\right)^2$$

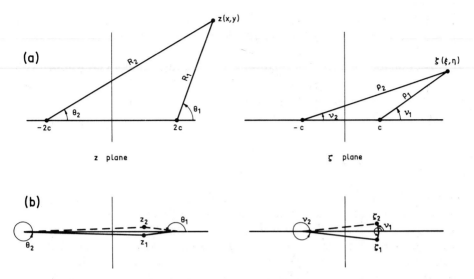

**FIGURE 4.13**
(a) Coordinate systems used to investigate the critical points of the Joukowski transformation and (b) the coordinate changes corresponding to a smooth curve passing through $\zeta = c$.

Thus, with reference to Fig. 4.13a,

$$\frac{R_1 e^{i\theta_1}}{R_2 e^{i\theta_2}} = \left(\frac{\rho_1 e^{i\nu_1}}{\rho_2 e^{i\nu_2}}\right)^2$$

or

$$\frac{R_1}{R_2} e^{i(\theta_1 - \theta_2)} = \left(\frac{\rho_1}{\rho_2}\right)^2 e^{i2(\nu_1 - \nu_2)}$$

Equating the modulus and the argument of each side of this equation shows that

$$\frac{R_1}{R_2} = \left(\frac{\rho_1}{\rho_2}\right)^2$$

and

$$\theta_1 - \theta_2 = 2(\nu_1 - \nu_2)$$

This last result shows that if a smooth curve passes through the point $\zeta = c$, the corresponding curve in the $z$ plane will form a knife-edge or cusp. This may be verified by considering a smooth curve to pass through the point $\zeta = c$. Two points on this curve are shown in Fig. 4.13b, from which it is seen that $\nu_1$ changes from $3\pi/2$ to $\pi/2$ and $\nu_2$ changes from $2\pi$ to $0$ as the critical point is passed. That is, the value of $\nu_1 - \nu_2$ changes from $-\pi/2$ to $\pi/2$, giving a difference of $\pi$. From the result $\theta_1 - \theta_2 = 2(\nu_1 - \nu_2)$, it follows that the corresponding difference in the value of $\theta_1 - \theta_2$ will be $2\pi$. This yields a knife-edge or cusp in the $z$ plane as shown in Fig. 4.13b. That is, if a smooth curve passes through either of the critical points $\zeta = \pm c$, the corresponding curve in the $z$ plane will contain a knife-edge at the corresponding critical point $z = \pm 2c$.

An example of a smooth curve which passes through both critical points is a circle centered at the origin of the $\zeta$ plane and whose radius is $c$, the constant which appears in the Joukowski transformation. Then, on this circle $\zeta = ce^{iv}$, so that the value of $z$ will be given by

$$z = ce^{iv} + ce^{-iv}$$

$$= 2c \cos v$$

That is, the circle in the $\zeta$ plane maps into the strip $y = 0$, $x = 2c \cos v$ in the $z$ plane. It is readily verified that all points which lie outside the circle $|\zeta| = c$ cover the entire $z$ plane. However, the points inside the circle $|\zeta| = c$ also cover the entire $z$ plane, so that the transformation is double-valued. This is readily verified by observing that for any value of $\zeta$ Eq. (4.20) yields the same value of $z$ for that value of $\zeta$ and also for $c^2/\zeta$. It will be noted that $c^2/\zeta$ is simply the image of the point $\zeta$ inside the circle of radius $c$.

This double-valued property of the Joukowski transformation is treated by connecting the two points $z = \pm 2c$ by a branch cut along the $x$ axis and creating two Riemann sheets. Then the mapping is single-valued if all the points outside the circle $|\zeta| = c$ are taken to fall on one of these sheets and all the points inside the circle to fall on the other sheet. In fluid mechanics, difficulties due to the double-valued behavior do not usually arise because the points $|\zeta| < c$ usually lie inside some body about which the flow is being studied, so that these points are not in the flow field in the $z$ plane.

## 4.14   FLOW AROUND ELLIPSES

Applications of the Joukowski transformation will be made in an inverse sense. That is, the simple geometry of the circle, the flow around which is known, will be placed in the $\zeta$ plane, and the corresponding body which results in the $z$ plane will be investigated by use of Eq. (4.20).

Consider, first, the constant $c$ in Eq. (4.20) to be real and positive, and consider a circle of radius $a > c$ to be centered at the origin of the $\zeta$ plane. The contour in the $z$ plane corresponding to this circle in the $\zeta$ plane may be identified by substituting $\zeta = ae^{iv}$ into Eq. (4.20).

$$z = ae^{iv} + \frac{c^2}{a}e^{-iv}$$

$$= \left(a + \frac{c^2}{a}\right)\cos v + i\left(a - \frac{c^2}{a}\right)\sin v$$

Equating real and imaginary parts of this equation gives

$$x = \left(a + \frac{c^2}{a}\right)\cos v$$

$$y = \left(a - \frac{c^2}{a}\right)\sin v$$

These are the parametric equations of the required curve in the $z$ plane. The equation of the curve may be obtained by eliminating $v$ by use of the identity $\cos^2 v + \sin^2 v = 1$. This gives

$$\left(\frac{x}{a + c^2/a}\right)^2 + \left(\frac{y}{a - c^2/a}\right)^2 = 1$$

which is the equation of an ellipse whose major semiaxis is of length $a + c^2/a$, aligned along the $x$ axis, and whose minor semiaxis is of length $a - c^2/a$. Then, in order to obtain the complex potential for a uniform flow of magnitude $U$ approaching this ellipse at an angle of attack $\alpha$, the same flow should be considered to approach the circular cylinder in the $\zeta$ plane. But it is shown in Prob. 4.5, Eq. (4.29), that the complex potential for a uniform flow of magnitude $U$ approaching a circular cylinder of radius $a$ at an angle $\alpha$ to the reference axis is

$$F(\zeta) = U\left(\zeta e^{-i\alpha} + \frac{a^2}{\zeta}e^{i\alpha}\right)$$

Then, by solving Eq. (4.20) for $\zeta$ in terms of $z$, the complex potential in the $z$ plane may be obtained. From Eq. (4.20),

$$\zeta^2 - z\zeta + c^2 = 0$$

$$\therefore \quad \zeta = \frac{z}{2} \pm \sqrt{\left(\frac{z}{2}\right)^2 - c^2}$$

Since it is known that $\zeta \to z$ for large values of $z$, the positive root must be chosen. Then the complex potential in the $z$ plane becomes

$$F(z) = U\left\{\left[\frac{z}{2} + \sqrt{\left(\frac{z}{2}\right)^2 - c^2}\right]e^{-i\alpha} + \frac{a^2 e^{i\alpha}}{z/2 + \sqrt{(z/2)^2 - c^2}}\right\}$$

$$= U\left\{\left[z - \frac{z}{2} + \sqrt{\left(\frac{z}{2}\right)^2 - c^2}\right]e^{-i\alpha} + \frac{a^2}{c^2}\left[\frac{z}{2} - \sqrt{\left(\frac{z}{2}\right)^2 - c^2}\right]e^{i\alpha}\right\}$$

where the last term has been rationalized. By writing $z/2$ as $z - z/2$ in the first

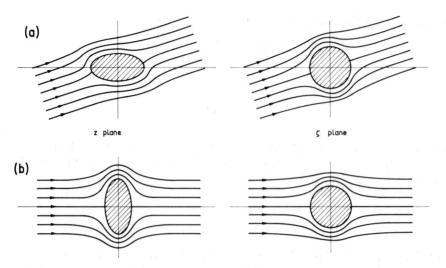

**(a)**

z plane

ζ plane

**(b)**

**FIGURE 4.14**
(*a*) Uniform flow approaching a horizontal ellipse at an angle of attack and (*b*) uniform parallel flow approaching a vertical ellipse.

term, two of the terms may be combined as follows:

$$F(z) = U\left[ze^{-i\alpha} + \left(\frac{a^2}{c^2}e^{i\alpha} - e^{-i\alpha}\right)\left(\frac{z}{2} - \sqrt{\left(\frac{z}{2}\right)^2 - c^2}\right)\right] \quad (4.21a)$$

Equation (4.21*a*) is the complex potential for a uniform flow of magnitude $U$ approaching an ellipse whose major semiaxis is $a + c^2/a$ and whose minor semiaxis is $a - c^2/a$ at an angle of attack $\alpha$ to the major axis. In this form it may be seen that the complex potential consists of that for a uniform flow at an angle $\alpha$ to the reference axis plus a perturbation which is large near the body but vanishes for large values of $z$. The flow field generated by the complex potential (4.21*a*) is shown in Fig. 4.14*a* together with that for the circular cylinder in the $\zeta$ plane.

The stagnation points in the $\zeta$ plane are located at $\zeta = ae^{i\alpha}$ and $\zeta = ae^{i(\alpha + \pi)}$, that is, at $\zeta = \pm ae^{i\alpha}$. Then, from Eq. (4.20), the corresponding points in the $z$ plane are

$$z = \pm ae^{i\alpha} \pm \frac{c^2}{a}e^{-i\alpha}$$

$$= \pm\left(a + \frac{c^2}{a}\right)\cos\alpha \pm i\left(a - \frac{c^2}{a}\right)\sin\alpha$$

This gives the coordinates of the stagnation points as

$$x = \pm \left( a + \frac{c^2}{a} \right) \cos \alpha$$

$$y = \pm \left( a - \frac{c^2}{a} \right) \sin \alpha$$

Equation (4.21a) includes two special cases within its range of validity. For $\alpha = 0$ it describes a uniform rectilinear flow approaching a horizontally oriented ellipse, and for $\alpha = \pi/2$ it describes a uniform vertical flow approaching the same horizontally oriented ellipse. However, it is of interest to note that the solution for a uniform rectilinear flow approaching a vertically oriented ellipse may be obtained directly from the Joukowski transformation with a slight modification. Substitute $c = ib$, where $b$ is real and positive, into Eq. (4.19).

$$z = \zeta - \frac{b^2}{\zeta}$$

Then, as with the horizontal ellipse, examining the mapping of the circle $\zeta = ae^{iv}$ gives the parametric equations of the mapped boundary.

$$x = \left( a - \frac{b^2}{a} \right) \cos v$$

$$y = \left( a + \frac{b^2}{a} \right) \sin v$$

Thus the equation of the contour in the $z$ plane is

$$\left( \frac{x}{a - b^2/a} \right)^2 + \left( \frac{y}{a + b^2/a} \right)^2 = 1$$

which is the equation of an ellipse whose major semiaxis is $a + b^2/a$ which is aligned along the $y$ axis. Then to obtain a uniform rectilinear flow approaching such an ellipse the same flow should approach the circle in the $\zeta$ plane. Thus the required complex potential, from Eq. (4.13), is

$$F(\zeta) = U \left( \zeta + \frac{a^2}{\zeta} \right)$$

But the inverted equation of the mapping for which $\zeta \to z$ as $z \to \infty$ is

$$\zeta = \frac{z}{2} + \sqrt{\left( \frac{z}{2} \right)^2 + b^2}$$

Hence the complex potential in the $z$ plane is

$$F(z) = U\left[\frac{z}{2} + \sqrt{\left(\frac{z}{2}\right)^2 + b^2} + \frac{a^2}{z/2 + \sqrt{(z/2)^2 + b^2}}\right]$$

(4.21b)

$$F(z) = U\left[z - \left(1 + \frac{a^2}{b^2}\right)\left(\frac{z}{2} - \sqrt{\left(\frac{z}{2}\right)^2 + b^2}\right)\right]$$

in which the same rationalization and simplification has been carried out as before. Again the complex potential is in the form of that for a uniform flow plus a perturbation which is large near the body and which vanishes at large distances from the body. Equation (4.21b) describes a uniform rectilinear flow of magnitude $U$ approaching a vertically oriented ellipse. The flow field for this situation is shown in Fig. 4.14b.

### 4.15   KUTTA CONDITION
### AND THE FLAT-PLATE AIRFOIL

It was observed in Sec. 4.6 that the potential flow solution for flow around a sharp edge contained a singularity at the edge itself. This singularity required an infinite velocity at the point in question which, of course, is physically impossible. The question arises, then, as to what the real flow situation would be in a physical experiment. Depending upon the actual physical configuration, one of two remedial situations will prevail. One possibility is that the fluid will separate from the solid surface at the knife-edge. The resulting free streamline configuration would be such that the radius of curvature at the edge becomes finite rather than being zero. As a consequence, the velocities there will remain finite. Examples of this type of solution will be discussed later in this chapter.

A second possibility is that a stagnation point exists at the sharp edge. For the flow around finite bodies, stagnation points exist, and it seems possible that a stagnation point could be induced by the flow field to move to the location of the sharp edge. This possibility leads to the so-called "Kutta condition," and it will be discussed below in the context of the flat-plate airfoil—that is, a flat plate which is at some angle of attack to the free stream.

In the previous section, the flow around an ellipse was obtained from the Joukowski transformation [Eq. (4.20)] by considering the flow around a circular cylinder of radius $a > c$ in the $\zeta$ plane. Now, if the constant $c$ is allowed to approach the magnitude of the radius $a$, the resulting ellipse in the $z$ plane degenerates to a flat plate defined by the strip $-2a \le x \le 2a$. The resulting flow field, as defined by Eq. (4.21a), is shown in Fig. 4.15a. Because of the angle of attack, the stagnation points do not coincide with the leading and trailing edges of the flat plate. Rather, the upstream stagnation point is located on the lower surface and the downstream stagnation point is located on the upper surface at the points $x = \pm 2a \cos \alpha$. Then, around both the leading and

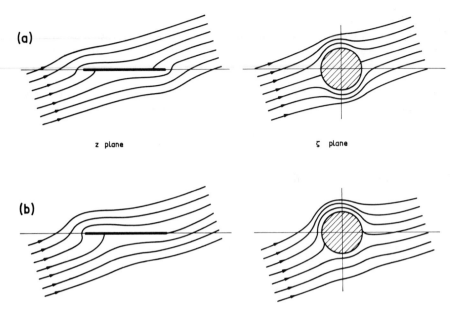

**FIGURE 4.15**
Flow around a flat plate at shallow angle of attack (*a*) without circulation and (*b*) satisfying Kutta condition.

trailing edges, the flow will be that associated with a sharp edge which was discussed in Sec. 4.6. In that section, it was observed that infinite velocity components existed at the edge itself, a situation which is physically impossible to realize.

The difficulty encountered above with the flat-plate airfoil does not occur at the leading edge of real airfoils because real airfoils have a finite thickness and so have a finite radius of curvature at the leading edge. However, the trailing edge of airfoils is usually quite sharp, so that the difficulty of infinite velocity components still exists there. However, this remaining difficulty would also be overcome if the stagnation point which is near the trailing edge was actually at the trailing edge. This would be accomplished if a circulation existed around the flat plate and the magnitude of this circulation was just the amount required to rotate the rear stagnation point so that its location coincides with the trailing edge. This condition is called the *Kutta condition*, and it may be restated as follows: For bodies with sharp trailing edges which are at small angles of attack to the free stream, the flow will adjust itself in such a way that the rear stagnation point coincides with the trailing edge.

The amount of circulation required to comply with the Kutta condition may be determined as follows: In the $\zeta$ plane of Fig. 4.15$a$, the rear stagnation point is located at the point $\zeta = ae^{i\alpha}$. But, according to the Kutta condition, the rear stagnation point should be located at the point $z = 2a$, which corresponds to the point $\zeta = a$. That is, the stagnation point on the downstream face of the circular cylinder in the $\zeta$ plane should be rotated clockwise through an angle $\alpha$. But from Eq. (4.16), the magnitude of the circulation which will do this is

$$\Gamma = 4\pi Ua \sin \alpha \qquad (4.22a)$$

in the clockwise direction (that is, negative circulation). Then the complex potential for the required flow in the $\zeta$ plane is, from Eqs. (4.14) and (4.29),

$$F(\zeta) = U\left(\zeta e^{-i\alpha} + \frac{a^2}{\zeta}e^{i\alpha}\right) + i2Ua \sin \alpha \, \log \frac{\zeta}{a}$$

But the equation of the mapping is

$$z = \zeta + \frac{a^2}{\zeta}$$

and the inverse which gives $\zeta \rightarrow z$ as $z \rightarrow \infty$ is

$$\zeta = \frac{z}{2} + \sqrt{\left(\frac{z}{2}\right)^2 - a^2}$$

Then the complex potential in the $z$ plane is

$$F(z) = U\left\{\left[\frac{z}{2} + \sqrt{\left(\frac{z}{2}\right)^2 - a^2}\right]e^{-i\alpha} + \frac{a^2 e^{i\alpha}}{z/2 + \sqrt{(z/2)^2 - a^2}}\right.$$

$$\left. + i2a \sin \alpha \, \log\left[\frac{1}{a}\left(\frac{z}{2} + \sqrt{\left(\frac{z}{2}\right)^2 - a^2}\right)\right]\right\} \qquad (4.22b)$$

The flow field corresponding to this complex potential is shown in Fig. 4.15$b$. Although the flow at the trailing edge is now regular, the singularity at the leading edge still exists. In an actual flow configuration the fluid would separate at the leading edge and reattach again on the top side of the airfoil. The streamline $\psi = 0$ would then correspond to a finite curvature, and the velocity components would remain finite at the leading edge.

The lift force generated by the flat-plate airfoil may be calculated from the Kutta-Joukowski law. Then, denoting the lift force by $Y$ and using the value of the circulation given by Eq. (4.22$a$),

$$Y = 4\pi\rho U^2 a \sin \alpha$$

It is usual to express lift forces in terms of the dimensionless *lift coefficient* $C_L$, which is defined as follows:

$$C_L = \frac{Y}{\frac{1}{2}\rho U^2 l}$$

where $l$ is the length or chord of the airfoil which, for the flat plate under consideration, equals $4a$. Then the value of the lift coefficient for the flat-plate airfoil is

$$C_L = 2\pi \sin \alpha \qquad (4.22c)$$

This result shows that the lift coefficient for the flat-plate airfoil increases with angle of attack, and for small values of $\alpha$, for which $\sin \alpha \approx \alpha$, the lift coefficient is proportional to the angle of attack with a constant of proportionality of $2\pi$. This result is very close to experimental observations, and so the Kutta condition appears to be well justified. If the Kutta condition were not valid, there would be no circulation around the flat plate, and consequently, no lift would be generated. This would mean that kites would not be able to fly.

## 4.16 SYMMETRICAL JOUKOWSKI AIRFOIL

A family of airfoils may be obtained in the $z$ plane by considering the Joukowski transformation in conjunction with a series of circles in the $\zeta$ plane whose centers are slightly displaced from the origin. These airfoils are known as the *Joukowski family of airfoils*.

Consider, first, the case where the center of the circle in the $\zeta$ plane is displaced from the origin along the real axis. It must then be decided in which direction the center should be moved and what radius should be employed, relative to the Joukowski constant $c$. From previous sections it is known that if the circumference of the circle passes through either of the two critical points of the Joukowski transformation, $\zeta = \pm c$, then a sharp edge or cusp is obtained in the $z$ plane. Then, if the leading edge of the airfoil is to have a finite radius of curvature and if there should be no singularities in the flow field itself, it follows that the point $\zeta = -c$ should be inside the circle in the $\zeta$ plane. Also, since the trailing edge of the airfoil should be sharp as opposed to being blunt, the circumference of the circle should pass through the point $\zeta = c$. These conditions will be satisfied by taking the center of the circle to be at $\zeta = -m$, where $m$ is real, and by choosing the radius of the circle to be $c + m$. Such a configuration is shown in Fig. 4.16$a$. The radius $a$ is given by

$$a = c + m = c(1 + \varepsilon)$$

where the parameter $\varepsilon = m/c$ will be assumed to be small compared with unity. When $\varepsilon = 0$, the flat-plat airfoil is recovered, so that for $\varepsilon \ll 1$ it may be anticipated that a thin airfoil will be obtained. The significance of the restriction $\varepsilon \ll 1$ will be that all the equations may be linearized in $\varepsilon$, which will permit a closed-form solution for the equation of the airfoil surface in the $z$ plane. Also shown in Fig. 4.16$a$ is the airfoil which is obtained in the $z$ plane and its principal parameters, the chord $l$ and the maximum thickness $t$. It is now required to relate these parameters to the free parameters $a$ and $m$ and to establish the equation of the airfoil surface in the $\zeta$ plane.

To establish the chord of the airfoil in terms of the chosen radius $a$ and offset $m$, it is only necessary to find the mapping of the points $\zeta = c$ and $\zeta = -(c + 2m)$, since these points correspond to the trailing and leading edges,

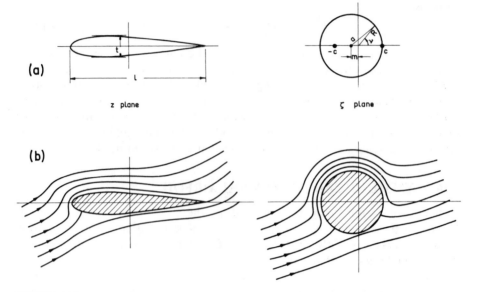

**FIGURE 4.16**
The symmetrical Joukowski airfoil; (*a*) the mapping planes and (*b*) uniform flow past the airfoil.

respectively. Using the Joukowski transformation, the mapping of the point $\zeta = c$ is $z = 2c$. Also, the mapping of the point $\zeta = -(c + 2m) = -c(1 + 2\varepsilon)$ is

$$z = -c(1 + 2\varepsilon) - \frac{c}{1 + 2\varepsilon}$$

Since it was decided to linearize all quantities in $\varepsilon$, the value of $z$ will be obtained to the first order in $\varepsilon$ only.

$$\therefore \quad z = -c(1 + 2\varepsilon) - c\left[1 - 2\varepsilon + O(\varepsilon^2)\right]$$
$$= -2c + O(\varepsilon^2)$$

That is, to the first order in $\varepsilon$ the leading edge of the airfoil is located at $z = -2c$ so that the chord length is

$$l = 4c$$

This means that, correct to the first order in $\varepsilon$, the length of the airfoil is unchanged by the shifting of the center of the circle in the $\zeta$ plane.

In order to determine the maximum thickness $t$, the equation of the airfoil surface must be obtained. This may be done by inserting the equation of the surface in the $\zeta$ plane into the Joukowski transformation. But in the $\zeta$ plane the polar radius $R$ to the circumference of the circle is a function of the angle $v$. In order to establish this dependence, the cosine rule will be applied to the triangle defined by the radius $a$, the coordinate $R$, and the real $\zeta$ axis, as shown in Fig. 4.16$a$. Thus

$$a^2 = R^2 + m^2 - 2Rm \cos(\pi - \nu)$$
$$= R^2 + m^2 + 2Rm \cos \nu$$

But $a = c + m$, so that the above equation may be written in the form

$$(c + m)^2 = R^2\left(1 + \frac{m^2}{R^2} + 2\frac{m}{R}\cos \nu\right)$$

Now since $R \geq c$, it follows that $m/R \leq m/c$ so that, to first order in $\varepsilon = m/c$, the term $m^2/R^2$ may be neglected. The equation for $R$ then becomes

$$c + m = R\left(1 + 2\frac{m}{R}\cos \nu\right)^{1/2}$$
$$= R\left[1 + \frac{m}{R}\cos \nu + O(\varepsilon^2)\right]$$

Thus to the first order in $\varepsilon$, this relation becomes

$$c(1 + \varepsilon) = R + m \cos \nu$$
$$\therefore \quad R = c[1 + \varepsilon(1 - \cos \nu)]$$

This is the required equation which gives the variation of the radius $R$ with the angle $\nu$ for points on the circumference of the circle in the $\zeta$ plane. Then, in order to determine the equation of the corresponding profile in the $\zeta$ plane, this result should be substituted into the Joukowski transformation [Eq. (4.20)]. Thus points on the surface of the airfoil will be defined by

$$z = c[1 + \varepsilon(1 - \cos \nu)]e^{i\nu} + \frac{ce^{-i\nu}}{1 + \varepsilon(1 - \cos \nu)}$$

This equation may also be linearized in $\varepsilon$ as follows:

$$z = c[1 + \varepsilon(1 - \cos \nu)]e^{i\nu} + c[1 - \varepsilon(1 - \cos \nu) + O(\varepsilon^2)]e^{-i\nu}$$
$$= c[2\cos \nu + i2\varepsilon(1 - \cos \nu)\sin \nu + O(\varepsilon^2)]$$

Then, by equating real and imaginary parts of this equation, the parametric equations of the airfoil are, to first order in $\varepsilon$,

$$x = 2c \cos \nu$$
$$y = 2c\varepsilon(1 - \cos \nu)\sin \nu$$

Using the first of these equations to eliminate $\nu$ from the second equation gives the following equation for the airfoil profile:

$$y = \pm 2c\varepsilon\left(1 - \frac{x}{2c}\right)\sqrt{1 - \left(\frac{x}{2c}\right)^2}$$

The location of the maximum thickness may now be obtained, and this is most readily done by using the parametric equation for the coordinate $y$ as derived above. Thus setting $dy/d\nu = 0$ for a maximum in $y$ gives the following equation for the value of $\nu$ at the maximum thickness:

$$\sin^2 \nu + (1 - \cos \nu) \cos \nu = 0$$

This relation reduces to

$$\cos 2\nu = \cos \nu$$

which is satisfied by $\nu = 0$, $\nu = 2\pi/3$, and $\nu = 4\pi/3$. The solution $\nu = 0$ corresponds to the trailing edge and so is the minimum thickness. The solutions $\nu = 2\pi/3$ and $\nu = 4\pi/3$ give the maximum thickness, and for these values of $\nu$ the coordinates of the airfoil surface are

$$x = -c$$

$$y = \pm \frac{3\sqrt{3}}{2} c\varepsilon$$

The maximum thickness $t$ will be twice the positive value of $y$, so that the thickness ratio $t/l$ of the airfoil will be

$$\frac{t}{l} = \frac{3\sqrt{3}}{4} \varepsilon$$

That is, the thickness-to-chord ratio of the airfoil is proportional to $\varepsilon$, which is the ratio of the offset of the center of the circle in the $\zeta$ plane to the radius $c$ of the critical points of the transformation. Since the thickness ratio of the airfoil is a parameter which may be thought of as being specified, it is useful to eliminate $\varepsilon$ in terms of this parameter. Hence

$$\varepsilon = \frac{4}{3\sqrt{3}} \frac{t}{l} = 0.77 \frac{t}{l}$$

Then the equation of the airfoil surface may be written in the form

$$\frac{y}{t} = \pm 0.385 \left(1 - 2\frac{x}{l}\right) \sqrt{1 - \left(2\frac{x}{l}\right)^2} \tag{4.23a}$$

where the maximum value of $y/t$ will be 0.5 and the minimum value will be $-0.5$, both of which occur at $x = -c$.

The magnitude of the circulation which is required to satisfy the Kutta condition is, from Eq. (4.16), $4\pi Ua \sin \alpha$, where $a = c + m$ and $m = c\varepsilon = 0.77tc/l$. Thus the required amount of circulation is

$$\Gamma = \pi Ul \left(1 + 0.77\frac{t}{l}\right) \sin \alpha \tag{4.23b}$$

where $c$ has been replaced by $l/4$. In this form the required circulation may be evaluated for the given free-stream velocity, angle of attack, and the chord and thickness of the airfoil. Using the Kutta-Joukowski law [Eq. (4.18)], the lift force acting on the airfoil may be evaluated as

$$Y = \pi \rho U^2 l \left(1 + 0.77\frac{t}{l}\right) \sin \alpha$$

Then the lift coefficient for the symmetrical Joukowski airfoil is

$$C_L = 2\pi \left(1 + 0.77\frac{t}{l}\right) \sin \alpha \qquad (4.23c)$$

It will be noticed that this result reduces to Eq. (4.22c) for the flat-plate airfoil as $t \rightarrow 0$. This indicates that the effect of thickness of an airfoil is to increase the lift coefficient. However, this fact cannot be used to produce high lift coefficients through thick airfoils, since the flow tends to separate from bluff bodies much more readily than it does from streamlined bodies. This separation of the flow is a viscous effect, and it will be discussed in the next part of the book. In the meantime, it is sufficient to say that separation of the flow results in a low-pressure wake which destroys the lift. The same result may occur for slender bodies, such as airfoils, which are at large angles of attack. In this context the separation is usually referred to as *stall*.

The center of the circle in the $\zeta$ plane is located at $\zeta = -m$ rather than $\zeta = 0$. Thus the complex potential for the flow in the $\zeta$ plane may be obtained from Eq. (4.29) by replacing $\zeta$ by $\zeta + m$ and adding circulation. The required complex potential then becomes

$$F(\zeta) = U\left[(\zeta + m)e^{-i\alpha} + \frac{a^2}{\zeta + m}e^{i\alpha}\right] + \frac{i\Gamma}{2\pi}\log\left(\frac{\zeta + m}{a}\right) \qquad (4.23d)$$

where

$$a = \frac{l}{4} + 0.77\frac{tc}{l}$$

and

$$m = 0.77\frac{tc}{l}$$

The magnitude of the circulation $\Gamma$ is given by Eq. (4.23b), and in the Joukowski transformation the parameter $c$ equals $l/4$. The flow field corresponding to this complex potential is shown in Fig. 4.16b.

## 4.17   CIRCULAR-ARC AIRFOIL

It was shown in the two previous sections that, using the Joukowski transformation, a circle of radius $c$ centered at the origin of the $\zeta$ plane produced a flat-plate airfoil while a slightly larger circle centered a small distance along the real axis from the origin produced a thin symmetrical airfoil. It will now be shown that a circle whose radius is slightly larger than $c$ and whose center is located on the imaginary axis of the $\zeta$ plane produces an airfoil which has no thickness but which has curvature or camber.

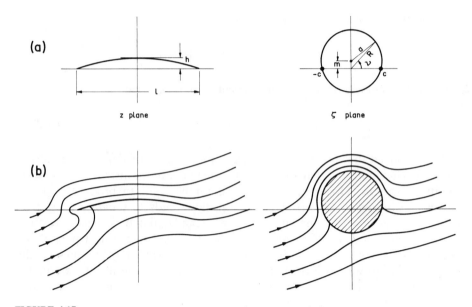

**FIGURE 4.17**
The circular-arc airfoil; (a) the mapping planes and (b) uniform flow past the airfoil.

Referring to Fig. 4.17a, consider a circle of radius $a > c$ in the $\zeta$ plane such that the center of the circle is located a distance $m$ along the positive imaginary axis. Since the trailing edge of the airfoil should be sharp, the circle should pass through the critical point $\zeta = c$ as before. Then, in this case, the circle will also pass through the other critical point, $\zeta = -c$.

The equation of the airfoil in the $z$ plane may be obtained by substituting $\zeta = Re^{i\nu}$ into the Joukowski transformation where, on the circumference of the circle in the $\zeta$ plane, $R$ is a function of $\nu$. This substitution gives

$$z = Re^{i\nu} + \frac{c^2}{R}e^{-i\nu}$$

$$= \left(R + \frac{c^2}{R}\right)\cos\nu + i\left(R - \frac{c^2}{R}\right)\sin\nu$$

Thus the parametric equations of the airfoil profile are

$$x = \left(R + \frac{c^2}{R}\right)\cos\nu$$

$$y = \left(R - \frac{c^2}{R}\right)\sin\nu$$

The variable $R$ may be eliminated from these equations as follows:

$$x^2 \sin^2 \nu - y^2 \cos^2 \nu = \left( R + \frac{c^2}{R} \right)^2 \sin^2 \nu \cos^2 \nu - \left( R - \frac{c^2}{R} \right)^2 \sin^2 \nu \cos^2 \nu$$

$$= 4c^2 \sin^2 \nu \cos^2 \nu$$

This is the equation of the airfoil surface in the $z$ plane, but it still contains the variable $\nu$. This variable may be eliminated by applying the cosine rule to the triangle defined by the radius $a$, the coordinate $R$, and the imaginary $\zeta$ axis. From this it follows that

$$a^2 = R^2 + m^2 - 2Rm \cos \left( \frac{\pi}{2} - \nu \right)$$

$$c^2 + m^2 = R^2 + m^2 - 2Rm \sin \nu$$

where the fact that $a^2 = c^2 + m^2$ has been used. Solving this equation for $\sin \nu$, it follows that

$$\sin \nu = \frac{R^2 - c^2}{2Rm}$$

But it was shown above that $y = [(R^2 - c^2) \sin \nu]/R$; hence it follows that

$$\sin \nu = \frac{y}{2m \sin \nu}$$

or

$$\sin^2 \nu = \frac{y}{2m}$$

and so

$$\cos^2 \nu = 1 - \frac{y}{2m}$$

Using these results to eliminate $\nu$, the equation of the airfoil surface becomes

$$x^2 \frac{y}{2m} - y^2 \left( 1 - \frac{y}{2m} \right) = 4c^2 \frac{y}{2m} \left( 1 - \frac{y}{2m} \right)$$

Collecting like terms, this equation may be put in the form

$$x^2 + y^2 + 2 \left( \frac{c^2}{m} - m \right) y = 4c^2$$

Completing the square in $y$ shows that the equation of the airfoil surface is

$$x^2 + \left[ y + c \left( \frac{c}{m} - \frac{m}{c} \right) \right]^2 = c^2 \left[ 4 + \left( \frac{c}{m} - \frac{m}{c} \right)^2 \right]$$

which is the equation of a circle. It should be noted that so far no approximations have been made. But to be consistent with the analysis in the previous section and to permit superposition in the next section, the parameter $\varepsilon = m/c$ will again be assumed to be small compared with unity. Then, linearizing in $\varepsilon$, the equation of the airfoil surface becomes

$$x^2 + \left(y + \frac{c^2}{m}\right)^2 = c^2\left(4 + \frac{c^2}{m^2}\right)$$

That is, correct to the first order in $\varepsilon$, the center of the circle in the $z$ plane is located at $y = -c^2/\mathrm{Im}$ and the radius of the circle is $c\sqrt{4 + c^2/m^2}$.

The characteristic parameters of the airfoil are the chord $l$ and the camber height $h$, and these are shown in Fig. 4.17a. Since the equation of the airfoil has now been established, it is possible to relate these parameters to those in the $\zeta$ plane, namely, $c$ and $m$. Since the ends of the circular-arc airfoil lie on the real axis $y = 0$, the foregoing equation of the airfoil shows that the corresponding values of $x$ are $\pm 2c$. That is, the chord of the airfoil is

$$l = 4c$$

This is the same chord length as for the two previous airfoils.

The simplest way of establishing the camber $h$ of the airfoil is to use the fact that, in view of the result that the center of the circular arc is at $x = 0$, the maximum value of $y$ will occur when $x = 0$. But the parametric equation $x = (R + c^2/R)\cos \nu$ shows that this corresponds to $\nu = \pi/2$. Then the other parametric equation, namely, $y = 2m \sin^2 \nu$, shows that the maximum value of $y$ is $2m$. That is,

$$h = 2m$$

Using the foregoing results, the $\zeta$-plane parameters $c$ and $m$ may be replaced by the $z$-plane parameters $l/4$ and $h/2$, respectively. Then the equation of the airfoil surface in the $z$ plane may be written in the form

$$x^2 + \left(y + \frac{l^2}{8h}\right)^2 = \frac{l^2}{4}\left(1 + \frac{l^2}{16h^2}\right) \qquad (4.24a)$$

In order to satisfy the Kutta condition, the rear stagnation point must rotate through an angle greater than $\alpha$, the angle of the free stream. By rotating through the angle $\alpha$, the rear stagnation point will be located on the surface of the circle in the $\zeta$ plane at a point which is in the same horizontal plane as the center of the circle. But the center of the circle is located a distance $m$ above the real $\zeta$ axis. Thus, in order to be located at the point $\zeta = c$, the rear stagnation point must rotate through a further angle given by

$$\tan^{-1}\frac{m}{c} = \tan^{-1}\varepsilon$$

$$= \varepsilon + O(\varepsilon^2)$$

That is, in order to comply with the Kutta condition, the rear stagnation point must rotate through the angle $\alpha + m/c$, to the first order in $\varepsilon$. Then, from Eq.

(4.16), the required circulation is

$$\Gamma = 4\pi Ua \sin\left(\alpha + \frac{m}{c}\right)$$

but $a = \sqrt{c^2 + m^2}$ so that, to first order in $\varepsilon$, $a = c$. Hence

$$\Gamma = 4\pi Uc \sin\left(\alpha + \frac{m}{c}\right)$$

Then, from the Kutta-Joukowski law, the lift force is

$$Y = 4\pi\rho U^2 c \sin\left(\alpha + \frac{m}{c}\right)$$

and the corresponding lift coefficient is

$$C_L = 8\pi\frac{c}{l} \sin\left(\alpha + \frac{m}{c}\right)$$

Using again the fact that $c = l/4$ and $m = h/2$, the lift coefficient becomes

$$C_L = 2\pi \sin\left(\alpha + \frac{2h}{l}\right) \tag{4.24b}$$

Comparing this result with Eq. (4.22c), the corresponding result for the flat plate, shows that the effect of positive camber in an airfoil is to increase its lift coefficient. As a consequence of this increased lift coefficient a nonzero lift exists at zero angle of attack.

Since the center of the circle in the $\zeta$ plane is at $\zeta = im$ rather than $\zeta = 0$, the complex potential in the $\zeta$ plane may be obtained by replacing $\zeta$ by $\zeta - im$ in Eq. (4.29) and adding circulation. Thus the required complex potential is

$$F(\zeta) = U\left[(\zeta - im)e^{-i\alpha} + \frac{a^2}{\zeta - im}e^{i\alpha}\right] + \frac{i\Gamma}{2\pi}\log\left(\frac{\zeta - im}{a}\right) \tag{4.24c}$$

where

$$a = \frac{l}{4}$$

and

$$m = \frac{h}{2}$$

The magnitude of the circulation $\Gamma$ is given by

$$\Gamma = \pi Ul \sin\left(\alpha + \frac{2h}{l}\right)$$

and the parameter $c$ in the Joukowski transformation is $l/4$. The flow field corresponding to this complex potential is shown in Fig. 4.17b. As was the case with the flatplate airfoil, this flow field has a singularity at the leading edge. This singularity would not exist for airfoils of finite nose radius and would not exist even for sharp leading edges because of separation of the flow at the nose. In spite of this local inaccuracy, the results derived above are representative of the flow around thin cambered airfoils.

## 4.18   JOUKOWSKI AIRFOIL

The results of the two previous sections suggest that a cambered airfoil of finite thickness may be obtained by considering the Joukowski transformation in conjunction with a circle in the $\zeta$ plane whose center is in the second quadrant. Such a configuration is shown in Fig. 4.18$a$ in which the center of the circle is displaced a distance $m$ from the origin at an angle $\delta$ from the reference axis. In order that the trailing edge of the corresponding airfoil may be sharp, the circumference of the circle passes through the critical point $\zeta = c$. The principal parameters of the airfoil in the $z$ plane are also shown in Fig. 4.18$a$. These are the chord $l$, the maximum thickness $t$, and the maximum camber of the centerline $h$.

From the previous two sections it follows that, to the first order in $\varepsilon$, the centerline of the airfoil will be a circular arc whose center is on the $y$ axis and the airfoil will be symmetrical about its centerline. Then the equation of the upper and lower surfaces of the airfoil may be obtained from the equation for the circular-arc centerline plus or minus, respectively, and thickness effect. Hence from Eqs. (4.23$a$) and (4.24$a$) the airfoil profile will be given by

$$y = \sqrt{\frac{l^2}{4}\left(1 + \frac{l^2}{16h^2}\right) - x^2} - \frac{l^2}{8h} \pm 0.385t\left(1 - 2\frac{x}{l}\right)\sqrt{1 - \left(2\frac{x}{l}\right)^2} \quad (4.25a)$$

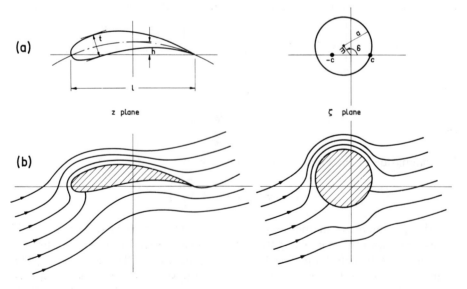

(a)

z plane

ς plane

(b)

**FIGURE 4.18**
The Joukowski airfoil; ($a$) the mapping planes and ($b$) uniform flow past the airfoil.

where the plus defines the upper surface and the minus defines the lower surface of the so-called "Joukowski airfoil."

It was observed that the effect of thickness on an airfoil was to increase its lift by an amount $0.77t/l$ and that the effect of camber was to increase the effective angle of attack to $\alpha + 2h/l$. The present airfoil has both these effects, so that, from Eqs. (4.23c) and (4.24b), the lift coefficient for the Joukowski airfoil will be

$$C_L = 2\pi\left(1 + 0.77\frac{t}{l}\right)\sin\left(\alpha + \frac{2h}{l}\right) \qquad (4.25b)$$

The complex potential in the $\zeta$ plane is

$$F(\zeta) = U\left[(\zeta - me^{i\delta})e^{-i\alpha} + \frac{a^2e^{i\alpha}}{\zeta - me^{i\delta}}\right] + \frac{i\Gamma}{2\pi}\log\left(\frac{\zeta - me^{i\delta}}{a}\right) \qquad (4.25c)$$

where $m\cos\delta = -0.77(tc/l)$, $m\sin\delta = h/2$, and $a = l/4 + 0.77(tc/l)$. These results follow from the observation that $-m\cos\delta$ replaces $m$ as used in the symmetrical Joukowski airfoil and $m\sin\delta$ replaces $m$ as used in the circular-arc airfoil. The magnitude of the circulation $\Gamma$ will include both the thickness and camber effects, and so it follows that

$$\Gamma = \pi Ul\left(1 + 0.77\frac{t}{l}\right)\sin\left(\alpha + \frac{2h}{l}\right)$$

The flow field corresponding to the foregoing complex potential is shown in Fig. 4.18b. It should be remembered that there is a limit to the amount of thickness and camber which may be introduced if the flow field is to remain as shown. As the thickness and/or camber of the airfoil increases, the bod departs more and more from a streamlined airfoil and approaches a bluff body. It was pointed out earlier that a consequence of this would be separation of the flow wich destroys the lift force and creates the so-called "stall condition."

## 4.19 SCHWARZ-CHRISTOFFEL TRANSFORMATION

Another conformal transformation which is of prime interest in the study of potential flows is the Schwarz-Christoffel transformation. This transformation is reviewed briefly in Appendix D, from which it will be seen that the mapping function is the solution to the following differential equation:

$$\frac{dz}{d\zeta} = K(\zeta - a)^{\alpha/\pi - 1}(\zeta - b)^{\beta/\pi - 1}(\zeta - c)^{\gamma/\pi - 1}\cdots$$

where $a$, $b$, $c$, etc., are the locations in the $\zeta$ plane of the vertices of a polygon in the $z$ plane which subtend the internal angles $\alpha$, $\beta$, $\gamma$, etc. The quantity $K$ is an arbitrary constant, and normally three of the quantities $a$, $b$, $c$, etc., may be chosen arbitrarily. The manner in which this transformation is used will be illustrated through its application to a simple problem whose solution may be

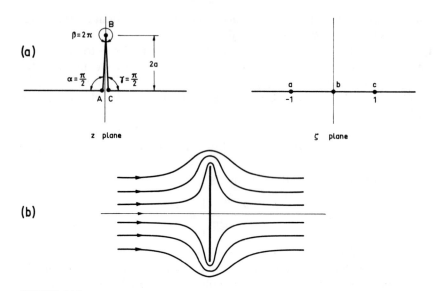

**FIGURE 4.19**
Flow around a vertical flat plate assuming nonseparated flow; (a) Schwarz-Christoffel mapping planes and (b) the flow field.

deduced from previously established results. This will permit a direct and independent check on the solution obtained through use of the Schwarz-Christoffel transformation.

The problem which will be considered is that of obtaining the complex potential for the flow around a flat plate of finite length which is oriented such that it is perpendicular to the oncoming flow; that is, the angle of attack is 90°. The solution to this problem may be deduced from that of a vertically oriented ellipse, which was treated in Sec. 4.14, by a limiting procedure. The length of the plate which will be so obtained is $4a$; such a plate will be considered here. The stagnation streamline will be a line of symmetry for this problem, so that only one half of the plate, say the top half, need be considered. The stagnation streamline $\psi = 0$ is shown in Fig. 4.19a for the top half of the vertical plate of length $4a$. The plate itself is considered to be made up of the line $ABC$ which folds back on itself. The location of the vertices $A$, $B$, and $C$ in the $z$ plane are shown as the points $a$, $b$, and $c$, respectively, on the $\zeta$ plane. The points chosen for $a$, $b$, and $c$ are $-1$, 0, and 1, respectively.

The equation of the Schwarz-Christoffel mapping function is the solution to

$$\frac{dz}{d\zeta} = K(\zeta + 1)^{-1/2}(\zeta - 0)^{1}(\zeta - 1)^{-1/2}$$

where the fact that $\alpha = \pi/2$, $\beta = 2\pi$, and $\gamma = \pi/2$, as indicated in Fig. 4.19a, has been used. That is, the equation of the mapping function is

$$\frac{dz}{d\zeta} = K \frac{\zeta}{\sqrt{\zeta^2 - 1}}$$

or

$$z = K\sqrt{\zeta^2 - 1} + D$$

where $D$ is a constant of integration which is, in general, complex. The constants $K$ and $D$ will now be evaluated such that the points $A, B, C$ and $a, b, c$ correspond to each other. The conditions to be satisfied are

1. When $\zeta = 1$, $z = 0$.
2. When $\zeta = -1$, $z = 0$.
3. When $\zeta = 0$, $z = i2a$.

The first two conditions are satisfied by taking $D = 0$, while the third condition is satisfied by choosing $K = 2a$. Then the required mapping function is

$$z = 2a\sqrt{\zeta^2 - 1}$$

The complex potential in the $\zeta$ plane is that of a uniform rectilinear flow, since the streamline $\psi = 0$ has been stretched out along the real $\zeta$ axis. To find the magnitude of the uniform velocity, it is observed from the mapping function that as $\zeta \to \infty$, $z \to 2a\zeta$. Then, from Eq. (4.19), $W(\zeta) \to 2aW(z)$, so that for a flow of magnitude $U$ in the $z$ plane the magnitude in the $\zeta$ plane should be $2aU$. Thus the required complex potential is

$$F(\zeta) = 2aU\zeta$$

But from the mapping function

$$\zeta = \pm \sqrt{\left(\frac{z}{2a}\right)^2 + 1}$$

In order that $\zeta \to +\infty$ as $z \to +\infty$, so that the direction of the flow is correct, the positive root must be chosen. Hence the required inverse of the mapping function is

$$\zeta = \frac{1}{2a}\sqrt{z^2 + 4a^2}$$

The complex potential in the $z$ plane then becomes

$$F(z) = U\sqrt{z^2 + 4a^2}$$

This result may be checked b using the result for the uniform flow of magnitude $U$ past an ellipse of major semiaxis $(a + b^2/a)$ and minor semiaxis $(a - b^2/a)$, the latter being along the $x$ axis. The resulting complex potential is given by Eq. (4.21b), which was obtained through use of the Joukowski transformation. As

$b \rightarrow a$ in this result, the ellipse degenerates to a vertical flat plate of length $4a$. Substituting $b = a$ in the complex potential confirms the result derived above by use of the Schwarz-Christoffel transformation. The corresponding flow field is shown in Fig. 4.19b. From this figure, or from inspection of the complex potential, it will be see that infinite velocities, of the type discussed in Sec. 4.6, exist at $y = \pm 2a$. Clearly the Kutta condition cannot be applied in such a case, and so the fluid will separate from the two edges of the plate. That is, the complex potential derived here does not represent the actual flow field accurately because the fluid does not remain in contact with the plate as was implicitly assumed here. A more representative flow configuration for this problem will be analyzed at the end of this chapter.

## 4.20   SOURCE IN A CHANNEL

The Schwarz-Christoffel transformation may be used to solve a sequence of problems which are related to that of the flow generated by a line source which is located in a two-dimensional channel. Then consider a channel of width $2l$ and of infinite length in which a source is located midway between the channel walls. If the origin of the coordinate system in the $z$ plane is taken to be at the location of the source, it is clear that the resulting flow field will be symmetrical about both the $x$ axis and the $y$ axis. Then the entire $x$ axis and the portion $-l \leq$ Iy $\leq l$ of the $y$ axis will be streamlines, so that only the first quadrant of the flow field need be considered; the remainder will follow from symmetry. Figure 4.20a shows the first quadrant of the flow field in the $z$ plane in which the source is located at $z = 0$.

Considering the region $0 \leq x$, $0 \leq y \leq l$ to be bounded by the polygon which is to be mapped, the vertices $A$ and $B$ will be chosen to correspond to the points $\zeta = -1$ and $\zeta = 1$ as shown in Fig. 4.20a. The interior angles corresponding to the vertices $A$ and $B$ in the $z$ plane are $\pi/2$, so that the differential equation of the mapping is

$$\frac{dz}{d\zeta} = K(\zeta + 1)^{-1/2}(\zeta - 1)^{-1/2}$$

$$= \frac{K}{\sqrt{\zeta^2 - 1}}$$

hence
$$z = K \cosh^{-1} \zeta + D$$

where $D$ is a constant of integration. The constants $K$ and $D$ will now be evaluated such that the point $A$ corresponds to the point $a$ and the point $B$ corresponds to the point $b$. The required conditions are

1. When $\zeta = 1$, $z = 0$.
2. When $\zeta = -1$, $z = il$.

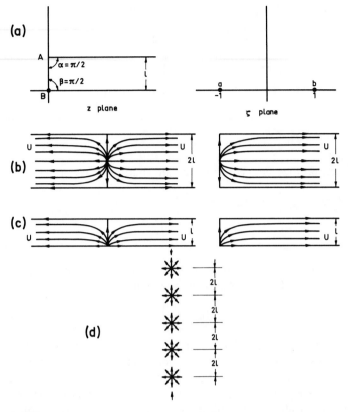

**FIGURE 4.20**
(*a*) Mapping planes for source in a channel, (*b*) the flow field for a full or semi-infinite channel, (*c*) the flow field for the source at the wall, and (*d*) an infinite array of sources.

The first condition is satisfied by setting $D = 0$, while the second condition is satisfied for $K = l/\pi$. Then the required mapping function is

$$z = \frac{l}{\pi} \cosh^{-1} \zeta$$

which has the inverse

$$\zeta = \cosh \frac{\pi z}{l}$$

The flow field in the $\zeta$ plane now corresponds to a source located at the point $\zeta = 1$. Hence the complex potential in the $\zeta$ plane is

$$F(\zeta) = \frac{m}{2\pi} \log(\zeta - 1)$$

so that the corresponding complex potential in the $z$ plane is

$$F(z) = \frac{m}{2\pi} \log \left( \cosh \frac{\pi z}{l} - 1 \right)$$

This result may be simplified slightly by using the identity

$$\cosh(X + Y) - \cosh(X - Y) = 2 \sinh X \sinh Y$$

for $X = Y = \pi z/(2l)$. Hence

$$\cosh \frac{\pi z}{l} - 1 = 2 \sinh^2 \frac{\pi z}{2l}$$

Thus the complex potential may be written in the alternative form

$$F(z) = \frac{m}{2\pi} \log \left( 2 \sinh^2 \frac{\pi z}{2l} \right)$$

$$= \frac{m}{\pi} \log \left( \sinh \frac{\pi z}{2l} \right) + \frac{m}{2\pi} \log 2$$

But the constant term may be neglected, since it does not affect the velocity components. That is, the complex potential may be taken to be

$$F(z) = \frac{m}{\pi} \log \left( \sinh \frac{\pi z}{2l} \right) \tag{4.26}$$

Equation (4.26) is the complex potential for the flow configurations shown in Fig. 4.20b, c, and d. Figure 4.20b shows the flow field due to a source which is located on the centerline of an infinitely long channel or at the center of the end of a semi-infinite channel. Figure 4.20c shows the flow field due to a source which is located on one wall of an infinite channel or at a corner of a semi-infinite channel. The foregoing flow configurations are clearly related to the largest flow field, shown in Fig. 4.20b, by symmetry. The total quantity of fluid leaving the source is $4lU$, so that the source strength $m$ in Eq. (4.26) should be $4lU$ in order that the channel velocity in the four configurations shown in Fig. 4.20b and c will be $U$.

Figure 4.20d depicts an infinite array of line sources spaced a distance $2l$ apart. The horizontal lines which pass midway between each pair of sources will obviously be streamlines for such an array of sources. It follows that the case of a source located in a horizontal channel may be thought of as only one component of an infinite number of such channels which are stacked on top of eac other in the vertical direction. Mathematically, the fact that Eq. (4.26) represents an infinite number of sources spread in the $y$ direction follows from the fact that the hyperbolic sine function repeats itself for imaginary values of its argument.

## 4.21  FLOW THROUGH AN APERTURE

One of the most impressive applications of the Schwarz-Christoffel transformation, in the field of fluid mechanics, is in the study of streaming motions which

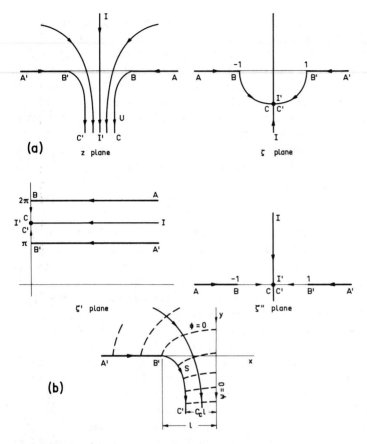

**FIGURE 4.21**
(*a*) Mapping planes for flow through a slit and (*b*) geometry of one of the free streamlines.

involve free streamlines. It is not usually known where these free streamlines lie, and this information must come out of the solution. The key to solving such problems is the so-called "hodograph plane," which uses the fact that along such free streamlines the pressure is constant. Two examples will be covered in this chapter, this first example being an application to the flow through a two-dimensional slit or aperture.

Figure 4.21*a* shows, in the *z* plane, a horizontal plate with an opening in it. The plate contains a semi-infinite expanse of fluid above it, and this fluid is draining through the aperture which is defined by the section *BB'* of the *x* axis. At the corners *B* and *B'* the flow will locally behave like that for the flow around a sharp edge, which was discussed in Sec. 4.6. It was pointed out in that section that if the fluid remained in contact with the solid boundary, infinite velocity components would result at the edge itself. Since this cannot be so physically, the fluid will not remain in contact with the solid boundary but will separate at the edge. In the case under discussion the bounding streamlines along the horizontal plate will curve toward the vertical as shown in Fig. 4.21a. The magnitude of the velocity in the resulting jet will reach some uniform magnitude $U$ downstream of all edge effects. The principal streamlines in the flow field have been labeled for identification purposes. These are the bounding streamline on the right, identified by the points $A$, $B$, and $C$, the bounding streamlines on the left, identified by the points $A'$, $B'$, and $C'$, and the central streamline $II'$. The free streamlines are $BC$ and $B'C'$.

The first transformation will be to the *hodograph plane*, which will be designated the $\zeta$ plane here. The transformation will be taken to be

$$\zeta = U\frac{dz}{dF} = \frac{U}{W} = \frac{U}{\sqrt{u^2 + v^2}}e^{i\theta} \tag{4.27a}$$

That is, the $\zeta$ plane is defined by the nondimensional reciprocal of the complex velocity and the last equality follows from the fact that $W = u - iv = \sqrt{u^2 + v^2}e^{-i\theta}$. The significance of this transformation is that the free streamlines, whose positions are unknown, are mapped onto the unit circle in the $\zeta$ plane, as will now be shown. In so doing, it should be noted that, by the above definition, $\theta$ is the angle subtended by the velocity vector in the $z$ plane.

Along the free streamlines $BC$ and $B'C'$ the pressure will be constant, typically atmospheric pressure, so that, from Bernoulli's equation, the quantity $u^2 + v^2$ will be constant. The value of this constant may be determined by noting that away from the edge effects, the velocity in the jet is $U$. Hence the value of $u^2 + v^2$ along the free streamlines is $U^2$. Then, along the free streamlines, $\zeta = e^{i\theta}$, which is the equation of the unit circle in the $\zeta$ plane. To find the portion of this unit circle which represents the free streamlines, it is observed that along the streamline $A'B'$ the angle $\theta$ of the velocity vector is 0 or $2\pi$ while the value of $\theta$ along $AB$ is $\pi$. Also, along the streamline $II'$ the angle $\theta$ of the velocity vector is $3\pi/2$. From these observations it is evident that the lower half of the unit circle in the $\zeta$ plane represents the streamlines $BC$ and $B'C'$, as shown in Fig. 4.21a. The other principal streamlines may be identified as follows: Along $A'B'$ the value of $\theta$ is 0 or $2\pi$, while $u^2 + v^2$ varies from 0 at $A'$ to $U^2$ at $B'$; hence $|\zeta|$ varies from infinity at $A'$ to unity at $B'$. Likewise, along $AB$ the value of $|\zeta|$ varies from infinity at $A$ to unity at $B$, with the value of $\theta$ being $\pi$. Finally, along the streamline $II'$ the value of $\theta$ is $3\pi/2$, while $u^2 + v^2$ varies from zero at $I$ to unity at $I'$, making $|\zeta|$ infinity at $I$ and unity at $I'$. This establishes the flow configuration shown in the $\zeta$ plane of Fig. 4.21a. Since the flow is toward the point $\zeta = -i$, which is identified by $C$, $C'$, and $I'$, there is a fluid sink there.

Since the principal streamlines in the $\zeta$ plane are either radial lines or the unit circle, the flow pattern may be mapped into a plane configuration by means of the logarithmic transformation. Then a second mapping is proposed to the $\zeta'$ plane, where $\zeta'$ is defined by

$$\zeta' = \log \zeta \tag{4.27b}$$

If a point in the $\zeta$ plane is represented by its polar coordinates $R$ and $\theta$, where

$R = U/(u^2 + v^2)^{1/2}$, then $\zeta = Re^{i\theta}$, so that

$$\zeta' = \log \zeta = \log R + i\theta$$

Thus the radial lines in the $\zeta$ plane become the horizontal lines defined by $\zeta' = \log R + i \times$ constant in the $\zeta'$ plane, while the unit circle $R = 1$ becomes the vertical line $\zeta' = i\theta$. Noting that the angle $\theta$ is the angle subtended by the velocity vector in the $z$ plane, it follows that the value of $\theta$ along $A'B'$ is 0 or $2\pi$. This gives the flow configuration shown in Fig. 4.21a in the $\zeta'$ plane, which corresponds to the flow in a semi-infinite channel due to a sink located at the center of the end of the channel. But it was seen in the previous section that such a configuration could be mapped into that of a simple source. Using the results of the previous section, a simple source flow will result in the $\zeta''$ plane through the mapping

$$\zeta'' = \cosh(\zeta' - i\pi)$$

i.e.,
$$\zeta'' = -\cosh \zeta' \qquad (4.27c)$$

Here the rectangle $ABCC'B'A'$ has been taken as the equivalent of the half channel of width $l$ which was considered in the previous section. Then the quantity $l$ which appeared in the transformation is $\pi$ in this case, and in order to bring the corner $B$ to the origin in the $\zeta'$ plane, the quantity $\zeta' - i\pi$ rather than $\zeta'$ is the appropriate variable.

The flow field in the $\zeta''$ plane is shown in Fig. 4.21a. The complex potential in this plane will be that for a simple sink which is located at $\zeta'' = 0$, so that

$$F(\zeta'') = -\frac{m}{2\pi} \log \zeta'' + K \qquad (4.27d)$$

where the constant $K$ has been added to permit the streamline $\psi = 0$ and the equipotential line $\phi = 0$ to correspond to a chosen streamline and equipotential line, respectively. Referring to Fig. 4.21b, it will now be specified that the streamline $\psi = 0$ be the streamline $II'$ and that the equipotential line $\phi = 0$ passes through the points $B'$ and $B$. Then, using the property of the stream function that the difference of the values of $\psi$ between two streamlines equals the volume of fluid flowing between these two streamlines, the value of $\psi$ along $A'B'C'$, which will be denoted by $\psi_{A'B'C'}$, may be identified. Considering the flow between the streamlines $II'$ and $A'B'C'$, it follows that

$$0 - \psi_{A'B'C'} = C_c lU$$

where $C_c$ is the contraction coefficient of the jet. Similarly, if $\psi_{ABC}$ is the value of $\psi$ along the streamline $ABC$, it follows by considering the flow between the streamlines $ABC$ and $II'$ that

$$\psi_{ABC} - 0 = C_c lU$$

That is,
$$\psi_{ABC} = C_c lU$$

and
$$\psi_{A'B'C'} = -C_c lU$$

Then, at the point $B'$, $\phi = 0$ and $\psi = -C_c lU$. Hence the value of the complex potential there is $0 - iC_c lU$. Applying this result to Eq. (4.27d) and noting that $\zeta'' = -1$ at the point $B'$ gives

$$0 - iC_c lU = -\frac{m}{2\pi} \log(-1) + K$$

$$-iC_c lU = -i\frac{m}{2} + K$$

Likewise at the point $B$ the complex potential is $0 + iC_c lU$ and the value of $\zeta''$ is unity; hence

$$0 + iC_c lU = -\frac{m}{2\pi} \log 1 + K$$

or

$$iC_c lU = K$$

These two equations show that $K = iC_c lU$ and $m = 4C_c lU$, so that the complex potential (4.27d) becomes

$$F(\zeta'') = -\frac{2C_c lU}{\pi} \log \zeta'' + iC_c lU$$

The corresponding complex potential in the $z$ plane may be obtained by use of the transformations (4.27a), (4.27b), and (4.27c). This gives

$$F(z) = -\frac{2C_c lU}{\pi} \log \left\{ \cosh\left[ \log\left( U\frac{dz}{dF} \right) - i\pi \right] \right\} + iC_c lU \quad (4.27e)$$

This result is an implicit expression for $F(z)$ rather than an explicit expression, since $dF/dz$ appears inside the expression for $F(z)$. However, the flow problem has, in principle, been solved, and it is possible t obtain useful information from the result. The quantity which is of prime interest in this problem is the value of the contraction coefficient $C_c$, so that this value will be determined below.

   In order to evaluate the contraction coefficient $C_c$, the equation of the free streamline $B'C'$ will be established. From this result the value of $x$ at the point $C'$ should be numerically equal to the half-jet dimension $C_c l$. This will enable the quantity $C_c$ to be evaluated. The equation of the free streamline $B'C'$ is most readily established in terms of a coordinate $s$ whose value is zero at the point $B'$ and whose magnitude increases along $B'C'$. Then, considering a small element of a curve, such as the streamline $B'C'$, whose slope is positive, it follows that

$$\frac{dx}{ds} = \cos\theta$$

$$\therefore \quad x = x_0 + \int_0^s \cos\theta \, ds$$

where the constant $x_0$ has been added to permit the condition $x = -l$ when $s = 0$ to be applied. The variation of $ds$ with $\theta$ is now required and, owing to the implicit nature of the mapping function (4.27a), this variation must be obtained by indirect methods as follows: The above expression for the lateral displacement $x$ of the jet surface may be written

$$x = x_0 + \int_{2\pi}^{\theta} \cos\theta \frac{ds}{d\zeta''} \frac{d\zeta''}{d\theta} d\theta$$

where the quantities $ds/d\zeta''$ and $d\zeta''/d\theta$ must be expressed in terms of $\theta$ before the integration may be performed. The value of $d\zeta''/d\theta$ on $B'C'$ will be obtained from the equations of the mappings, while $ds/d\zeta''$ will be obtained from the complex potential $F(\zeta'')$. Considering first the value of $ds/d\zeta''$, it may be stated that, on the streamline $B'C'$

$$1 = |\zeta| = \left|\frac{U}{W}\right| = \left|U\frac{dz}{dF}\right|$$

$$= \left|U\frac{dz}{d\zeta''}\frac{d\zeta''}{dF}\right|$$

But from Eq. (4.27d) with $m = 4C_c lU$,

$$\frac{dF}{d\zeta''} = -\frac{2C_c lU}{\pi}\frac{1}{\zeta''}$$

$$\therefore \quad 1 = \left|U\frac{dz}{d\zeta''}\frac{\pi}{2C_c lU}\zeta''\right|$$

On $B'C'$, $\zeta'' > 0$, so that

$$\left|\frac{dz}{d\zeta''}\right| = \frac{2C_c l}{\pi}\frac{1}{\zeta''}$$

Now on the streamline $B'C'$ the value of $dz$ may be represented by $ds\, e^{i\theta}$, where $ds$ is an element of the coordinate $s$ which was previously introduced. Also, along $B'C'$, $\zeta''$ is decreasing, so that $d\zeta'' < 0$. Hence

$$\frac{ds}{d\zeta''} = -\frac{2C_c l}{\pi}\frac{1}{\zeta''}$$

The equations of the various mappings may now be used to evaluate $\zeta''$ in terms of $\theta$. On the streamline $B'C'$ the value of $\zeta$ is

$$\zeta = e^{i\theta}$$

$$\therefore \quad \zeta' = \log\zeta = i\theta$$

and

$$\zeta'' = -\cosh\zeta' = -\cos\theta$$

$$\therefore \quad \frac{d\zeta''}{d\theta} = \sin\theta$$

and

$$\frac{ds}{d\zeta''} = \frac{2C_c l}{\pi}\frac{1}{\cos\theta}$$

Using these last two equations, the expression for the lateral displacement $x$ of the free streamline becomes

$$x = x_0 + \int_{2\pi}^{\theta} \cos\theta \, \frac{2C_c l}{\pi \cos\theta} \sin\theta \, d\theta$$

$$= x_0 + \frac{2C_c l}{\pi} \int_{2\pi}^{\theta} \sin\theta \, d\theta$$

$$= x_0 + \frac{2C_c l}{\pi} (1 - \cos\theta)$$

But when $\theta = 2\pi$, that is, at the point $B'$, $x = -l$. Hence $x_0 = -l$, so that

$$x = -l + \frac{2C_c l}{\pi} (1 - \cos\theta)$$

Also, the value of $x$ at the point $C'$ is $-C_c l$, while the value of $\theta$ is $3\pi/2$. Thus

$$-C_c l = -l + \frac{2C_c l}{\pi}$$

$$\therefore \quad C_c = \frac{\pi}{\pi + 2} \tag{4.27f}$$

Equation (4.27f) predicts that the free jet which emerges from the aperture will assume a width which is .611 of the width of the slit. This result is well established experimentally, and the figure of 0.611 has been confirmed for openings under deep liquids.

## 4.22   FLOW PAST A VERTICAL FLAT PLATE

In Sec. 4.19 the complex potential for the flow around a flat plate which is oriented perpendicular to the free stream was obtained. However, it was pointed out that the result obtained at that time was unrealistic because it required infinite velocity components at the two edges of the plate. It would therefore appear that the assumption of attached flow, which was implicitly made at that time, is not valid. The same problem will be treated here, but this time it will be assumed that the flow separates from the surface of the plate at the two edges. The resulting free streamlines will be treated in a manner which is similar to that in which the free jet was treated in the previous section.

Figure 4.22 shows, in the $z$ plane, the assumed flow configuration for a uniform rectilinear flow of magnitude $U$ approaching a vertically oriented flat plate of height $2l$. The stagnation streamline $II'$ splits upon reaching the plate and forms the bounding streamlines $ABC$ and $A'B'C'$, where $BC$ and $B'C'$ are the free streamlines. The region downstream of the plate between the two free streamlines is interpreted as being a cavity which has a uniform pressure throughout.

As was the case in the previous section, the free streamlines may be handled by use of the hodograph plane. Hence a transformation is made to the

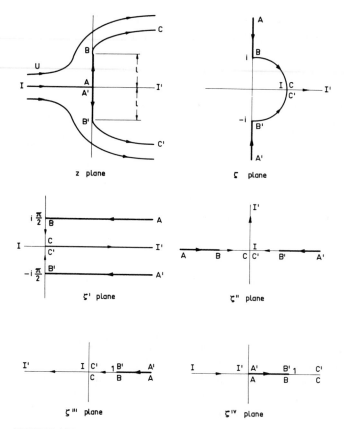

**FIGURE 4.22**
Mapping planes for flow over a flat plate which is oriented perpendicular to a uniform flow.

$\zeta$ plane, where

$$\zeta = U\frac{dz}{dF} = \frac{U}{W} = \frac{U}{\sqrt{u^2 + v^2}}e^{i\theta} \qquad (4.28a)$$

The boundaries of the flow field in the $\zeta$ plane are shown in Fig. 4.22. The free streamlines $BC$ and $B'C'$ again become part of the unit circle in the $\zeta$ plane. Since the value of $\theta$ along $ABC$ lies between $\pi/2$ and 0 and the value of $\theta$ along $A'B'C'$ lies between $-\pi/2$ and 0, the appropriate portion of the unit circle is that which lies in the first and fourth quadrants. Since the flow boundary crosses the positive portion of the real $\zeta$ axis, it is no longer convenient to consider the range of $\theta$ to be $0 \le \theta \le 2\pi$. If this range were adopted, the multivalued functions, which require branch cuts, would divide the flow boundary. This difficulty may be simply overcome by considering the branch cut to lie along the negative real $\zeta$ axis so that the principal value of the multivalued functions will correspond to $-\pi \le \theta \le \pi$.

The geometry of radial lines and the circular contour is next converted to that of a plane figure by means of the logarithmic transformation. That is, a mapping to the $\zeta'$ plane is made where

$$\zeta' = \log \zeta \qquad (4.28b)$$

This maps the flow boundary into that of a rectangular channel as shown in Fig. 4.22. Since the range of $\theta$ is now $-\pi \leq \theta \leq \pi$, the lower wall of this channel corresponds to the imaginary part of $\zeta'$ being $-\pi/2$, and the upper wall corresponds to $-\pi/2$. That is, the centerline of the channel corresponds to the real $\zeta'$ axis. Then the flow field may be stretched out using the same transformation which was used in Sec. 4.20 for a source in a channel. Here the corner $B'$ is located at $\zeta' = -i\pi/2$ rather than $\zeta' = 0$ and the channel half width is $\pi$ instead of $l$. Hence the required transformation is

$$\zeta'' = \cosh\left(\zeta' + i\frac{\pi}{2}\right) \qquad (4.28c)$$

The principal flow lines in the $\zeta''$ plane ma be made collinear by means of the mapping

$$\zeta''' = (\zeta'')^2 \qquad (4.28d)$$

This doubles the angles subtended by the principal streamlines, so that the flow in the $\zeta'''$ plane is unidirectional along the principal streamlines, as shown in Fig. 4.22. However, the flow is still not that of a uniform flow as the principal streamlines might suggest. This may be confirmed by observing that, in the $z$ plane, there is a source of fluid at $I$ which flows toward a sink at $CC'$. But in the $\zeta'''$ plane $CC'$ and $I$ are at the same location, so that the flow in the $\zeta'''$ plane is probably that of a doublet. Rather than prove this is so, one final transformation will be made to the $\zeta''''$ plane where

$$\zeta'''' = \frac{1}{\zeta'''} \qquad (4.28e)$$

The effect of this transformation is to map the origin to infinity, and vice versa, as shown in Fig. 4.22. Fluid emanates from $I$ and flows toward $CC'$, as was the case in the $z$ plane. That is, the flow in the $\zeta''''$ plane is that of a uniform rectilinear flow so that the complex potential is

$$F(\zeta''') = K\zeta''''$$

The value of the constant $K$, which represents the magnitude of the uniform flow, would normally be obtained by relating complex velocities through Eq. (4.19). However, the implicit nature of the hodograph transformation prohibits this to be done directly, so that indirect methods must be used, as was the case in the previous section. The hodograph transformation involves $dF/dz$, but $F$ is known only as a function of $\zeta''''$. Hence it is proposed to start with the hodograph transformation and express the variables in terms of $\zeta''$. Also, $F(\zeta'''')$

is known, so that $F(\zeta'')$ may be calculated, and so an identity will be established in the $\zeta''$ plane. The details now follow.

From Eq. (4.28a) the following identity is established:

$$U \frac{dz}{dF} = \zeta$$

$$\therefore \quad U \frac{dz}{d\zeta''} \frac{d\zeta''}{dF} = \zeta$$

$dF/d\zeta''$ may be evaluated from the complex potential $F(\zeta'''')$ and the mapping functions, while $\zeta$ may be evaluated from the mapping functions. Both quantities will be expressed in terms of $\zeta''$.

$$F(\zeta'''') = K\zeta''''$$

$$\therefore \quad F(\zeta''') = \frac{K}{\zeta'''}$$

and

$$F(\zeta'') = \frac{K}{(\zeta'')^2}$$

$$\therefore \quad \frac{dF}{d\zeta''} = -\frac{2K}{(\zeta'')^3}$$

also

$$\zeta = e^{\zeta'}$$

$$= e^{(\cosh^{-1}\zeta'' - i\pi/2)}$$

$$= -ie^{\cosh^{-1}\zeta''}$$

But $\cosh^{-1} x = \log(x + \sqrt{x^2 - 1})$, so that

$$\zeta = -i\left(\zeta'' + \sqrt{(\zeta'')^2 - 1}\right)$$

Substituting the above expressions for $dF/d\zeta''$ and $\zeta$ into the identity established from the hodograph transformation gives

$$-U \frac{dz}{d\zeta''} \frac{(\zeta'')^3}{2K} = -i\left(\zeta'' + \sqrt{(\zeta'')^2 - 1}\right)$$

or

$$U\,dz = i2K \frac{\zeta'' + \sqrt{(\zeta'')^2 - 1}}{(\zeta'')^3} d\zeta''$$

In order to establish an algebraic identity which will permit the constant $K$ to be evaluated, the above expression must be integrated. It is proposed to integrate over the region $B'$ to $A'$ so that

$$U\int_{-il}^{0} dz = i2K \int_{1}^{\infty} \frac{\zeta'' + \sqrt{(\zeta'')^2 - 1}}{(\zeta'')^3} d\zeta''$$

where the upper limits of integration correspond to the point $A'$ and the lower limits to the point $B'$. The integral on the right-hand side may be conveniently evaluated by use of the substitution $\zeta' = 1/\sin v$. Then

$$U \int_{-il}^{0} dz = -i2K \int_{\pi/2}^{0} (1 + \cos v) \cos v \, dv$$

$$\therefore \quad iUl = i2K\left(1 + \frac{\pi}{4}\right)$$

or

$$K = \frac{2Ul}{\pi + 4}$$

Then the complex potential in the $\zeta''''$ plane becomes

$$F(\zeta'''') = \frac{2Ul}{\pi + 4} \zeta''''$$

The corresponding complex potential in the $z$ plane may be obtained by using the mapping equations (4.28a), (4.28b), (4.28c), (4.28d), and (4.28e). Hence, using the fact that $\cosh(\zeta' + i\pi/2) = i \sinh \zeta'$, the corresponding expression for $F(z)$ is

$$F(z) = -\frac{2Ul}{\pi + 4} \frac{1}{\sinh^2\{\log[U(dz/dF)]\}} \tag{4.28f}$$

This result is again an implicit expression for $F(z)$ rather than an explicit expression. However, since the flow problem has been solved, results may be deduced from the solution. Here the result which is of interest is the drag force acting on the plate. Thus if $X$ is the drag force acting on plate in the positive $x$ direction and $P$ is the pressure in the cavity, it follows that

$$X = 2\int_{-l}^{0} (p - P) \, dy$$

where the symmetry of the flow field about the $x$ axis has been invoked. But from the Bernoulli equation

$$\frac{p}{\rho} + \tfrac{1}{2}(u^2 + v^2) = \frac{P}{\rho} + \tfrac{1}{2}U^2$$

where the Bernoulli constant has been evaluated on the free streamlines at a position well downstream of the plate. Thus the expression for the drag force $X$ becomes

$$X = 2\int_{-l}^{0} \tfrac{1}{2}\rho\left[U^2 - (u^2 + v^2)\right] dy$$

$$= \rho U^2 \int_{-1}^{0} dy - \rho \int_{-1}^{0} (u^2 + v^2) \, dy$$

The first integral may be evaluated explicitly, while the integrand of the second integral may be written as $v^2$, since $u = 0$ on the surface of the plate. Then $W = dF/dz = -iv$ on the surface of the plate, so that $v^2 = -W^2$ there.

That is,

$$X = \rho U^2 l + \rho \int_{-l}^{0} \left( \frac{dF}{dz} \right)^2 dy$$

Also, $x = 0$ on the surface of the plate, so that $z = iy$ there. That is, $dz = i\,dy$ on the plate, so that

$$X = \rho U^2 l - i\rho \int_{-il}^{0} \left( \frac{dF}{dz} \right)^2 dz$$

Now since $F(z)$ is an implicit expression, it is proposed to evaluate $F(\zeta'')$ and to perform the indicated integration in the $\zeta''$ plane rather than the $z$ plane. This may be done as follows:

$$X = \rho U^2 l - i\rho \int_{1}^{\infty} \left( \frac{dF}{d\zeta''} \right)^2 \frac{d\zeta''}{dz} \, d\zeta''$$

where it has been noted that as $z$ varies from $-il$ to 0, $\zeta''$ varies from unity to

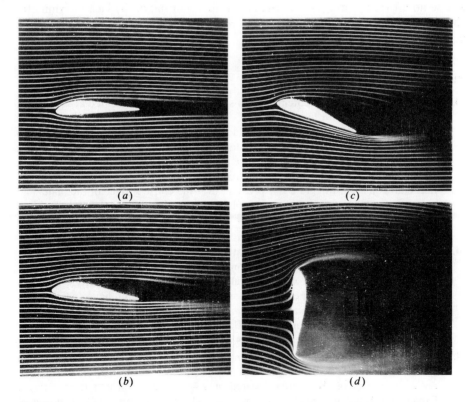

**PLATE 1**
Airfoil in uniform flow at angle of attack of $(a)$ $-2°$, $(b)$ $8°$, $(c)$ $20°$, and $(d)$ $90°$.

infinity. But expressions for $dF/d\zeta''$ and $dz/d\zeta''$ were obtained earlier in terms of $\zeta''$ and $K$. Then using these expressions and the fact that $K = 2Ul/(\pi + 4)$, the expression for the drag force becomes

$$X = \rho U^2 l - i\rho \int_1^\infty \left[ -\frac{4Ul}{\pi + 4} \frac{1}{(\zeta'')^3} \right]^2 \frac{(\pi + 4)(\zeta'')^3}{i4l\left[ \zeta'' + \sqrt{(\zeta'')^2 - 1} \right]} \, d\zeta$$

$$= \rho U^2 l - \frac{4\rho U^2}{\pi + 4} \int_1^\infty \frac{d\zeta''}{(\zeta'')^3 \left[ \zeta'' + \sqrt{(\zeta'')^2 - 1} \right]}$$

The integral may now be readily evaluated by means of the substitution $\zeta'' = 1/\sin \nu$. This gives

$$X = \rho U^2 l \left[ 1 + \frac{4}{\pi + 4} \int_{\pi/2}^0 (1 - \cos \nu) \cos \nu \, d\nu \right]$$

$$X = \frac{2\pi}{\pi + 4} \rho U^2 l$$

(4.28 g)

From the symmetry of the flow field about the $x$ axis it may be stated that there will be no lift force acting on the plate. On the other hand, the lack of symmetry about the $y$ axis implies the existence of a drag force, and Eq. (4.28 g) gives the magnitude of this drag force.

## PROBLEMS

**4.1.** Write down the complex potential for a source of strength $m$ located at $z = ih$ and a source of strength $m$ located at $z = -ih$. Show that the real axis is a streamline in the resulting flow field, and so deduce that the complex potential for the two sources is also the complex potential for a flat plate located along $y = 0$ with a source of strength $m$ located a distance $h$ above it.

Obtain the pressure on the surface of the plate mentioned above from the Bernoulli equation. Integrate this pressure over the entire surface of the plate, and so show that the force acting on the plate, due to the presence of the source, is $\rho m^2/(4\pi h)$. Take the pressure below the plate to be equal to the stagnation pressure in the fluid.

**4.2.** Consider a source of strength $m$ located at $z = -b$ and a sink of strength $m$ located at $z = b$. Write down the complex potential for the resulting flow field, adding a constant term $-im/2$ to make the streamline $\psi = 0$ correspond to a certain position. Expand the result for small values of $z/b$ and hence show that if $b \to \infty$ and $m \to \infty$ in such a way that $m/b \to \pi U$, the resulting complex potential is that of a uniform flow of magnitude $U$. That is, a uniform flow may be thought of as consisting of a source located at $-\infty$ and a sink at $+\infty$.

**4.3.** Write down the complex potential for a source of strength $m$ located at $z = -b$, a source of strength $m$ located at $z = -a^2/b$, a sink of strength $m$ located at $z = a^2/b$, a sink of strength $m$ located at $z = b$, and a constant term $-im/2$. Expand the result for small values of $z/b$ and $a^2/(bz)$, and hence show that if

$b \to \infty$ and $m \to \infty$ in such a way that $m/b \to \pi U$, the resulting complex potential is that of a uniform flow of magnitude $U$ flowing past a circular cylinder of radius $a$.

**4.4.** Consider a system of singularities consisting of a source of strength $m$ located at $z = -b$, a source of strength $m$ located at $z = -a^2/b$, a sink of strength $m$ located at $z = a^2/l$, and a sink of strength $m$ located at $z = l$. Write down the complex potential for these sources and sinks, and add a constant term $-[m/(2\pi)] \log b$. Let $b \to \infty$, and show that the result represents the complex potential for a circular cylinder of radius $a$ with a sink of strength $m$ located a distance $l$ to the right of the axis of the cylinder. This may be done by showing that the circle of radius $a$ is a streamline.

Use the Blasius integral theorem for a contour of integration which includes the cylinder but excludes the sink, and hence show that the force acting on the cylinder is

$$X = \frac{\rho m^2 a^2}{2\pi l (l^2 - a^2)}$$

**4.5.** Obtain the complex potential for a source of strength $m$ located at $z = be^{i(\alpha + \pi)}$, a source of strength $m$ located at $z = (a^2/b)e^{i(\alpha + \pi)}$, a sink of strength $m$ located at $z = (a^2/b)e^{i\alpha}$, a sink of strength $m$ located at $z = be^{i\alpha}$, and a constant term of magnitude $-im/2$. Expand this result for small values of $z/b$ and $a^2/(bz)$, and hence show that as $b \to \infty$ and as $m \to \infty$ such that $m/b \to \pi U$, the circle of radius $a$ is a streamline. Hence show that the complex potential for a uniform flow of magnitude $U$ approaching a circular cylinder of radius $a$ at an angle of attack $\alpha$ to the horizontal is

$$F(z) = U\left( ze^{-i\alpha} + \frac{a^2}{z} e^{i\alpha} \right) \tag{4.29}$$

**4.6.** Determine the complex potential for a circular cylinder of radius $a$ in a flow field which is produced by a counterclockwise vortex of strength $\Gamma$ located a distance $l$ from the axis of the cylinder. This may be done by writing the complex potential for a clockwise vortex of strength $\Gamma$ located at $z = a^2/l$, a counterclockwise vortex of strength $\Gamma$ located at $z = l$, and a constant of magnitude $-[i\Gamma/(2\pi)] \log b$. Then let $b \to \infty$ and show that the circle of radius $a$ is a streamline.

Obtain the value of the force acting on the cylinder by applying the Blasius law to a contour which includes the cylinder but excludes the vortex at $z = l$.

**4.7.** From the Bernoulli equation and Eq. (4.14), obtain an expression for the pressure $p(a, \theta)$ on the surface of a circular cylinder of radius $a$ in a uniform flow of magnitude $U$ in which the cylinder has a bound clockwise vortex of strength $\Gamma$ around it. Integrate the quantity $-p(a, \theta)a \sin \theta$ around the surface of the cylinder, and hence confirm the validity of the Kutta-Joukowski law for this particular flow.

**4.8.** Figure 4.23($a$) shows the assumed configuration for separated flow of magnitude $U$ approaching a circular cylinder of radius $a$. Figure 4.23($b$) shows a system of flow elements with which it is proposed to model this particular flow. The model consists of a uniform flow of magnitude $U$, a pair of sources of strength $m_1$, one being located at $z = ae^{i\theta_1}$ and the other at $z = ae^{-i\theta_1}$, a pair of sinks of strength

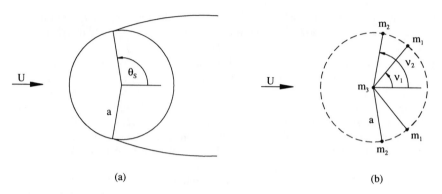

**FIGURE 4.23**
(*a*) Assumed flow configuration and (*b*) flow model.

$m_2$, one being located at $z = ae^{i\theta_2}$ and the other at $z = ae^{-i\theta_2}$, and a sink of strength $m_3$ located at $z = 0$. Write down the complex potential for this system of flow elements.

Find the strength of the sink $m_3$ which makes the circle of radius $R = a$ a streamline.

Let $q(R, \theta)$ denote the magnitude of the velocity vector at the point $(R, \theta)$. Obtain an expression for the value of the velocity magnitude on the circle of radius $a$; that is, obtain an expression for $q(a, \theta)$. The result should be expressed in terms of the parameters $U$, $a$, $m_1$, $m_2$, $\theta_1$, and $\theta_2$.

In order to determine the parameters $m_2$ and $\delta_2$, two conditions are postulated to be valid at the separation point $(a, \theta_s)$, where $\theta_s$ is the polar angle corresponding to the location of the separation points. These two conditions are given below in which $c$ is an experimentally determined constant and $R_e$ is the Reynolds number:

$$\frac{dU}{d\theta}(a, \theta_s) = 0$$

and

$$\frac{d^2U}{d\theta^2}(a, \theta_s) = cUR_e^{1/2}$$

Express the result in the functional form:

$$\delta_2 = \delta_2(\theta_s, c, R_e, m_1, \delta_1)$$

and

$$m_2 = m_2(\theta_s, m_1, \delta_1, \delta_2)$$

**4.9.** A mapping function is defined by the following equation:

$$z = \zeta + \frac{c^n}{(n-1)\zeta^{n-1}}$$

where

$$z = x + iy = Re^{i\theta}$$

and

$$\zeta = \xi + i\eta = \rho e^{i\nu}$$

In the above, $c$ is a real constant and $n$ is an integer. Find the location of the critical points of this mapping function in the $\zeta$ plane, and sketch their locations for $n = 2$, for $n = 3$ and for $n = 4$.

Using the cartesian representation of $z$ and the polar representation of $\zeta$, find the equations of the mapping in the form:

$$x = x(\rho, \nu) \quad \text{and} \quad y = y(\rho, \nu)$$

From these equations obtain expressions for the polar coordinates $R$ and $\theta$ of the form:

$$R = R(\rho, \nu) \quad \text{and} \quad \theta = \theta(\rho, \nu)$$

Consider a circle in the $\zeta$ plane whose radius $\rho$ is large so that

$$\frac{c}{\rho} = \varepsilon \ll 1$$

Use the results obtained above for $R$ and $\theta$ to find the equations for the mapping of this circle in the $z$ plane, working to the first order only in the small parameter $\varepsilon$. Sketch the resulting shape in the $z$ plane for $n = 3$.

**4.10.** Consider the Joukowski transformation as defined by Eq. (4.20). Suppose that the figure which exists in the $\zeta$ plane is a circle whose center lies in the second quadrant. It is required to construct from an accurately prepared drawing the corresponding figure in the $z$ plane as follows.

Using either (a) or (b) below, depending on the preferred system of units, draw the circle specified in the $\zeta$ plane. From this drawing, prepare a table of values of the cartesian coordinates for points on the circle using 15° increments in the polar angle between adjacent points. Next, use the Joukowski transformation to prepare a second table of values for the cartesian points in the $z$ plane corresponding to the measured values in the $\zeta$ plane. Finally, draw the figure which is produced in the $z$ plane.

(a) SI Units:

| | | |
|---|---|---|
| Joukowski parameter, $c$ | = | 60.0 mm |
| Center of circle in $\zeta$ plane | = | $(-5.0 \text{ mm}, +7.5 \text{ mm})$ |
| Circle to pass through point | = | $(+60.0 \text{ mm}, 0.0 \text{ mm})$ |

(b) English Units:

| | | |
|---|---|---|
| Joukowski parameter, $c$ | = | 2.4″ |
| Center of circle in $\zeta$ plane | = | $(-0.2″, +0.3″)$ |
| Circle to pass through point | = | $(+2.4″, 0.0″)$ |

**4.11.** Using either (a) or (b) below, depending on the preferred system of units, determine the lift force generated by a short span of an aircraft wing whose cross section is the same shape as that of problem 4.10. Take the air properties to be defined by the standard atmosphere at sea level.

(a) SI Units:

| | | |
|---|---|---|
| Length of wing element | = | 1.0 m |
| Chord of wing element | = | 3.0 m |
| Flight speed | = | 250 m/second |

(b) English Units:

| | | |
|---|---|---|
| Length of wing element | = | 3.0 feet |
| Chord of wing element | = | 9.0 feet |
| Flight speed | = | 750 feet/second |

**4.12.** Find the transformation which maps the interior of the sector $0 \le \theta \le \pi/n$ in the z plane onto the upper half of the $\zeta$ plane. Thus by considering a uniform flow in the $\zeta$ plane obtain the complex potential for the flow in the sector $0 \le \theta \le \pi/n$ in the z plane.

**4.13.** Use the Schwarz-Christoffel transformation to find the mapping which transforms the interior of the 90° bend shown in the z plane of Fig. 4.24 onto the upper half of the $\zeta$ plane as shown. Hence obtain the complex potential for the flow around in a right-angled bend in terms of the channel width $l$ and the approach velocity $U$.

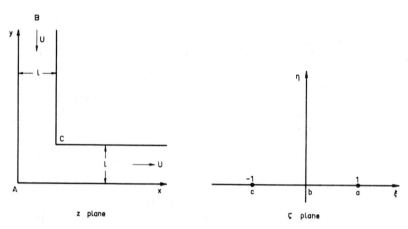

**FIGURE 4.24**
Mapping planes for flow in a channel having a 90° bend.

**4.14.** The z plane of Fig. 4.25 shows one half of a symmetric expansion device, or diffuser. It is assumed that the angle $\phi$ is not large and that the flow remains in contact with the wall. In the $\zeta$ plane the points specified by lower case letters correspond to the points indicated by the capitalized points in the z plane. The

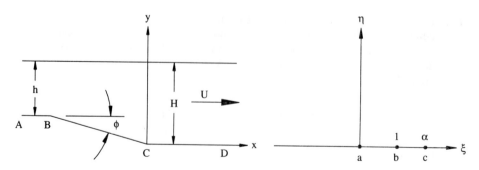

**FIGURE 4.25**
Mapping planes for diffuser.

location of the point $e$, indicated by the value $\alpha$, is undefined at this time. Using the correspondence indicated, find the differential equation of the mapping function if the angle $\phi$ is taken to be the ratio of two integers times $90°$; that is:

$$\phi = \frac{r}{n}\frac{\pi}{2}$$

where $r$ and $n$ are integers. Express the result in terms of the parameters $r$, $n$, $\alpha$, and $K$, where $K$ is the scale parameter in the Schwarz-Christoffel transformation.

Noting that the strength of the source in the $\zeta$ plane is $m = 2UH$, obtain the complex potential for the flow field in the $\zeta$ plane. From this result obtain an expression for the complex velocity in the $z$ plane, expressing the result in terms of the variable $\zeta$ and the parameters $U$, $\alpha$, $r$, $n$, and $K$.

Use the result obtained above to evaluate the parameters $K$ and $\alpha$, expressing them in terms of the remaining parameters $h$, $H$, $r$, and $n$.

**4.15.** Figure 4.26 shows a channel with a step in it in the $z$ plane. Show that the mapping function which maps the interior of this channel onto the upper half of the $\zeta$ plane is

$$z = \frac{H}{\pi}\left[\log\left(\frac{1+s}{1-s}\right) - \frac{h}{H}\log\left(\frac{H/h+s}{H/h-s}\right)\right]$$

where

$$s^2 = \frac{\zeta - (H/h)^2}{\zeta - 1}$$

Let the points $A$, $B$, and $C$ in the $z$ plane correspond to the points 0, 1, and $\alpha$ respectively, in the $\zeta$ plane. The quantity $\alpha$ may not be specified priori, but it should be determined from the mapping function after the points $A$ and $B$ have been located as desired. Note that the streamline $ABCD$ may be considered to be the streamline dividing two symmetrical regions, so that this mapping function also applies to a channel of width $2H$ with an obstacle of width $2(H - h)$ located along the centerline of the channel.

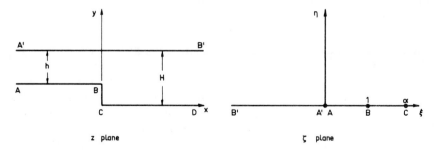

**FIGURE 4.26**
Mapping planes for a channel with a step in it.

# CHAPTER
# 5

# THREE-DIMENSIONAL POTENTIAL FLOWS

Although there are no significant phenomena associated with three-dimensional flows which do not exist in two-dimensional flows, the method of analyzing flow problems is completely different. The method of employing analytic functions of complex variables cannot be used here in view of the three-dimensionality of the problems. Then, we must resort to solving the partial differential equations which govern the variables of the flow field. These partial differential equations were reviewed at the beginning of Part II and, for irrotational motion, the equation governing the velocity potential is given by Eq. (II.5). Having solved the flow problem for $\phi$, the pressure may be obtained from the Bernoulli equation, which is expressed in Eq. (II.6).

The chapter begins by reviewing the equation which is to be satisfied by the velocity potential in spherical coordinates. Then it is shown that for axisymmetric flows a stream function exists, called the *Stokes stream function*. Although the flow fields may be solved through the velocity potential, the stream function is useful for interpreting the flow fields. Fundamental solutions are then established by solving the Laplace equation for $\phi$ by separation of variables. These fundamental solutions are then superimposed to establish the flow around a few three-dimensional bodies, including the sphere and a family of solid bodies known as *Rankine solids*. A study of the forces which act on three-dimensional bodies is then made, which leads to *d'Alembert's paradox*. This paradox shows that for any body immersed in a potential flow no forces exist on the body, in spite of the fact that forces are observed to exist experimentally. The chapter ends by introducing the notion of an apparent mass

138

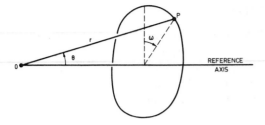

**FIGURE 5.1**
Definition sketch of spherical coordinates.

for a body in a potential flow. This concept allows the fluid to be ignored if a certain additional mass is associated with the body when its dynamics are considered.

Since bodies of interest, such as airship hulls and submarine vehicle hulls, have an axis of symmetry, this chapter will consider only three-dimensional bodies which are axisymmetric. In so doing, it will be found convenient to work in the spherical coordinates $(r, \theta, \omega)$. These coordinates are shown in Fig. 5.1. Since the axis of symmetry of bodies is invariably in the streamwise direction and since the approaching flow is normally taken to be horizontal, the reference axis of the coordinate system is also taken to be horizontal. Then, in terms of the spherical coordinates $(r, \theta, \omega)$, a point $P$ may be represented by its radius $r$ from the origin, the angle $\theta$ between the reference axis and the radius vector $r$, and the angle $\omega$ subtended by the perpendicular to the reference axis which passes through $P$. For axisymmetric flows, there will be no variation in the fluid properties as $\omega$ varies from 0 to $2\pi$ while $r$ and $\theta$ are held constant.

## 5.1  VELOCITY POTENTIAL

Although the topic of discussion is three-dimensional flows, these flow fields are supposed to be potential. That is, the fluid motion is assumed to be irrotational so that a velocity potential exists, irrespective of the dimensionality of the flow field. Then, the equation to be satisfied by the velocity potential is Laplace's equation, given by Eq. (II.5). Hence, expanding the Laplacian in spherical coordinates and using the fact that $\partial/\partial\omega = 0$ for axisymmetric flows, the equation to be satisfied by $\phi$ is

$$\frac{1}{r^2}\frac{\partial}{\partial r}\left(r^2\frac{\partial\phi}{\partial r}\right) + \frac{1}{r^2\sin\theta}\frac{\partial}{\partial\theta}\left(\sin\theta\frac{\partial\phi}{\partial\theta}\right) = 0 \tag{5.1}$$

The velocity components are related to the velocity potential by Eq. (II.4), which in spherical coordinates, gives

$$u_r = \frac{\partial\phi}{\partial r} \tag{5.2a}$$

$$u_\theta = \frac{1}{r}\frac{\partial\phi}{\partial\theta} \tag{5.2b}$$

and the third velocity component $u_\omega$ is zero for axisymmetric flows.

## 5.2   STOKES' STREAM FUNCTION

In the previous chapter a stream function was introduced which, by its definition, satisfied the two-dimensional continuity equation. In three dimensions it is
not possible, in general, to satisfy the continuity equation by a single scalar
function. However, in the case of axisymmetric flows such a function does exist.
The continuity equation for the incompressible case under consideration is, for
axisymmetric flows,

$$\frac{1}{r^2}\frac{\partial}{\partial r}\left(r^2 u_r\right) + \frac{1}{r\sin\theta}\frac{\partial}{\partial\theta}\left(u_\theta\sin\theta\right) = 0$$

Now consider the velocity components to be related to a function $\psi$ in the
following way:

$$u_r = \frac{1}{r^2\sin\theta}\frac{\partial\psi}{\partial\theta} \tag{5.3a}$$

$$u_\theta = -\frac{1}{r\sin\theta}\frac{\partial\psi}{\partial r} \tag{5.3b}$$

Direct substitution shows that if the velocity components are defined in this
way, the continuity equation will be identically satisfied for all functions $\psi$. It
will now be shown that the quantity $2\pi\psi$ corresponds to the volume of fluid
crossing the surface of revolution which is formed by rotating the position vector
$OP$, in Fig. 5.1, around the reference axis. This statement will be proved in the
following way. First, the statement will be assumed to be true and to constitute
the definition of a quantity $\psi$. Then it will be shown that as a result of this
definition the velocity components must be related to this function $\psi$ by Eqs.
(5.3a) and (5.3b).

Let a function $\psi$ be defined such that if the position vector $OP$ is rotated
around the reference axis, that is, if the coordinate $\omega$ is varied through $2\pi$
while $r$ and $\theta$ are held fixed, the quantity of fluid which crosses the surface of
revolution formed by the vector $OP$ will be $2\pi\psi$. Now apply this definition to
two points $P$ and $P'$ which are close together as shown in Fig. 5.2. Then if the
line element $PP'$ is rotated about the reference axis, the resulting surface will
have a quantity of fluid $2\pi\,d\psi$ crossing it per unit time. But reference to Fig. 5.2
shows that a quantity of fluid $u_r r\,d\theta - u_\theta\,dr$ crosses a unit area of this surface

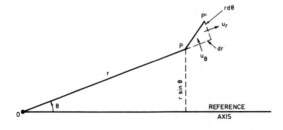

**FIGURE 5.2**
Velocity components and flow areas
defined by a reference point $P$ and
neighboring point $P'$.

so that

$$2\pi \, d\psi = 2\pi r \sin \theta (u_r r \, d\theta - u_\theta \, dr)$$

$$\therefore \quad d\psi = u_r r^2 \sin \theta \, d\theta - u_\theta r \sin \theta \, dr$$

But if $\psi$ is a function of both $r$ and $\theta$, it follows from differential calculus that

$$d\psi = \frac{\partial \psi}{\partial \theta} \, d\theta + \frac{\partial \psi}{\partial r} \, dr$$

Comparing these two expressions for $d\psi$ shows that

$$\frac{\partial \psi}{\partial \theta} = u_r r^2 \sin \theta$$

$$\frac{\partial \psi}{\partial r} = -u_\theta r \sin \theta$$

This confirms Eqs. (5.3a) and (5.3b), so that the definition of $\psi$ agrees with the requirements for satisfying the continuity equation. Then it may be concluded that, for the stream function defined by Eqs. (5.3a) and (5.3b), the volume of fluid crossing an element of surface generated by revolving a line element about the reference axis is $2\pi \, d\psi$.

The stream function defined above is known as the *Stokes stream function*. It will be used here in a auxiliary way only, since flow solutions will be obtained through solutions for the velocity potential $\phi$. The equation which must be satisfied by $\psi$ is therefore referred to the problems at the end of the chapter. It should be pointed out, however, that for rotational flows the velocity potential does not exist, and the stream function then offers the only mechanism for reducing the vector equations of motion to a single scalar equation.

## 5.3 SOLUTION OF THE POTENTIAL EQUATION

The equation to be satisfied by the velocity potential $\phi$ has been established. Rather than attempt to solve this equation as part of a boundary-value problem for various physical situations that may arise, it is proposed to obtain here a general form of solution by separation of variables. The fundamental solutions so obtained will subsequently be superimposed to produce more complex solutions in a manner similar to that which was used in the previous chapter.

The velocity potential will be a function of $r$ and $\theta$ only for axisymmetric flows, and so a separable solution will be sought of the form

$$\phi(r, \theta) = R(r)T(\theta)$$

Substituting this assumed form of solution into Eq. (5.1) gives

$$\frac{T}{r^2} \frac{d}{dr}\left(r^2 \frac{dR}{dr}\right) + \frac{R}{r^2 \sin \theta} \frac{d}{d\theta}\left(\sin \theta \frac{dT}{d\theta}\right) = 0$$

This equation may now be reduced to a separated form by multiplying it by $r^2/(RT)$.

$$\frac{1}{R}\frac{d}{dr}\left(r^2\frac{dR}{dr}\right) = -\frac{1}{T\sin\theta}\frac{d}{d\theta}\left(\sin\theta\frac{dT}{d\theta}\right)$$

The usual argument of separation of variables is now invoked. The left-side of this equation is a function of $r$ only and the right-hand side is a function of $\theta$ only. Hence, if either $r$ or $\theta$ alone is changed, one side of the equation will change while the other does not. Then the only way the equation can remain valid is for each side to be equal to a constant, say $l(l + 1)$. Then

$$\frac{1}{R}\frac{d}{dr}\left(r^2\frac{dR}{dr}\right) = l(l + 1)$$

and

$$-\frac{1}{T\sin\theta}\frac{d}{d\theta}\left(\sin\theta\frac{dT}{d\theta}\right) = l(l + 1)$$

The significance of choosing the separation constant as $l(l + 1)$, rather than simply $\beta$, is that with this choice the resulting ordinary differential equation for $T(\theta)$ appears in standard form, and so a subsequent transformation becomes unnecessary. For the time being there is no implication that the quantity $l$ need be an integer. The differential equation to be satisfied by $R(r)$ is

$$\frac{d}{dr}\left(r^2\frac{dR}{dr}\right) - l(l + 1)R = 0$$

This is an equidimensional equation, and so its solution will be of the form

$$R(r) = Kr^\alpha$$

Substituting this form of solution into the differential equation gives

$$\alpha(\alpha + 1)Kr^\alpha - l(l + 1)Kr^\alpha = 0$$

which will be satisfied by $\alpha = l$ and $\alpha = -(l + 1)$. Then the general solution for $R(r)$ will be a linear combination of these two solutions. That is,

$$R_l(r) = A_l r^l + \frac{B_l}{r^{l+1}}$$

Since this is a valid solution for any choice of $l$, the arbitrary constants $A_l$ and $B_l$ have been assigned subscripts to indicate which value of $l$ is being considered. Likewise, the solution $R_l(r)$ has been assigned a subscript to indicate which solution is being considered.

The equation for $T(\theta)$ is

$$\frac{1}{\sin\theta}\frac{d}{d\theta}\left(\sin\theta\frac{dT}{d\theta}\right) + l(l + 1)T = 0$$

This is Legendre's equation, and it may be reduced to its standard form by the transformation $x = \cos \theta$, which yields

$$\frac{d}{dx}\left[(1 - x^2)\frac{dT}{dx}\right] + l(l + 1)T = 0$$

The solutions to this equation are Legendre's function of the first kind, denoted by $P_l(x)$, and Legendre's function of the second kind, denoted by $Q_l(x)$. Thus the general solution for $T(\theta)$ is

$$T_l(\theta) = C_l P_l(\cos \theta) + D_l Q_l(\cos \theta)$$

But the $Q_l(\cos \theta)$ diverge for $\cos \theta = \pm 1$ for all values of $l$. The coefficient $D_l$ must then be specified as being zero, since there should be no singularities in the flow field. Also, $P_l(\cos \theta)$ diverges for $\cos \theta = \pm 1$ unless $l$ is an integer. Then it must be specified that the quantity $l$ be an integer, so that the continuous spectrum of separation constants $l(l + 1)$ now becomes a discrete spectrum.

Combining the solution for $R_l(r)$ with that for $T_l(\theta)$ gives the following solution for $\phi_l(\cos \theta)$:

$$\phi_l(r, \theta) = \left(A_l r^l + \frac{B_l}{r^{l+1}}\right) P_l(\cos \theta)$$

where the arbitrary constant $C_l$ has been absorbed into the two other arbitrary constants $A_l$ and $B_l$. This solution is valid for any integer $l$. Then, since the partial differential equation being solved is linear, all such possible solutions may be superimposed to yield a more general type of solution. That is, $\phi(r, \theta)$ may be considered to be the sum of all possible solutions $\phi_l(r, \theta)$. Hence

$$\phi(r, \theta) = \sum_{l=0}^{\infty} \left(A_l r^l + \frac{B_l}{r^{l+1}}\right) P_l(\cos \theta) \tag{5.4}$$

The Legendre function of the first kind which appears in Eq. (5.4) is defined by

$$P_l(x) = \frac{1}{2^l l!}\frac{d^l}{dx^l}(x^2 - 1)^l$$

In view of the nature of this function, it is frequently referred to as *Legendre's polynomial of order l*. The first three Legendre polynomials are written out explicitly below for reference purposes.

$$P_0(x) = 1$$

$$P_1(x) = x$$

$$P_2(x) = \tfrac{1}{2}(3x^2 - 1)$$

Equation (5.4) contains certain fundamental solutions which are useful for superimposing to establish additional solutions. These fundamental solutions will now be studied.

## 5.4   UNIFORM FLOW

One of the solutions contained in Eq. (5.4) corresponds to a uniform flow. It may be obtained by setting

$$B_l = 0 \qquad \text{for all } l$$

$$A_l = \begin{cases} 0 & \text{for } l \neq 1 \\ U & \text{for } l = 1 \end{cases}$$

Using the fact that $P_1(\cos \theta) = \cos \theta$, the solution given by Eq. (5.4) then becomes

$$\phi(r, \theta) = Ur \cos \theta \qquad (5.5a)$$

The simplest way of confirming that the velocity potential given by Eq. (5.5a) corresponds to a uniform flow is to note that the cartesian coordinate $x$ is related to the spherical coordinates $r$ and $\theta$ by the relation $x = r \cos \theta$. Thus Eq. (5.5a) states that $\phi = Ux$, which is the velocity potential for a uniform flow of magnitude $U$.

The stream function for a uniform flow may be deduced from Eq. (5.5a) as follows: Using the result (5.5a), it follows from Eq. (5.2a) that

$$u_r = \frac{\partial \phi}{\partial r} = U \cos \theta$$

Hence from Eq. (5.3a),

$$\frac{1}{r^2 \sin \theta} \frac{\partial \psi}{\partial \theta} = U \cos \theta$$

$$\therefore \quad \psi = \tfrac{1}{2} Ur^2 \sin^2 \theta + f(r)$$

where $f(r)$ is any function of $r$. Likewise, the velocity component $u_\theta$ may be evaluated from $\phi$ and expressed as a derivative of $\psi$, giving

$$u_\theta = -\frac{1}{r} \frac{\partial \phi}{\partial \theta} = -U \sin \theta$$

$$\therefore \quad -\frac{1}{r \sin \theta} \frac{\partial \psi}{\partial r} = -U \sin \theta$$

or

$$\psi = \tfrac{1}{2} Ur^2 \sin^2 \theta + g(\theta)$$

where $g(\theta)$ is any function of $\theta$. Then, comparing these two expressions for $\psi$, it follows that $f(r) = g(\theta) = 0$ and

$$\psi(r, \theta) = \tfrac{1}{2} Ur^2 \sin^2 \theta \qquad (5.5b)$$

An alternative way of evaluating $\psi(r, \theta)$ is simply to invoke its definition. Then, considering an arbitrary point $P$ in the fluid as shown in Fig. 5.3, the amount of fluid crossing the surface generated by $OP$ due to the uniform flow will be $2\pi\psi$. But the flow area perpendicular to the velocity vector is $\pi(r \sin \theta)^2$.

**FIGURE 5.3**
Geometry for evaluating the stream
function for a uniform flow.

Hence it follows from the definition of $\psi$ that

$$2\pi\psi = U\pi(r \sin \theta)^2$$

or $$\psi(r,\theta) = \tfrac{1}{2}Ur^2 \sin^2 \theta$$

This agrees with the result obtained by the other method.

Both the methods outlined above for evaluating the stream function are useful, and each will be used in the following sections. The particular method which is employed will depend upon the complexity of the problem, and it is evident that the second method can be conveniently employed only for very simple flow fields.

## 5.5 SOURCE AND SINK

The velocity potential corresponding to a three-dimensional source or sink is obtained from Eq. (5.4) through the term whose coefficient is $B_0$. Then let

$$A_l = 0 \qquad \text{for all } l$$

$$B_l = \begin{cases} 0 & \text{for } l \neq 0 \\ B_0 \neq 0 & \text{for } l = 0 \end{cases}$$

Then, from Eq. (5.4), using the fact that $P_0(\cos \theta) = 1$,

$$\phi(r,\theta) = \frac{B_0}{r}$$

The velocity components for the resulting flow field are

$$u_r = -\frac{B_0}{r^2}$$

$$u_\theta = 0$$

Hence the velocity is purely radial, its magnitude increases as the origin is approached, and there is a singularity at the origin. Clearly there is a source or sink of fluid at $r = 0$, and the quantity of fluid leaving or entering this singularity may be evaluated by enclosing it with a spherical control surface or radius $r$. Then if $Q$ is the volume of the fluid leaving the control surface per unit time, it follows that

$$Q = \int_s \mathbf{u} \cdot \mathbf{n} \, ds$$

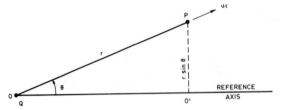

**FIGURE 5.4**
Geometry for evaluating the stream function for flow due to a source.

But the velocity vector is radial, and so $\mathbf{u} \cdot \mathbf{n} = |\mathbf{u}| = -B_0/r^2$ and $ds = r^2 \sin\theta \, d\theta \, d\omega$. Hence

$$Q = \int_0^{2\pi} d\omega \int_0^{\pi} \left(\frac{B_0}{r^2}\right) r^2 \sin\theta \, d\theta$$

$$= -4\pi B_0$$

Then, for a source of strength $Q$, the constant $B_0$ should be set equal to $-Q/(4\pi)$. That is, the velocity potential for a source of strength $Q$ located at $r = 0$ is

$$\phi(r,\theta) = -\frac{Q}{4\pi r} \tag{5.6a}$$

It should be noted that the minus sign is associated with the source, and so a positive sign would be associated with a sink. For a source of strength $Q$ located at $r = r_0$ rather than $r = 0$, the quantity $r$ should be replaced by $(r - r_0)$.

In order to establish the stream function corresponding to Eq. (5.6a), the definition of $\psi$ will be invoked. Referring to Fig. 5.4, a source of strength $Q$ is shown at the origin. At any arbitrary point $P$ the velocity will be radial and is indicated by $u_r$. The quantity of fluid which crosses the surface generated by revolving the line $OP$ about the reference axis will depend upon whether the source $Q$ is considered to be slightly to the left of the origin or slightly to the right of it. Here the source $Q$ will be considered to be slightly to the right of $O$, so that the quantity of fluid crossing the surface generated by $OP$ will be $2\pi\psi + Q$. Then from Fig. 5.4 it follows that

$$2\pi\psi + Q = \int_0^{\theta} u_r \cos\theta \, 2\pi r \sin\theta \, \frac{r \, d\theta}{\cos\theta}$$

where $u_r \cos\theta$ is the component of the velocity vector which is perpendicular to $O'P$ and $r \, d\theta/\cos\theta$ is the element of surface area along $O'P$. Performing the integration yields

$$\psi(r,\theta) = -\frac{Q}{4\pi}(1 + \cos\theta) \tag{5.6b}$$

It now becomes evident that if the source $Q$ had been considered to be slightly to the left of the origin, the constant term in Eq. (5.6b) would have been different. However, the velocity components would be the same.

## 5.6   FLOW DUE TO A DOUBLET

As was the case in two dimensions, the flow due to a doublet may be obtained by superimposing a source and sink of equal strength and letting the distance separating the source and the sink shrink to zero. Figure 5.5 shows a source of strength $Q$ located at the origin and a sink of strength $Q$ located a distance $\delta x$ along the positive portion of the reference axis. The distance from the source to some point $P$ in the fluid will be $r$, and the corresponding distance to the sink will be $r - \delta r$.

From Eq. (5.6a), the velocity potential for the flow due to this source and sink will be

$$\phi(r,\theta) = -\frac{Q}{4\pi r} + \frac{Q}{4\pi(r - \delta r)}$$

$$= -\frac{Q}{4\pi r}\left(1 - \frac{1}{1 - \delta r/r}\right)$$

If the source and sink are close together, the quantity $\delta r/r$ will be small, so that the expression for the velocity potential may be expanded as follows:

$$\phi(r,\theta) = -\frac{Q}{4\pi r}\left\{1 - \left[1 + \frac{\delta r}{r} + O\left(\frac{\delta r}{r}\right)^2\right]\right\}$$

$$= \frac{Q}{4\pi r}\left[\frac{\delta r}{r} + O\left(\frac{\delta r}{r}\right)^2\right]$$

The quantity $\delta r$ may be eliminated in favor of $\delta x$ by applying the cosine rule to the triangle defined by the vectors $r$ and $r - \delta r$, and the distance $\delta x$ separating the source and the sink. Thus

$$(r - \delta r)^2 = r^2 + (\delta x)^2 - 2r\,\delta x\cos\theta$$

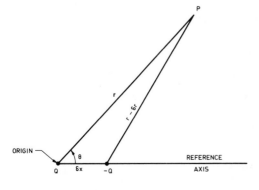

**FIGURE 5.5**
Superposition of a source and sink which become a doublet as $\delta x \to 0$.

Solving this equation for $\cos \theta$ gives

$$\cos \theta = \frac{r^2 + (\delta x)^2 - (r - \delta r)^2}{2r\, \delta x}$$

$$= \frac{\delta r}{\delta x} - \frac{\delta r}{2r} \frac{\delta r}{\delta x} + \frac{\delta x}{2r}$$

$$= \frac{\delta r}{\delta x}\left[1 + O\left(\frac{\delta r}{r}\right)\right]$$

$$\therefore \quad \delta r = \delta x \cos \theta \left[1 + O\left(\frac{\delta r}{r}\right)\right]$$

Using this result, the expression for $\phi(r, \theta)$ becomes

$$\phi(r, \theta) = \frac{Q}{4\pi r}\left\{\frac{\delta x}{r}\cos \theta\left[1 + O\left(\frac{\delta r}{r}\right)\right]\right\}$$

Now let the distance $\delta x \to 0$ and the source strength $Q \to \infty$ such that the product $Q\,\delta x \to \mu$. Then

$$\phi(r, \theta) = \frac{\mu}{4\pi r^2}\cos \theta \qquad (5.7a)$$

Equation (5.7a) is the velocity potential for a positive doublet of strength $\mu$, that is, a doublet which expels fluid along the negative portion of the reference axis and absorbs fluid along the positive portion.

The stream function corresponding to Eq. (5.7a) will be obtained by using the equivalent expressions for the velocity components given by Eqs. (5.2) and (5.3). Thus

$$u_r = \frac{\partial \phi}{\partial r} = -\frac{\mu}{2\pi r^3}\cos \theta = \frac{1}{r^2 \sin \theta}\frac{\partial \psi}{\partial \theta}$$

$$\therefore \quad \frac{\partial \psi}{\partial \theta} = -\frac{\mu}{2\pi r}\sin \theta \cos \theta$$

and $\qquad \psi(r, \theta) = -\frac{\mu}{4\pi r}\sin^2 \theta + f(r)$

Likewise, the two expressions for $u_\theta$ give

$$u_\theta = -\frac{1}{r}\frac{\partial \phi}{\partial \theta} = -\frac{\mu}{4\pi r^3}\sin \theta = -\frac{1}{r \sin \theta}\frac{\partial \psi}{\partial r}$$

$$\therefore \quad \frac{\partial \psi}{\partial r} = \frac{\mu}{4\pi r^2}\sin^2 \theta$$

and $\qquad \psi(r, \theta) = -\frac{\mu}{4\pi r}\sin^2 \theta + g(\theta)$

Comparing these two expressions for $\psi(r, \theta)$ shows that $f(r) = g(\theta) = 0$ and

$$\psi(r, \theta) = -\frac{\mu}{4\pi r} \sin^2 \theta \tag{5.7b}$$

Equation (5.7b) gives the stream function for a doublet which discharges fluid along the negative portion of the reference axis and attracts fluid along the positive part of the reference axis.

## 5.7 FLOW NEAR A BLUNT NOSE

By superimposing the solutions for a uniform flow and a source, the solution corresponding to a long cylinder with a blunt nose is obtained. From Eqs. (5.5b) and (5.6b), the stream function for a uniform flow of magnitude $U$ and a source of strength $Q$ located at the origin is

$$\psi(r, \theta) = \tfrac{1}{2} U r^2 \sin^2 \theta - \frac{Q}{4\pi}(1 + \cos \theta)$$

In order to interpret the flow field which this solution represents, consider $\psi$ to be constant and solve the above equation for $r$ in terms of $\theta$:

$$r = \sqrt{\frac{2\psi}{U \sin^2 \theta} + \frac{Q}{2\pi U}\frac{1 + \cos \theta}{\sin^2 \theta}}$$

$$= \sqrt{\frac{2\psi}{U \sin^2 \theta} + \frac{Q}{4\pi U \sin^2 (\theta/2)}}$$

where the fact that $1 + \cos \theta = 2 \cos^2 (\theta/2)$ and the fact that $\sin(\theta) = 2 \sin(\theta/2) \cos(\theta/2)$ has been used. Then, denoting the value of $r$ for which $\psi = 0$ by $r_0$, the radius to the surface corresponding to $\psi = 0$ is

$$r_0 = \sqrt{\frac{Q}{4\pi U}\frac{1}{\sin(\theta/2)}}$$

Thus the radius $r_0$ corresponding to the principal values of $\theta$ are as follows:

When $\theta = 0$, $\qquad r_0 = \infty$

When $\theta = \dfrac{\pi}{2}$, $\qquad r_0 = \sqrt{\dfrac{Q}{2\pi U}}$

When $\theta = \pi$, $\qquad r_0 = \sqrt{\dfrac{Q}{4\pi U}}$

This defines the stream surface $\psi = 0$ as shown in Fig. 5.6.

Although the polar radius $r_0$ is infinite for $\theta = 0$, the cylindrical radius $R_0$ is finite. This may be verified by noting that $R = r \sin \theta$, so that

$$R_0 = \sqrt{\frac{Q}{4\pi U}\frac{\sin \theta}{\sin(\theta/2)}}$$

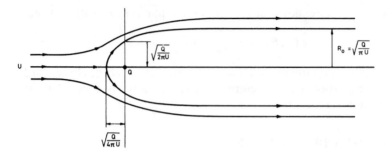

**FIGURE 5.6**
Flow around an axisymmetric body created by a source in a uniform flow.

Then, as $\theta \to 0$

$$\frac{\sin \theta}{\sin (\theta/2)} = \frac{\theta - \theta^3/3! + \cdots}{\theta/2 - 1/3!(\theta/2)^3 + \cdots} \to 2$$

Hence the cylindrical radius far from the source becomes

$$R_0 = \sqrt{\frac{Q}{\pi U}}$$

The fluid which emanates from the source located at the origin does not mix with the fluid which constitutes the uniform flow. Then a shell could be fitted to the shape of the surface corresponding to $\psi = 0$ and the source could be removed without disturbing the outer flow. That is, the stream function for the semi-infinite body shown in Fig. 5.6 is

$$\psi(r, \theta) = \tfrac{1}{2}Ur^2 \sin^2 \theta - \frac{Q}{4\pi}(1 + \cos \theta) \tag{5.8a}$$

The corresponding velocity potential may be obtained from Eqs. (5.5a) and (5.6a), giving

$$\phi(r, \theta) = Ur \cos \theta - \frac{Q}{4\pi r} \tag{5.8b}$$

Equations (5.8) may be used to deduce the velocity and pressure distribution in the vicinity of the nose of a blunt axisymmetric body such as an aircraft fuselage or a submarine hull.

## 5.8   FLOW AROUND A SPHERE

The stream function for a uniform flow past a sphere may be obtained by superimposing the solution for a uniform flow and that for a doublet. From Eqs.

(5.5b) and (5.7b), the stream function for such a superposition is

$$\psi(r,\theta) = \tfrac{1}{2}Ur^2 \sin^2 \theta - \frac{\mu}{4\pi r} \sin^2 \theta$$

Then the equation defining the surface which corresponds to $\psi = 0$ is

$$0 = \tfrac{1}{2}Ur_0^2 \sin^2 \theta - \frac{\mu}{4\pi r_0} \sin^2 \theta$$

where $r$ is the value of the polar radius $r$ which defines the surface on which $\psi = 0$. Solving this equation for $r_0$ gives

$$r_0 = \left(\frac{\mu}{2\pi U}\right)^{1/3}$$

Since $r_0 = $ constant, the surface which corresponds to $\psi = 0$ is that of a sphere. If the doublet strength is chosen to be $\mu = 2\pi Ua^3$, the radius of this spherical surface will be $r_0 = a$. Then, by choosing $\mu = 2\pi Ua^3$, the stream function for a uniform flow of magnitude $U$ approaching a sphere of radius $a$ is

$$\psi(r,\theta) = \tfrac{1}{2}U\left(r^2 - \frac{a^3}{r}\right) \sin^2 \theta \qquad (5.9a)$$

The corresponding velocity potential may be obtained from Eqs. (5.5a) and (5.7a), in which the doublet strength $\mu = 2\pi Ua^3$ is used. This gives

$$\phi(r,\theta) = U\left(r + \frac{1}{2}\frac{a^3}{r^2}\right) \cos \theta \qquad (5.9b)$$

## 5.9 LINE-DISTRIBUTED SOURCE

The stream function and the velocity potential for a source which is distributed over a finite strip will be established in this section. The result is useful as one element in superpositions which lead to additional solutions to flow problems.

Figure 5.7 shows a source which is uniformly distributed over the section $0 \le x \le L$ of the reference axis. The source strength, which is constant, is $q$ per unit length, so that $qL$ is the total volume of fluid which emanates from the source per unit time. An arbitrary field point $P$ is shown whose coordinates are $r$, $\theta$, and $\omega$. One end of the line source, which is at the origin, is a distance $r$ from this point and subtends an angle $\theta$ to the $x$ axis. The other end of the line source is a distance $\eta$ from the point $P$ and subtends an angle $\alpha$ to the $x$ axis. Also, an element of the line source of length $d\xi$, which is a distance $\xi$ from the origin, subtends an angle $\nu$ to the $x$ axis. But the strength of this element of the source is $q\,d\xi$, so that, from Eq. (5.6b), the stream function for the line source will be

$$\psi = -\int_0^L \frac{q\,d\xi}{4\pi}(1 + \cos \nu)$$

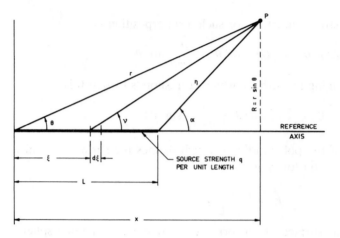

**FIGURE 5.7**
Geometry connecting a field point $P$ and line source of length $L$ distributed uniformly along the reference axis.

where the angle $\nu$ is a function of $\xi$ and so will be a variable in the integration. Rather than express $\nu$ as a function of $\xi$, the variable of integration will be changed from $\xi$ to $\nu$. Referring to Fig. 5.7, it will be observed that the cylindrical radius $R = r \sin \theta = \eta \sin \alpha$ remains constant throughout the integration. Also, it may be observed that

$$x - \xi = R \cot \nu$$

$$\therefore \quad -d\xi = -R \csc^2 \nu \, d\nu$$

Hence the expression for $\psi$ may be written in the form

$$\psi(r, \theta) = -\frac{qR}{4\pi} \int_\theta^\alpha \csc^2 \nu (1 + \cos \nu) \, d\nu$$

$$= -\frac{qR}{4\pi} \left( \cot \theta - \cot \alpha + \frac{1}{\sin \theta} - \frac{1}{\sin \alpha} \right)$$

But from Fig. 5.7 the following relations may be established:

$$x = R \cot \theta$$

$$x - L = R \cot \alpha$$

$$r = \frac{R}{\sin \theta}$$

$$\eta = \frac{R}{\sin \alpha}$$

Using these relations, the expression for the stream function for a line source of strength $q$ per unit length and of length $L$ is

$$\psi = -\frac{q}{4\pi}(L + r - \eta) \tag{5.10a}$$

The velocity potential corresponding to Eq. (5.10a) may be obtained in an analogous way. From Eq. (5.6a) it follows that

$$\phi = -\int_0^L \frac{q\,d\xi}{4\pi(R/\sin\nu)}$$

where it has been observed that the distance from the point at $\xi$ on the line source to the field point $P$ is $R/\sin\nu$. As before it is observed that

$$x - \xi = R\cot\nu$$

and so

$$-d\xi = -R\csc^2\nu\,d\nu$$

Then the expression for $\phi$ becomes

$$\phi = -\frac{q}{4\pi}\int_\theta^\alpha \sin\nu\,\csc^2\nu\,d\nu$$

$$= -\frac{q}{4\pi}\int_\theta^\alpha \frac{d\nu}{\sin\nu} \tag{5.10b}$$

$$\phi = -\frac{q}{4\pi}\log\left(\frac{\tan\alpha/2}{\tan\theta/2}\right)$$

Although the result for the stream function was more compact when expressed in terms of lengths, the result for the velocity potential is more compact in terms of angles, so that Eq. (5.10b) will be considered to be the final result.

## 5.10  SPHERE IN THE FLOW FIELD OF A SOURCE

In the problems at the end of Chap. 4 it was established that the solution for a circular cylinder in a source flow could be obtained from the solutions for two sources and two sinks, all of which have the same strength. It will be shown here that the solution for a sphere in a source flow may be obtained in an analogous manner, although the singularities which must be imposed are two sources, of unequal strength, and a line sink.

Figure 5.8 shows the connection between a field point $P$ and certain singularities. The singularities are a source of strength $Q$, which is located at the point $Q$, which is a distance $l$ along the reference axis, a source of strength $Q^*$, which is located at the point $Q^*$, which is at the image point $a^2/l$ of the point $Q$ in the sphere of radius $a$, and a uniformly distributed line sink of strength $q$ per unit length along the section $OQ^*$ of the reference axis. It will be shown that for an appropriate choice of source and sink strengths the sphere $r = a$ corresponds to $\psi = 0$.

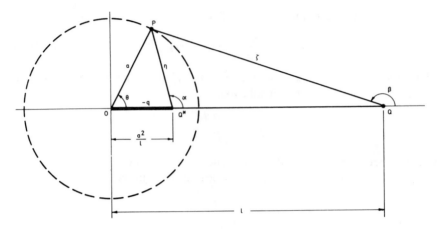

**FIGURE 5.8**
Superposition of a line sink of strength $q$ per unit length, a source of strength $Q^*$, and a source of strength $Q$ outside a sphere of radius $a$.

If the spherical surface $r = a$ is to be a stream surface, the total sink strength inside this region must equal the total source strength there. That is, $qa^2/l = Q^*$, which establishes the sink strength $q$ in terms of the source strength $Q^*$. Then, using Eqs. (5.6b) and (5.10a), the stream function for the singularities shown in Fig. 5.8 will be

$$\psi(r,\theta) = -\frac{Q}{4\pi}(1 + \cos\beta) - \frac{Q^*}{4\pi}(1 + \cos\alpha) + \frac{Q^*}{4\pi}\frac{l}{a^2}\left(\frac{a^2}{l} + r - \eta\right)$$

where $\alpha$, $\beta$, and $\eta$ are functions of $r$ and $\theta$. Then for points on the surface of the sphere $r = a$ the value of $\psi$ will be

$$\psi(a,\theta) = -\frac{q}{4\pi}(1 + \cos\beta) - \frac{Q^*}{4\pi}(1 + \cos\alpha) + \frac{Q^*}{4\pi}\left(1 + \frac{1}{a} - \frac{l\eta}{a^2}\right)$$

Now if the point $P$ lies on the spherical surface $r = a$, a relationship will exist among the parameters $\eta$, $a$, $\beta$, $\alpha$, and $\theta$. To establish this relationship, it will be noted that

$$\frac{a^2/l}{a} = \frac{a}{l}$$

But the numerator and denominator of each side of this equation represent the lengths of one of the vectors shown in Fig. 5.8. Thus it follows that

$$\frac{OQ^*}{OP} = \frac{OP}{OQ}$$

But these lengths represent corresponding sides of the triangles $OPQ^*$ and $OQP$. Then, since the angle $\theta$ is common to both these triangles, it follows that the two triangles are similar. Then the angle $O\hat{P}Q^*$ must equal the angle $O\hat{Q}P$, which, in turn, equals $\pi - \beta$. Hence the length $\eta$ may be written as

$$\eta = \frac{a^2}{l} \cos(\pi - \alpha) + a \cos(\pi - \beta)$$

$$= -\frac{a^2}{l} \cos \alpha - a \cos \beta$$

Substituting this result into the expression for $\psi(a, \theta)$ gives

$$\psi(a, \theta) = -\frac{Q}{4\pi}(1 + \cos \beta) - \frac{Q^*}{4\pi}(1 + \cos \alpha)$$

$$+ \frac{Q^*}{4\pi}\left(1 + \frac{l}{a} + \cos \alpha + \frac{l}{a} \cos \beta\right)$$

$$= (1 + \cos \beta)\left(-\frac{Q}{4\pi} + \frac{Q^*}{4\pi}\frac{l}{a}\right)$$

Thus by choosing the source strength $Q^*$ to be equal to $aQ/l$, the surface $r = a$ will correspond to the stream surface $\psi = 0$. Then the stream function for a sphere of radius $a$ whose center is at the origin and which is exposed to a point source of strength $Q$ located a distance $l$ along the positive reference axis is

$$\psi(r, \theta) = -\frac{Q}{4\pi}(1 + \cos \beta) - \frac{Q}{4\pi}\frac{a}{l}(1 + \cos \alpha) + \frac{Q}{4\pi}\left(\frac{a}{l} + \frac{r}{a} - \frac{\eta}{a}\right)$$

$$(5.11a)$$

The velocity potential corresponding to Eq. (5.11$a$) may be obtained from Eqs. (5.6$a$) and (5.10$b$). This gives

$$\phi(r, \theta) = -\frac{Q}{4\pi\zeta} - \frac{Q^*}{4\pi\eta} + \frac{q}{4\pi} \log\left(\frac{\tan \alpha/2}{\tan \theta/2}\right) \qquad (5.11b)$$

Using the fact that $Q^* = aQ/l$ and $q = lQ^*/a^2 = Q/a$, the expression for the velocity potential for a sphere of radius $a$ in the presence of a source of strength $Q$ becomes

$$\phi(r, \theta) = -\frac{Q}{4\pi\zeta} - \frac{Qa}{4\pi\eta l} + \frac{Q}{4\pi a} \log\left(\frac{\tan \alpha/2}{\tan \theta/2}\right) \qquad (5.11c)$$

The quantity $\zeta$ is the distance from the field point $P$ to the source $Q$ as shown in Fig. 5.8.

## 5.11  RANKINE SOLIDS

The solution for the flow around a family of bodies, which are known as Rankine solids, is obtained by superimposing a source and a sink of equal strength in a uniform flow field. Let the magnitude of the uniform flow be $U$

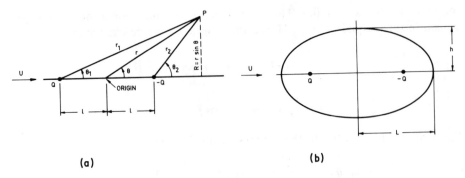

**FIGURE 5.9**
(a) Superposition of uniform flow, source and sink, and (b) uniform flow approaching a Rankine solid.

and the strength of the source and the sink be $Q$. Consider the source and the sink to be located equal distances $l$ from the origin as shown in Fig. 5.9a.

From Eqs. (5.5b) and (5.6b), the stream function for the configuration shown in this figure is

$$\psi(r,\theta) = \tfrac{1}{2}Ur^2 \sin^2 \theta - \frac{Q}{4\pi}(\cos\theta_1 - \cos\theta_2) \qquad (5.12a)$$

Then if $r_0$ is the radius to the surface on which $\psi = 0$, the radius $r_0$ must satisfy the equation

$$0 = \tfrac{1}{2}Ur_0^2 \sin^2 \theta - \frac{Q}{4\pi}(\cos\theta_1 - \cos\theta_2)$$

Working with the cylindrical radius $R = r \sin\theta$ rather than the polar radius $r$, it follows that the cylindrical radius $R_0$ which corresponds to the surface $\psi = 0$ will be

$$R_0^2 = \frac{Q}{2\pi U}(\cos\theta_1 - \cos\theta_2)$$

Then when $\theta_1 = \theta_2 = 0$, and when $\theta_1 = \theta_2 = \pi$, the value of $R_0$ is zero. Also, the maximum value of $R_0$ occurs when $\cos\theta_1 = -\cos\theta_2$, which corresponds to $\theta = \pi/2$ or $\theta = 3\pi/2$. Thus the stream surface which corresponds to $\psi = 0$ defines a body as shown in Fig. 5.9b. The principal dimensions of this body are the half width $L$ and the half height $h$. Both these parameters depend upon the free-stream velocity $U$, the source and sink strength $Q$, and the distance $l$.

The value of $L$ may be obtained from the equation which results from the observation that the velocity at the downstream stagnation point is zero. But the velocity at that point is the superposition of a uniform flow of magnitude $U$, a source of strength $Q$ a distance $L + l$ away, and a sink of strength $Q$ a distance

$L - l$ away. Hence

$$U + \frac{Q}{4\pi(L+l)^2} - \frac{Q}{4\pi(L-l)^2} = 0$$

Rearranging this equation gives the following equation to be satisfied by $L$ in terms of the parameters $U$, $Q$, and $l$:

$$(L^2 - l^2)^2 - \frac{Qi}{\pi U}L = 0 \qquad (5.12b)$$

An analogous expression for the half height $h$ may be obtained by noting that the value of the cylindrical radius $R_0$ is $h$ when $\cos \theta_1 = -\cos \theta_2$, where $\tan \theta_1 = h/l$. Hence

$$h^2 = \frac{Q}{2\pi U} \left( \frac{l}{\sqrt{h^2 + l^2}} + \frac{l}{\sqrt{h^2 + l^2}} \right)$$

Rearranging this expression shows that $h$ must satisfy the following equation:

$$h^2\sqrt{h^2 + l^2} - \frac{Ql}{\pi U} = 0 \qquad (5.12c)$$

For various values of the parameters $U$, $Q$, and $l$, Eqs. (5.12b) and (5.12c) define a family of bodies of revolution for which the stream function is given by Eq. (5.12a). The corresponding velocity potential is

$$\phi(r, \theta) = Ur \cos \theta - \frac{Q}{4\pi r_1} + \frac{Q}{4\pi r_2} \qquad (5.12d)$$

## 5.12   D'ALEMBERT'S PARADOX

It will be shown in this section that if an arbitrary three-dimensional body is immersed in a uniform flow, the equations of hydrodynamics predict that there will be no force exerted on the body by the fluid. Experimentally it is known that a drag force exists on a body which is in a fluid flow, so that this theoretical result is known as d'Alembert's paradox.

Figure 5.10 shows a body of arbitrary shape whose center of gravity is located at the origin of a coordinate system. The surface of the body is denoted by $S$, and the unit outward normal to $S$, locally, is denoted by **n**. The hydrodynamic force, which may act on the body, is denoted by the force vector **F**. A spherical control surface $S_0$ is set up around the body under consideration, and $\mathbf{n}_0$ is the unit outward normal to $S_0$. That is, $\mathbf{n}_0 = \mathbf{e}_r$, where $\mathbf{e}_r$ is the unit radial vector.

The equation of force equilibrium will now be written for the body of fluid which is contained between the surfaces $S$ and $S_0$. The fluid force acting on the body through the surface $S$ is **F**; hence the force acting on the fluid through that surface is $-\mathbf{F}$. There is no transfer of momentum across the surface $S$, since that surface is a stream surface. Around the surface $S_0$ there will be a force due

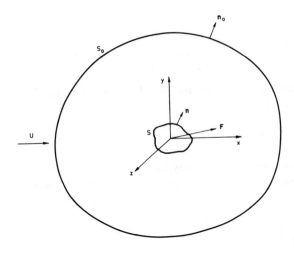

**FIGURE 5.10**
Spherical control surface $S_0$ enclosing an arbitrary body of surface $S$. The force acting on the body is $\mathbf{F}$.

to the pressure distribution. The magnitude of this force will be $-p\mathbf{n}_0$ per unit of surface area, so that the total pressure force will be the surface integral of this quantity. Across the surface $S_0$ there will be a momentum flux corresponding to the mass flux $\rho\mathbf{u} \cdot \mathbf{n}_0$ per unit area. Then the momentum flux will be $\rho\mathbf{u}(\mathbf{u} \cdot \mathbf{n}_0)$ per unit area, so that the inertia force per unit surface area will be $-\rho\mathbf{u}(\mathbf{u} \cdot \mathbf{n}_0)$. Thus the equation of force equilibrium for the fluid which is bounded by the surfaces $S$ and $S_0$ is

$$0 = -\mathbf{F} - \int_{S_0} \left[ p\mathbf{n}_0 + \rho\mathbf{u}(\mathbf{u} \cdot \mathbf{n}_0) \right] dS$$

The pressure may be eliminated from this equation through use of the Bernoulli equation, which for the case of steady irrotational motion under consideration, may be written in the form

$$p + \tfrac{1}{2}\rho\mathbf{u} \cdot \mathbf{u} = B$$

where $B$ is the Bernoulli constant. Then the force $\mathbf{F}$ acting on the body will be given by the following integral:

$$\mathbf{F} = \rho\int_{S_0} \left[ \tfrac{1}{2}(\mathbf{u} \cdot \mathbf{u})\mathbf{n}_0 - \mathbf{u}(\mathbf{u} \cdot \mathbf{n}_0) \right] dS$$

Here, it has been observed that the surface integral of $B\mathbf{n}_0$ is zero for any closed surface.

It is now proposed to write the velocity vector $\mathbf{u}$ as the sum of the free-stream velocity vector $\mathbf{U} = U\mathbf{e}$ and a perturbation $\mathbf{u}'$. The perturbation velocity $\mathbf{u}'$ will be large near the body, but it will tend to zero far from the body. Then, writing

$$\mathbf{u} = \mathbf{U} + \mathbf{u}'$$

the expression for the force **F** becomes

$$F = \rho \int_{S_0} \{ (\tfrac{1}{2}U^2 + U \cdot u' + \tfrac{1}{2}u' \cdot u')n_0 - (U + u')[(U + u') \cdot n_0] \} \, dS$$

Expand the integrand now and note that $\int_{S_0} U^2 \, ds = 0$ since $U^2$ is a constant, $\int_{S_0} U \cdot n_0 \, ds = 0$, since **U** is a constant vector and $\int_{S_0} u' \cdot n_0 \, ds = 0$ from the continuity equation. Hence

$$F = \rho \int_{S_0} \{ (U \cdot u' + \tfrac{1}{2}u' \cdot u')n_0 - [u'(U \cdot n_0) + u'(u' \cdot n_0)] \} \, ds$$

The first and third terms in the integrand may be replaced by $-U \times (u \times n_0)$ in view of the vector identity

$$U \times (u' \times n_0) = u'(U \cdot n_0) - n_0(U \cdot u')$$

Hence the expression for the hydrodynamic force **F** may be written in the form

$$F = \rho \int_{S_0} [ -U \times (u' \times n_0) + \tfrac{1}{2}(u' \cdot u')n_0 - u'(u' \cdot n_0)] \, ds$$

It will now be shown that each of these terms is zero.

Let $\phi'$ be the velocity potential corresponding to the perturbation velocity $u'$. Then, from Eq. (5.4), $\phi'$ must be of the form

$$\phi' = \sum_{l=0}^{\infty} A_l \frac{P_l(\cos \theta)}{r^{l+1}}$$

$$= \frac{Q}{4\pi r} + \frac{\mu \cos \theta}{4\pi r^2} + O\left(\frac{1}{r^3}\right)$$

where the first two terms, which correspond to a source and a doublet, have been written out explicitly and any remaining terms must vary as $1/r^3$ or some greater power of $1/r$. Then, since $u' = \nabla \phi'$, it follows that

$$|u'| = O\left(\frac{1}{r^2}\right)$$

That is, the perturbation velocity varies, at most, as $1/r^2$. Also

$$u' \times n_0 = \nabla \left[ \frac{Q}{4\pi r} - \frac{\mu \cos \theta}{4\pi r^2} + O\left(\frac{1}{r^3}\right) \right] \xi e_r$$

$$= \left[ -\frac{Q}{4\pi r^2} e_r + O\left(\frac{1}{r^3}\right) \right] \times e_r$$

That is, since $n_0 = e_r$ and since $e_r \times e_r = 0$, it follows that

$$|u' \times n_0| = O\left(\frac{1}{r^3}\right)$$

Finally, since an element of surface area $dS$ equals $r^2 \sin \theta \, d\theta \, d\omega$, it is evident that

$$dS = O(r^2)$$

Using the foregoing results, it is possible to establish the order of magnitude of each of the integrals which appears in the expression for **F**. Thus

$$\int_{S_0} \mathbf{U} \times (\mathbf{u}' \times \mathbf{n}_0) \, dS = O\left(\frac{1}{r}\right)$$

$$\int_{S_0} (\mathbf{u}' \cdot \mathbf{u}') \mathbf{n}_0 \, dS = O\left(\frac{1}{r^2}\right)$$

$$\int_{S_0} \mathbf{u}' (\mathbf{u}' \cdot \mathbf{n}_0) \, dS = O\left(\frac{1}{r^2}\right)$$

That is, if the radius of the spherical surface $S_0$ is taken to be very large, each of these integrals will be vanishingly small. Thus in the limit,

$$\mathbf{F} = 0 \tag{5.13}$$

Since it is known that any body which is immersed in a flow field experiences a drag force, Eq. (5.13) poses a paradox which is known as d'Alembert's paradox. The resolution of this paradox lies in the fact that viscous effects have been omitted from the equations which led to Eq. (5.13). It will be seen in Part III that there is a thin fluid layer around such a body in which viscous effects cannot be neglected. This fluid layer, or boundary layer, exerts a shear stress on the body which gives rise to a drag force. In addition, the boundary layer may separate from the surface of the body, creating a low-pressure wake. This, in turn, will induce an additional drag, called the *form drag*, owing to the pressure differential around the surface of the body. However, for streamlined bodies Eq. (5.13) is approached because of the absence of form drag, although the viscous-shear drag will still exist.

## 5.13 FORCES INDUCED BY SINGULARITIES

It was established in the previous section that, according to the equations of hydrodynamics, no force exists on a body which is in a uniform flow field. This agrees with the results of the last chapter, since the Kutta-Joukowski law shows that in the absence of circulation around a body there are no forces acting on two-dimensional bodies. In view of the fact that $\nabla \cdot \boldsymbol{\omega} = 0$, it is very difficult to establish an appreciable circulation around short bodies—that is, around three-dimensional bodies. However, it was established in the problems at the end of Chap. 4 that a force will exist on a cylinder which is exposed to a singularity in the flow such as a source, a sink, or a vortex. Likewise, it will be shown here that a force exists on a three-dimensional body if it is exposed to a point singularity in the fluid.

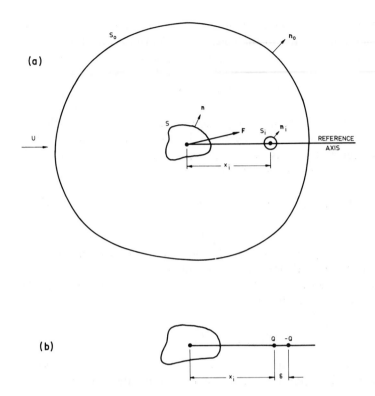

**FIGURE 5.11**
(*a*) Control surfaces for a body located at the origin and a point singularity at $x = x_i$ and (*b*) a source and a sink close together near the body.

Figure 5.11*a* shows an arbitrary body whose center of gravity coincides with the origin of a coordinate system. The surface of the body is denoted by $S$, and **n** is the outward unit normal to $S$. A singularity is assumed to exist at the point $x = x_i$, since the polar axis may be made to pass through the singularity without loss of generality. A small spherical control surface denoted by $S_i$ and of radius $\varepsilon$ is established around the singularity. The unit outward normal to the surface $S_i$ is denoted by $\mathbf{n}_i$. A large spherical surface, denoted by $S_0$, is drawn around both the body and the singularity. The unit normal to this surface is denoted by $\mathbf{n}_0$. The hydrodynamic force which acts on the body and whose magnitude is sought is denoted by **F**.

For equilibrium of the forces which act on the body of fluid which is inside $S_0$ but outside $S$ and $S_i$, the sum of the forces must be zero. Hence

$$0 = -\mathbf{F} - \int_{S_0} \left[ p\mathbf{n}_0 + \rho\mathbf{u}(\mathbf{u} \cdot \mathbf{n}_0) \right] dS + \int_{S_i} \left[ p\mathbf{n}_i + \rho\mathbf{u}(\mathbf{u} \cdot \mathbf{n}_i) \right] dS$$

The first two terms on the right-hand side of this equation are identical with those which appeared in the previous section, and the third term represents the pressure and momentum integral for the new surface $S_i$. But it was shown in the previous section that the integral around $S_0$ which appears above is zero. Then

$$\mathbf{F} = \int_{S_i} [\,p\mathbf{n}_i + \rho\mathbf{u}(\mathbf{u} \cdot \mathbf{n}_i)] \, dS$$

Since the integral which appears in this equation represents the force acting on the surface $S_i$, whose radius is arbitrarily small, it follows that if a force $\mathbf{F}$ acts on the body $S$, the reaction of this force must act on the singularity. From the Bernoulli equation, $p = \mathbf{B} - \rho(\mathbf{u} \cdot \mathbf{u})/2$, so that

$$\mathbf{F} = \rho\int_{S_i} [\,-\tfrac{1}{2}(\mathbf{u} \cdot \mathbf{u})\mathbf{n}_i + \mathbf{u}(\mathbf{u} \cdot \mathbf{n}_i)] \, dS \tag{5.14a}$$

In order to further reduce the integral in Eq. (5.14a), it is necessary to specify the nature of the singularity which is located inside the surface $S_i$. The case of a source, or sink, and that of a doublet will be examined.

Consider, first, the singularity at $x = x_i$ to be a source of strength $Q$. Then, from Eq. (5.6a), the velocity on the surface $S_i$ will be

$$\mathbf{u} = \frac{Q}{4\pi\varepsilon^2}\mathbf{e}_\varepsilon + \mathbf{u}_i$$

where $\mathbf{e}_\varepsilon$ is the unit vector radial from the point $x = x_i$ and $\mathbf{u}_i$ is the velocity induced by all means other than the source under consideration. Then

$$\mathbf{u} \cdot \mathbf{u} = \frac{Q^2}{16\pi^2\varepsilon^4} + \frac{Q}{2\pi\varepsilon^2}\mathbf{e}_\varepsilon \cdot \mathbf{u}_i + \mathbf{u}_i \cdot \mathbf{u}_i$$

and

$$\mathbf{u} \cdot \mathbf{n}_i = \mathbf{u} \cdot \mathbf{e}_\varepsilon$$

$$= \frac{Q}{4\pi\varepsilon^2} + \mathbf{u}_i \cdot \mathbf{e}_\varepsilon$$

Hence from Eq. (5.14a),

$$\mathbf{F} = \rho\int_{S_i}\left[-\frac{1}{2}\left(\frac{Q^2}{16\pi^2\varepsilon^4} + \frac{Q}{2\pi\varepsilon^2}\mathbf{e}_\varepsilon \cdot \mathbf{u}_i + \mathbf{u}_i \cdot \mathbf{u}_i\right)\mathbf{e}_\varepsilon\right.$$

$$\left. + \left(\frac{Q}{4\pi\varepsilon^2}\mathbf{e}_\varepsilon + \mathbf{u}i\right)\left(\frac{Q}{4\pi\varepsilon^2} + \mathbf{u}_i \cdot \mathbf{e}_\varepsilon\right)\right] dS$$

$$= \rho\int_{S_i}\left[\frac{Q^2}{32\pi^2\varepsilon^4}\mathbf{e}_\varepsilon - \tfrac{1}{2}(\mathbf{u}_i \cdot \mathbf{u}_i)\mathbf{e}_\varepsilon + \frac{Q}{4\pi\varepsilon^2}\mathbf{u}_i + (\mathbf{u}_i \cdot \mathbf{e}_\varepsilon)\mathbf{u}_i\right] dS$$

Of these four integrals, the first is zero, since it involves a constant times $\mathbf{e}_\varepsilon$ integrated around a closed surface. Since the radius $\varepsilon$ is arbitrarily small, the quantity $\mathbf{u}_i \cdot \mathbf{u}_i$ may be considered to be constant over the surface $S_i$, and so the second integral will likewise be zero in the limit as $\varepsilon \to 0$. The last term in the integrand will likewise involve a quantity $\mathbf{u}_i$, which will be constant, and a quantity $\mathbf{e}_\varepsilon$, which will change direction around $S_i$. Thus the product $\mathbf{u}_i \cdot \mathbf{e}_\varepsilon$ will have equal positive and negative regions over the surface $S_i$, so that the integral of $(\mathbf{u}_i \cdot \mathbf{e}_\varepsilon)\mathbf{u}_i$ over $S_i$ will be zero. Then the expression for $\mathbf{F}$ becomes

$$\mathbf{F} = \rho \int_{S_i} \frac{Q}{4\pi\varepsilon^2} \mathbf{u}_i \, dS$$

where, again, $\mathbf{u}_i$ may be considered to be constant throughout the integration for vanishingly small values of $\varepsilon$. Hence

$$\mathbf{F} = \frac{\rho Q}{4\pi} \mathbf{u}_i \int_0^{2\pi} d\omega \int_0^\pi \sin\theta \, d\theta \tag{5.14b}$$

$$\mathbf{F} = \rho Q \mathbf{u}_i$$

That is, the force on the body, and on the source, is proportional to the source strength and to the magnitude of the velocity $\mathbf{u}_i$ induced at the location of the source by all mechanisms other than the source itself. The direction of the force coincides with that of the velocity vector $\mathbf{u}_i$. For a sink, $Q$ should be replaced by $-Q$ in Eq. (5.14b).

Consider now the case when the singularity is a doublet. It was shown in Sec. 5.6 that a doublet may be obtained by superimposing a source and a sink of equal strength. Hence, consider a source of strength $Q$ to be located at $x = x_i$ and a sink of strength $Q$ to be located at $x = x_i + \delta$, as shown in Fig. 5.11b, where $\delta$ is a vanishingly small distance. Then if $\mathbf{u}_i$ is the fluid velocity at $x = x_i$ due to all components of the flow except the source and the sink under consideration, the velocity at $x = x_i$, less that due to the source itself, will be

$$\frac{Q}{4\pi\delta^2} \mathbf{e}_x + \mathbf{u}_i$$

and the velocity at $x = x_i + \delta$, less that due to the sink, will be

$$\frac{Q}{4\pi\delta^2} \mathbf{e}_x + \mathbf{u}_i + \delta \frac{\partial \mathbf{u}_i}{\partial x} + \cdots$$

Then, from Eq. (5.14b), the force acting on the body due to the source will be

$$\rho Q \left( \frac{Q}{4\pi\delta^2} \mathbf{e}_x + \mathbf{u}_i \right)$$

and the force acting on the body due to the sink will be

$$-\rho Q \left( \frac{Q}{4\pi\delta^2} \mathbf{e}_x + \mathbf{u}_i + \delta \frac{\partial \mathbf{u}_i}{\partial x} + \cdots \right)$$

where the minus sign results from the fact that a sink is being considered. The net force which will act on the body due to the combined source and sink will then be

$$-\rho Q \delta \frac{\partial \mathbf{u}_i}{\partial x}$$

Now if $\delta$ is allowed to shrink to zero and $Q$ is allowed to become infinite such that $Q\delta \to \mu$, the force acting on the doublet of strength $\mu$ which will result at $x = x_i$ will be

$$-\rho \mu \frac{\partial \mathbf{u}_i}{\partial x}$$

Hence the force acting on the body due to a doublet of strength $\mu$ will be

$$\mathbf{F} = -\rho \mu \frac{\partial \mathbf{u}_i}{\partial x} \qquad (5.14c)$$

As an example of an application of the foregoing results, consider a sphere in the presence of a source which was discussed in Sec. 5.10. The flow field was found to consist of the source of strength $Q$ which was located at $x = l$, an image source of strength $Qa/l$ located at $x = a^2/l$, and a line sink of strength $Q/a$ extending over the region $x = 0$ to $x = a^2/l$. Then the velocity $\mathbf{u}_i$ at the point $x = l$ due to all causes except the source of strength $Q$ will be

$$
\begin{aligned}
\mathbf{u}_i &= \frac{Qa/l}{4\pi} \frac{1}{\left(l - a^2/l\right)^2} \mathbf{e}_x - \int_0^{a^2/l} \frac{Q/a}{4\pi} \frac{\mathbf{e}_x}{\left(l - x\right)^2} \, dx \\
&= \frac{Qa/l}{4\pi} \frac{1}{\left(l - a^2/l\right)^2} \mathbf{e}_x - \frac{Q/a}{4\pi} \left[ \frac{1}{\left(l - a^2/l\right)} - \frac{1}{l} \right] \mathbf{e}_x \\
&= \frac{Qa^3}{4\pi l \left(l^2 - a^2\right)^2} \mathbf{e}_x
\end{aligned}
$$

Then, from Eq. (5.14b), the force $\mathbf{F}$ acting on the sphere due to the source will be

$$\mathbf{F} = \frac{\rho Q^2 a^3}{4\pi l \left(l^2 - a^2\right)^2} \mathbf{e}_x \qquad (5.14d)$$

That is, the sphere is attracted to the source with a force which is proportional to $Q^2$.

## 5.14   KINETIC ENERGY OF A MOVING FLUID

It is sometimes of interest to calculate the kinetic energy associated with a fluid disturbance. An example of the utility of this quantity in the context of flow around immersed bodies will be given in the next section, and an application to free-surface flows will be made in the next chapter.

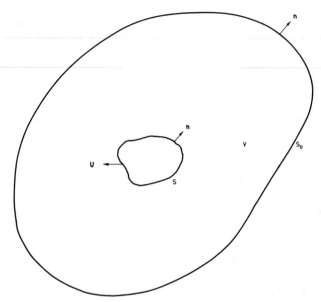

**FIGURE 5.12**
Control surface for arbitrary body moving through a quiescent fluid.

The kinetic energy associated with the fluid in the uniform flow around a stationary body will be infinite if the flow field is infinite in extent. However, the kinetic energy induced in a quiescent fluid by the passage of a body through it will be finite, even if the flow field is infinite in extent. For this reason, discussions of kinetic-energy considerations are based on a frame of reference in which the fluid far from the body is at rest and the body is moving.

Referring to Fig. 5.12, we consider an arbitrary body of surface area $S$ which is moving with velocity $\mathbf{U}$ through a stationary fluid. An arbitrarily shaped control surface $S_0$ is constructed around the body. The unit outward normals to the surfaces $S$ and $S_0$, denoted by $\mathbf{n}$, are indicated. If $V$ is the volume of fluid contained between the surfaces $S$ and $S_0$, the kinetic energy of $T$ of this volume of fluid will be

$$T = \int_V \tfrac{1}{2}\rho(\mathbf{u} \cdot \mathbf{u}) \, dV$$

$$= \tfrac{1}{2}\rho \int_V \nabla\phi \cdot \nabla\phi \, dV$$

where $\phi$ is the velocity potential corresponding to the motion induced in the fluid by the moving body. This volume integral may be converted to a surface integral by use of Green's theorem in the form given in Appendix A. Thus, since $\nabla^2\phi = 0$, it follows from Green's theorem that

$$T = \tfrac{1}{2}\rho \int_\Sigma \phi \frac{\partial\phi}{\partial n} \, dS$$

where $\Sigma$ is the surface which encloses $V$ and so consists of the surfaces $S$ and

$S_0$. But on the surface $S$ the unit normal points away from the surface and into the volume $V$. Using this fact, the surface integral above may be expanded to give

$$T = \tfrac{1}{2}\rho \int_{S_0} \phi \frac{\partial \phi}{\partial n} \, dS - \tfrac{1}{2}\rho \int_S \phi \frac{\partial \phi}{\partial n} \, dS$$

The first integral which appears in this expression is zero, which will now be shown.

From the continuity equation, it follows that

$$\int_V \nabla \cdot \mathbf{u} \, dV = 0$$

This volume integral may be converted into two surface integrals by use of Gauss' theorem to yield the following:

$$\int_{S_0} \mathbf{u} \cdot \mathbf{n} \, dS - \int_S \mathbf{u} \cdot \mathbf{n} \, dS = 0$$

But $\mathbf{u} \cdot \mathbf{n} = \partial \phi / \partial n$ and on the surface $S$, $\mathbf{u} = \mathbf{U}$. Hence

$$\int_{S_0} \frac{\partial \phi}{\partial n} \, dS - \int_S \mathbf{U} \cdot \mathbf{n} \, dS = 0$$

The second integral in this identity is zero, since $\mathbf{U}$ is a constant vector, so that for any constant $C$ it follows that

$$\int_{S_0} \mathbf{C} \frac{\partial \phi}{\partial n} \, dS = 0$$

Subtracting this quantity from the right-hand side of the expression for $T$ gives

$$T = \tfrac{1}{2}\rho \int_{S_0} (\phi - C) \frac{\partial \phi}{\partial n} \, dS - \tfrac{1}{2}\rho \int_S \phi \frac{\partial \phi}{\partial n} \, dS$$

Now since the fluid velocity far from the body is zero, the value of $\phi$ there can at most be a constant. Thus by considering the surface $S_0$ to be large and by choosing $C$ to be the value of $\phi$ far from the body, the first integral may be made to vanish. That is, the kinetic energy induced in the fluid by the movement of the body is

$$T = -\tfrac{1}{2}\rho \int_S \phi \frac{\partial \phi}{\partial n} \, dS \qquad (5.15)$$

where, it should be recalled, the velocity potential corresponds to the body moving through a stationary fluid.

## 5.15   APPARENT MASS

When a body moves through a quiescent fluid, a certain mass of the fluid is induced to move to some greater or lesser extent. A question which may then be asked is, what equivalent mass of fluid, if it moved with the same velocity as the body, would exhibit the same kinetic energy as the actual case? If the fluid may be considered as being ideal, the mass of fluid referred to above is found to depend upon the body shape only, and this mass of fluid is called the apparent mass.

We define the apparent mass of a fluid $M'$ as that mass of fluid which, if it were moving with the same velocity as the body, would have the same kinetic energy as the entire fluid. That is,

$$\tfrac{1}{2}M'U^2 = -\tfrac{1}{2}\rho \int_S \phi \frac{\partial \phi}{\partial n}\, dS$$

$$M' = -\frac{\rho}{U^2}\int_S \phi \frac{\partial \phi}{\partial n}\, dS$$

(5.16)

For arbitrarily shaped bodies the velocity potential will depend upon the direction of the flow. That is, the apparent mass of fluid associated with a given body will be a property of the shape of that body, and like inertia, there will in general be three principal axes of the apparent mass. For axisymmetric bodies there will be two principal values of $M'$, while for the sphere there will be only one.

As an example of an application of Eq. (5.16), the apparent mass for the sphere will be worked out here. The velocity potential (5.9$b$) corresponds to a stationary sphere of radius $a$ with a uniform flow of magnitude $U$ approaching it. Then the required velocity potential may be obtained from Eq. (5.9$b$) by adding the velocity potential for a uniform flow of magnitude $U$ in the negative $x$ direction. This gives

$$\phi(r,\theta) = U\left(r + \frac{1}{2}\frac{a^3}{r^2}\right)\cos\theta - Ur\cos\theta$$

$$= \tfrac{1}{2}U\frac{a^3}{r^2}\cos\theta$$

$$\therefore \quad \frac{\partial\phi}{\partial n}(r,\theta) = \frac{\partial\phi}{\partial r}(r,\theta) = -U\frac{a^3}{r^3}\cos\theta$$

Hence on the surface $S$ where $r = a$

$$\phi\frac{\partial\phi}{\partial n} = -\tfrac{1}{2}U^2 a \cos^2\theta$$

Then, from Eq. (5.16), the apparent mass for the sphere is

$$M' = -\frac{\rho}{U^2}\int_0^{2\pi} d\omega \int_0^{\pi}\left(-\tfrac{1}{2}U^2 a \cos^2\theta\right)a^2\sin\theta\, d\theta$$

$$M' = \tfrac{2}{3}\pi a^3 \rho$$

(5.17)

That is, the apparent mass for a sphere is one-half of the mass of the same volume of fluid. This apparent mass may be added to the actual mass of the sphere, and the total mass may be used in the dynamic equations of the sphere. That is, the existence of the fluid may be ignored if the apparent mass of fluid is added to the actual mass of the body.

## PROBLEMS

**5.1.** Use the definition of the Stokes stream function and the $\omega$ component of the condition of irrotationality to show that the equation to be satisfied by $\psi(r, \theta)$ for axisymmetric flows is

$$r^2 \frac{\partial^2 \psi}{\partial r^2} + \sin\theta \frac{\partial}{\partial \theta}\left(\frac{1}{\sin\theta}\frac{\partial\psi}{\partial\theta}\right) = 0 \qquad (5.18)$$

**5.2.** Show, by direct substitution, that the stream functions obtained for a uniform flow, a source, and a doublet, as given by Eqs. (5.5b), (5.6b), and (5.7b), respectively, satisfy Eq. (5.18).

**5.3.** Look for a separable solution to Eq. (5.18) of the form

$$\psi(r, \theta) = R(r)T(\theta)$$

Hence show that the finite solutions for $R(r)$ are of the form

$$R_n(r) = A_n r^{-n}$$

and that the equation to be satisfied by $T(\theta)$ is

$$(1 - \eta^2)\frac{d^2 T}{d\eta^2} + n(n + 1)T = 0$$

where $\eta = \cos\theta$. Show that the substitution $T = (1 - \eta^2)^{1/2}\tau$ transforms the equation to the form

$$(1 - \eta^2)\frac{d^2\tau}{d\eta^2} - 2\eta\frac{d\tau}{d\eta} + \left[n(n + 1) - \frac{1}{1 - \eta^2}\right]T = 0$$

which is the associated Legendre equation. Show that the nonsingular solution to this equation is

$$\tau_n(\eta) = (1 - \eta^2)^{1/2}\frac{dP_n(\eta)}{d\eta}$$

where $P_n(\eta)$ is Legendre's polynomial of order $n$. Thus deduce that general solutions to Eq. (5.18) will be of the form

$$\psi(r, \theta) = \sum_{n=1}^{\infty} A_n \frac{\sin\theta}{r^n}\frac{d}{d\theta}[P_n(\cos\theta)] \qquad (5.19)$$

**5.4.** Show that setting $A_n = 0$ for $n \neq 1$ in Eq. (5.19) yields the solution for a doublet.

**5.5.** Figure 5.13 shows a doublet of strength $\mu$ located at $x = l$ and a doublet of strength $\mu^*$ located at $x = a^2/l$. Show that the surface $r = a$ corresponds to $\psi = 0$ if $\mu^* = -a^3\mu/l^3$, and hence deduce that the stream function for a doublet of strength

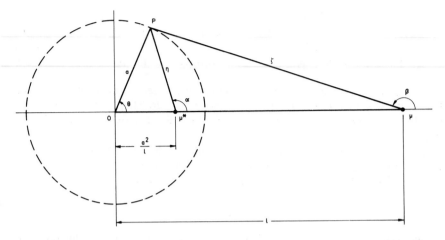

**FIGURE 5.13**
Superposition of doublet of strength $\mu$ and doublet of strength $\mu^*$ which leads to doublet of strength $\mu$ outside a sphere of radius $a$.

$\mu$ located a distance $l$ from the center of a sphere of radius $a$ is

$$\psi(r,\theta) = -\frac{\mu}{4\pi\xi}\sin^2\beta + \frac{\mu a^3}{4\pi l^3\eta}\sin^2\alpha \qquad (5.20a)$$

Also deduce that the corresponding velocity potential is

$$\phi(r,\theta) = -\frac{\mu}{4\pi\xi^2}\cos\beta - \frac{\mu a^3}{4\pi l^3\eta^2}\cos\alpha \qquad (5.20b)$$

**5.6.** Show that the force which acts on a sphere of radius $a$ owing to a doublet of strength $\mu$ located a distance $l$ from the center of the sphere along the $x$ axis is

$$\mathbf{F} = \frac{3\rho\mu^2 a^3 l}{2\pi(l^2-a^2)^4}\mathbf{e}_x$$

**5.7.** A spherical gas bubble of radius $R(t)$ exists in a liquid; that is, the radius of the bubble is changing with time. The liquid is quiescent, apart from any motion which is caused by the bubble. It is assumed that the fluid motion does not involve any viscous or compressible effects, and it may therefore be represented by a time-dependent velocity potential which satisfies the following conditions:

$$\nabla^2\psi = \frac{1}{r}\frac{\partial}{\partial r}\left(r^2\frac{\partial\phi}{\partial r}\right) = 0$$

$$\frac{\partial\phi}{\partial r}(r\to\infty,t) = 0$$

$$\frac{\partial\phi}{\partial r}(r=R,t) = \dot{R}$$

where $\dot{R}$ is the derivative of $R$ with respect to time. Obtain an expression for the radial velocity at the surface of the bubble and for the velocity potential $\phi(r, t)$, both of these expressions being in terms of $R$, $\dot{R}$, and $r$. Also obtain an expression for the pressure at the surface of the bubble $p(R, t)$ if the pressure far from the bubble is $p_0$ which is a constant.

Suppose that at time $t = 0$ the pressure at the surface of the bubble is $p_0$, the radius of the bubble is $R_0$ and its initial velocity is $-\dot{R}$. That is, the radius of the bubble is decreasing as time increases. Find the time required for the radius of the bubble to shrink to zero.

**5.8.** A sphere of radius $a$ moves along the $x$ axis with velocity $U(t)$ which is varying with time. A fixed-origin coordinate system is defined by the location of the sphere at $t = 0$, so that its location at any subsequent time is given by

$$x_0(t) = \int_0^t U(\tau)\, d\tau$$

as depicted in Fig. 5.14. Then if $P$ is any fixed field point, its coordinates $(r, \theta)$ relative to the sphere will change with time. Obtain the velocity potential for a sphere in a stationary fluid, first in terms of $r$ and $\theta$, then in terms of $x$, $R$, and $x_0$. If the undisturbed pressure is $p_\infty$, find the pressure at the field point $P$ in terms of $r$ and $\theta$. Hence, by integrating the pressure over the surface of the sphere, find the force acting on the sphere. Compare the result so obtained with that obtained by using the apparent-mass concept in conjunction with Newton's second law.

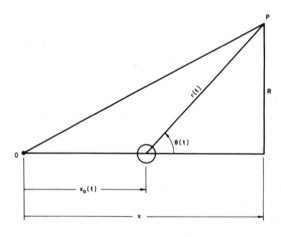

**FIGURE 5.14**
Coordinate systems for a sphere moving through a stationary fluid.

# CHAPTER

# 6

# SURFACE WAVES

The effect of gravity on liquid surfaces is treated in this chapter. The flows associated with surface waves will be assumed to be potential in nature, which is a valid approximation for many free-surface phenomena. Most of the flows which will be treated here will be two-dimensional. However, the treatment of surface waves has been separated from the other two-dimensional potential flows because of the different nature of the problems and the different approaches to their solutions.

The formulation of surface-wave problems is discussed first. The linearized version of this formulation is then presented, and this version is used throughout most of the remainder of the chapter. The propagation speed of small-amplitude waves is established, and the effect of surface tension on this result is investigated. Waves on shallow liquids are discussed next, and the manner in which waves of arbitrary form and amplitude propagate is established.

The complex potential for traveling waves is calculated, and this result is used to establish the pathlines for fluid particles in a body of liquid on the surface of which surface waves are propagating. A superposition of traveling waves is then used to introduce the topic of standing waves. The particle paths for this type of wave are also established. The topic of standing waves leads, quite naturally, to the question of what type of waves may exist on the free surface of liquids which are contained in vessels of finite dimensions. In particular, rectangular and cylindrical vessels are discussed. The response of the free surface to arbitrary motions of such vessels is obtained. Finally, the behavior of waves at the interface of the two different fluid streams is investigated. This leads to the topics of Helmholtz or Rayleigh instability and Taylor instability.

## 6.1   THE GENERAL SURFACE-WAVE PROBLEM

When a quiescent body of liquid experiences gravity waves on its free surface, the motion induced by the surface waves may be considered to be irrotational in most instances. Then the velocity vector may be expressed as the gradient of a velocity potential, which in turn, must satisfy Laplace's equation. That is, the governing equation is the same as that for each of the two previous chapters, so that surface-wave theory introduces no new difficulties with respect to the governing equation. The boundary conditions which are to be satisfied will now be established.

Figure 6.1 shows a body of liquid on a flat surface in which waves exist on the free surface of the liquid. The $x$ axis of a coordinate system is located at the mean level of the free surface, which is defined by the equation $y = \eta(x, z, t)$ and the mean depth of the liquid is $h$. Two boundary conditions must be imposed on the free surface $y = \eta$. The first of these conditions is called the *kinematic condition*, and it states that a particle of fluid which is at some time on the free surface will always remain on the free surface. Then, since the equation of the free surface is $y - \eta = 0$, it follows that

$$\frac{D}{Dt}(y - \eta) = 0$$

In terms of eulerian coordinates this boundary condition becomes

$$\frac{\partial}{\partial t}(y - \eta) + \mathbf{u} \cdot \nabla(y - \eta) = 0$$

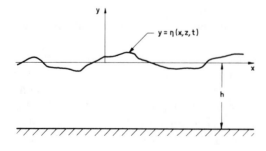

**FIGURE 6.1**
Coordinate system for surface-wave problems.

But in the eulerian frame of reference the coordinates $x$, $y$, $z$, and $t$ are independent. Also, the function $\eta$ depends on $x$, $z$, and $t$ only. Hence the above equation may be expanded to give

$$-\frac{\partial \eta}{\partial t} - u\frac{\partial \eta}{\partial x} + v - w\frac{\partial \eta}{\partial z} = 0$$

where it has been noted that $\partial x_i/\partial x_j = \delta_{ij}$. Finally, expressing the velocity components in terms of the velocity potential $\phi$, the kinematic surface condition becomes

$$\frac{\partial \eta}{\partial t} + \frac{\partial \phi}{\partial x}\frac{\partial \eta}{\partial x} + \frac{\partial \phi}{\partial z}\frac{\partial \eta}{\partial z} = \frac{\partial \phi}{\partial y}$$

The other boundary condition which must be imposed on the free surface is a dynamic one. Typically, the statement amounts to specifying that the pressure is constant, but in general it may be stated that $p = P(x, z, t)$ on $y = \eta$. This condition is implemented through the Bernouli equation. The appropriate form of the Bernoulli equation is that for unsteady, irrotational motion. Since gravitational forces are intrinsically important in free-surface waves, gravity must be included in the body-force term. Thus, from Eq. (II.6) with $G = -gy$, the boundary condition $p = P$ on $y = \eta$ becomes

$$\frac{\partial \phi}{\partial t} + \frac{P}{\rho} + \tfrac{1}{2}\nabla\phi \cdot \nabla\phi + g\eta = F(t)$$

Finally, the boundary condition at the bed must be imposed. For the case of an inviscid fluid which is under consideration, this amounts to specifying that the velocity component normal to the boundary be zero. For a flat bed as shown in Fig. 6.1, this simply amounts to specifying that $\partial \phi/\partial y = 0$ on $y = -h$.

To summarize, in terms of the velocity potential $\phi$, the conditions to be satisfied for surface-wave motions are the following:

$$\nabla^2\phi = 0 \qquad\qquad\qquad (6.1a)$$

$$\frac{\partial \eta}{\partial t} + \frac{\partial \phi}{\partial x}\frac{\partial \eta}{\partial x} + \frac{\partial \phi}{\partial z}\frac{\partial \eta}{\partial z} = \frac{\partial \phi}{\partial y} \qquad \text{on } y = \eta \qquad (6.1b)$$

$$\frac{\partial \phi}{\partial t} + \frac{P}{\rho} + \tfrac{1}{2}\nabla\phi \cdot \nabla\phi + g\eta = F(t) \qquad \text{on } y = \eta \qquad (6.1c)$$

$$\frac{\partial \phi}{\partial y} = 0 \qquad \text{on } y = -h \qquad (6.1d)$$

The difficulty in solving surface-wave problems may be seen to be in the boundary conditions rather than the differential equation. Equation (6.1c) is nonlinear, and both it and Eq. (6.1b) are to be imposed on the surface $y = \eta$. In many real situations this surface may not be known a priori and may be one of the quantities which comes out of the solution itself. However, many interesting features of surface-wave flows do not depend upon these complex features of the problem. That is, by linearizing the problem, the difficulties discussed above may be avoided while the basic features of the flow are not destroyed. Such a linearization will be carried out in the next section.

## 6.2   SMALL-AMPLITUDE PLANE WAVES

For simplicity we consider plane waves, that is, two-dimensional flow fields with waves on the surface. Then, without any further approximation, the differential equation to be satisfied by the velocity potential in the $xy$ plane that is

$$\frac{\partial^2 \phi}{\partial x^2} + \frac{\partial^2 \phi}{\partial y^2} = 0 \tag{6.2a}$$

In order to make the surface boundary conditions more tractable, small-amplitude waves will be considered. That is, only waves for which the amplitude is small compared with the other characteristic length scales will be considered. The other characteristic length scales are the liquid depth $h$ and the wavelength of the waves. But if $\eta$ is small compared with the wavelength, the quantity $\partial \eta / \partial x$, which is the slope of the free surface, will be small. Furthermore, the quantity $\partial \phi / \partial x$, which is a velocity component, will be small, since surface waves do not involve high frequencies and since the amplitude of the motion has been assumed to be small. Then the product of $\partial \phi / \partial x$ and $\partial \eta / \partial x$, which appears in Eq. (6.1b), will be quadratically small and hence may be neglected to first order. The kinematic boundary condition on the free surface then becomes

$$\frac{\partial \eta}{\partial t}(x, t) = \frac{\partial \phi}{\partial y}(x, \eta, t)$$

Although this equation is free from quadratic terms, it still contains the difficulty that it must be imposed on $y = \eta$. However, in our present approximation $\eta$ is small, so that a Taylor expansion may be written for the quantity $\partial \phi / \partial y$ at $y = \eta$ about the line $y = 0$. Thus

$$\frac{\partial \phi}{\partial y}(x, \eta, t) = \frac{\partial \phi}{\partial y}(x, 0, t) + \eta \frac{\partial^2 \phi}{\partial y^2}(x, 0, t) + O(\eta^2)$$

The second term in this expansion is quadratically small and so, to the first order, may be neglected. That is, to the first order in small quantities, the boundary condition (6.1*b*) may be written in the form

$$\frac{\partial \phi}{\partial y}(x,0,t) = \frac{\partial \eta}{\partial t}(x,t) \tag{6.2b}$$

The dynamic boundary condition on the free surface may be treated in the same way. Since the fluid is essentially quiescent and any fluid motion is induced by the waves, the nonlinear term $\mathbf{u} \cdot \mathbf{u} = \nabla \phi \cdot \nabla \phi$ may be neglected as being quadratically small. Thus Eq. (6.1*c*) becomes

$$\frac{\partial \phi}{\partial t}(x,\eta,t) + \frac{P(x,t)}{\rho} + g\eta(x,t) = F(t)$$

The quantity $\partial \phi / \partial t$ may be expanded in a Taylor series about the line $y = 0$, and only the first term in this expansion need be retained. This gives

$$\frac{\partial \phi}{\partial t}(x,0,t) + \frac{P(x,t)}{\rho} + g\eta(x,t) = F(t)$$

The quantity $F(t)$ may be absorbed into the velocity potential $\phi(x,y,t)$ by considering $\phi(x,y,t)$ to be replaced by

$$\phi(x,y,t) + \int F(t)\, dt$$

Thus the linearized version of Eq. (6.1*c*) may be written in the form

$$\frac{\partial \phi}{\partial t}(x,0,t) + \frac{P(x,t)}{\rho} + g\eta(x,t) = 0$$

If the time derivative of this equation is formed, the term $\partial \eta / \partial t$ may be eliminated in favor of $\partial \phi / \partial y$ from Eq. (6.2*b*). Thus the preferred form of the dynamic boundary condition on the free surface is

$$\frac{\partial^2 \phi}{\partial t^2}(x,0,t) + \frac{1}{\rho}\frac{\partial P(x,t)}{\partial t} + g\frac{\partial \phi}{\partial y}(x,0,t) = 0 \tag{6.2c}$$

The boundary condition on the bed is unaffected by the linearization and requires

$$\frac{\partial \phi}{\partial y}(x,-h,t) = 0 \tag{6.2d}$$

Equations (6.2) represent a much more tractable set than the general equations which were presented in the previous section. However, as was mentioned earlier, they correctly predicted many of the features of surface waves, and so they will form the basis of most of the remaining sections of this chapter.

## 6.3  PROPAGATION OF SURFACE WAVES

Consider a quiescent body of water or other liquid of depth $h$ as shown in Fig. 6.2. A small-amplitude plane wave is traveling along the surface of this liquid with velocity $c$. The form of the wave is taken to be sinusoidal in which the amplitude of the wave is $\varepsilon$ and its wavelength is $\lambda$. Thus the equation of the free surface will be $y = \eta(x, t)$, where

$$\eta(x, t) = \varepsilon \sin \frac{2\pi}{\lambda}(x - ct)$$

This corresponds to the wave traveling in the positive $x$ direction with velocity $c$.

The question we ask is the following: Given the wave amplitude $\varepsilon$ and wavelength $\lambda$ and given the depth $h$, what will be the propagation speed $c$? The answer to this question may presumably be obtained by solving the flow problem for the velocity potential. For the time being, surface-tension effects will be neglected; so the pressure on the surface of the liquid will be constant and equal to, say, atmospheric pressure. That is, $P(x, t) =$ constant in this instance. Then, from Eqs. (6.2), the problem to be solved for $\phi(x, y, t)$ is

$$\frac{\partial^2 \phi}{\partial x^2} + \frac{\partial^2 \phi}{\partial y^2} = 0$$

$$\frac{\partial \phi}{\partial y}(x, 0, t) = -\varepsilon \frac{2\pi c}{\lambda} \cos \frac{2\pi}{\lambda}(x - ct)$$

$$\frac{\partial^2 \phi}{\partial t^2}(x, 0, t) + g \frac{\partial \phi}{\partial y}(x, 0, t) = 0$$

$$\frac{\partial \phi}{\partial y}(x, -h, t) = 0$$

Here the equation for $\eta(x, t)$ has been used in the kinematic condition on the free surface. The appropriate solution to the Laplace equation by separation of variables will be trigonometric in $x$, and hence it will be exponential or

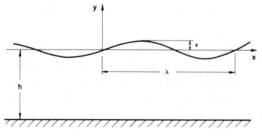

**FIGURE 6.2**
Parameters for pure sinusoidal waveform.

hyperbolic in $y$. This deduction follows from the nature of the value of $\partial\phi/\partial y$, which is prescribed on $y = 0$ by the kinematic boundary condition. In fact, inspection of this boundary condition yields even stronger information. Since $\partial\phi/\partial y$ must vary as $\cos 2\pi(x - ct)/\lambda$, then so must $\phi$. That is, the nature of the time dependence is brought in through this boundary condition, as is the nature of the $x$ dependence. Furthermore, since the separation constant in the $x$ direction must be $2\pi/\lambda$, the separation constant in the $y$ direction must also be $2\pi/\lambda$. Hence the appropriate form of solution to the Laplace equation is

$$\phi(x, y, t) = \cos \frac{2\pi}{\lambda}(x - ct)\left(C_1 \sinh \frac{2\pi y}{\lambda} + C_2 \cosh \frac{2\pi y}{\lambda}\right)$$

Here, the hyperbolic form of solution in $y$ has been used in preference to the exponential form, since the region in $y$ is finite rather than infinite. This facilitates application of the boundary conditions. Having used the form of the first boundary condition, the third boundary condition will now be imposed. Thus the condition that $\partial\phi/\partial y$ must vanish on $y = -h$ gives

$$\cos \frac{2\pi}{\lambda}(x - ct)\left(\frac{2\pi}{\lambda}C_1 \cosh \frac{2\pi h}{\lambda} - \frac{2\pi}{\lambda}C_2 \sinh \frac{2\pi h}{\lambda}\right) = 0$$

Since this condition is to be satisfied for all values of $x$ and $t$, the quantity inside the second parentheses must be zero. This gives

$$C_1 = C_2 \tanh \frac{2\pi h}{\lambda}$$

Hence the solution for $\phi(x, y, t)$ becomes

$$\phi(x, y, t) = C_2 \cos \frac{2\pi}{\lambda}(x - ct)\left(\tanh \frac{2\pi h}{\lambda} \sinh \frac{2\pi y}{\lambda} + \cosh \frac{2\pi y}{\lambda}\right)$$

Finally, the second boundary condition, corresponding to the dynamic condition on the free surface, will be imposed. This gives

$$C_2 \cos \frac{2\pi}{\lambda}(x - ct)\left[-\left(\frac{2\pi c}{\lambda}\right)^2 + g\frac{2\pi}{\lambda} \tanh \frac{2\pi h}{\lambda}\right] = 0$$

Again this equation is to be satisfied for all values of $x$ and $t$, so that the quantity inside the brackets must vanish. But the only unknown quantity inside the brackets is the wave speed $c$. That is, imposing this final boundary condition determines the speed $c$ with which the wave train is traveling. In nondimensional form the result is

$$\frac{c^2}{gh} = \frac{\lambda}{2\pi h} \tanh \frac{2\pi h}{\lambda} \tag{6.3a}$$

Equation (6.3a) was obtained using a small-amplitude approximation, which means that it is valid provided $\varepsilon \ll \lambda$ and $\varepsilon \ll h$.

As a special case, we consider deep liquids, that is, liquids for which $h \gg \lambda$. Then the parameter $2\pi h/\lambda$ will be large, so that $\tanh(2\pi h/\lambda)$ will be approximately unity. Then, for deep liquids Eq. (6.3a) may be approximated by

$$\frac{c^2}{gh} = \frac{\lambda}{2\pi h} \tag{6.3b}$$

Equation (6.3b) will be valid for $\varepsilon \ll \lambda \ll h$.

The other obvious limit is that of shallow liquids, that is, liquids for which $h \ll \lambda$. In this case the parameter $2\pi h/\lambda$ will be small so that

$$\tanh \frac{2\pi h}{\lambda} \approx \frac{2\pi h}{\lambda}$$

Then Eq. (6.3a) becomes

$$\frac{c^2}{gh} = 1 \tag{6.3c}$$

which will be valid for $\varepsilon \ll h \ll \lambda$.

The foregoing results are presented schematically in Fig. 6.3, in which the general solution [Eq. (6.3a)] is shown by a solid line and the two asymptotic limits are shown dotted.

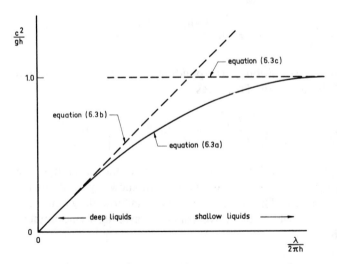

**FIGURE 6.3**
Propagation speed $c$ for small-amplitude surface waves of sinusoidal form.

An arbitrary shaped wave train may be considered to be a superposition of sinusoidal waves of the type just treated. That is, waves of arbitrary form may be Fourier-analyzed and so decomposed into a number of pure sinusoidal waves. Thus the foregoing results show that such waves will not, in general, propagate in an undisturbed way. That is because the propagation speed $c$, or celerity, as it is sometimes called, depends upon the wavelength $\lambda$ of its sinusoidal components. Only in the case of shallow liquids [Eq. (6.3$c$)] is the propagation speed independent of the wavelength. That is, unless the shallow-liquid conditions apply, the different Fourier components of an arbitrary shaped wave will all travel at different speeds so that the waveform will continuously change. This process is usually referred to as *dispersion*.

## 6.4  EFFECT OF SURFACE TENSION

In the previous section it was assumed that the pressure along the topmost layer of the liquid was constant corresponding to atmospheric pressure. However, if surface-tension effects are included, the pressure along the edge of the liquid will be different from the pressure outside the liquid unless the surface is flat. To establish the effect of this pressure differential, an element of the surface is isolated in Fig. 6.4 and the forces due to surface tension are indicated.

At the reference position $x$ the value of the surface tension is $\sigma$, and the slope of the surface there is $\partial \eta / \partial x$. Then, a short distance $\Delta x$ farther from the origin the value of the surface tension will be $\sigma + (\partial \sigma / \partial x)\,\Delta x$, and the slope of the surface will be $\partial \eta / \partial x + (\partial^2 \eta / \partial x^2)\,\Delta x$. Only the first-order corrections have been written down here, since the unwritten terms in the Taylor series will contain terms of order $(\Delta x)^2$ or smaller. Then, if $p_0$ is the pressure above the liquid and if $P(x, t)$ is the pressure at the edge of the liquid, vertical equilibrium of the element of surface shown in Fig. 6.4 requires that the following equation be satisfied:

$$(P - p_0)\,\Delta x + \left(\sigma + \frac{\partial \sigma}{\partial x}\,\Delta x\right)\left(\frac{\partial \eta}{\partial x} + \frac{\partial^2 \eta}{\partial x^2}\,\Delta x\right) - \sigma \frac{\partial \eta}{\partial x} = 0$$

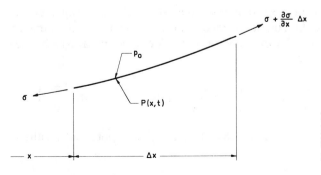

**FIGURE 6.4**
Element of liquid surface showing forces due to surface tension.

Expanding the terms in this equation and neglecting terms which are quadratic in the length $\Delta x$ gives

$$(P - p_0) + \sigma \frac{\partial^2 \eta}{\partial x^2} + \frac{\partial \sigma}{\partial x} \frac{\partial \eta}{\partial x} = 0$$

The length $\Delta x$ may now be permitted to shrink to zero, so that the neglected terms become identically zero and the above equation becomes exact. The last term in the above equation will be zero if $\sigma$ is constant and will be quadratically small if $\sigma$ is almost constant. Thus to the first order in small quantities the pressure $P(x, t)$ at the edge of the liquid becomes

$$P(x, t) = p_0 - \sigma \frac{\partial^2 \eta}{\partial x^2}$$

In the dynamic boundary condition on the free surface [Eq. (6.2c)] the pressure enters through the term $\partial P / \partial t$. But, if the pressure $p_0$ outside the liquid is constant, the expression for $\partial P / \partial t$ is

$$\frac{\partial P}{\partial t} = -\sigma \frac{\partial^2}{\partial x^2} \left( \frac{\partial \eta}{\partial t} \right)$$

$$= -\sigma \frac{\partial^3 \phi}{\partial x^2 \partial y} (x, 0, t)$$

where the order of differentiation has been interchanged in the first equation and $\partial \eta / \partial t$ has been eliminated in favor of derivatives of $\phi$ using the kinematic boundary conditions (6.2b). Using the above result, the dynamic boundary condition on the free surface [Eq. (6.2c)] becomes

$$\frac{\partial^2 \phi}{\partial t^2} (x, 0, t) - \frac{\sigma}{\rho} \frac{\partial^3 \phi}{\partial x^2 \partial y} (x, 0, t) + g \frac{\partial \phi}{\partial y} (x, 0, t) = 0 \qquad (6.4)$$

This revised form of the dynamic boundary condition will be used to recalculate the propagation speed of a sinusoidal wave.

The existence of surface tension does not affect the governing partial differential equation, the kinematic surface condition, or the bed boundary condition. Hence, from the previous section, the velocity potential which satisfies these unchanged equations is

$$\phi(x, y, t) = C_2 \cos \frac{2\pi}{\lambda} (x - ct) \left( \tanh \frac{2\pi h}{\lambda} \sinh \frac{2\pi y}{\lambda} + \cosh \frac{2\pi y}{\lambda} \right)$$

Application of the boundary condition (6.4) to this velocity potential results in the requirement

$$C_2 \cos \frac{2\pi}{\lambda}(x - ct)\left[-\left(\frac{2\pi c}{\lambda}\right)^2 + \frac{\sigma}{\rho}\left(\frac{2\pi}{\lambda}\right)^3 \tanh \frac{2\pi h}{\lambda} + g\frac{2\pi}{\lambda}\tanh \frac{2\pi h}{\lambda}\right] = 0$$

The general solution to this equation requires that the quantity inside the brackets vanish, which gives, in nondimensional form,

$$\frac{c^2}{gh} = \frac{\lambda}{2\pi h}\left[1 + \frac{\sigma}{\rho g}\left(\frac{2\pi}{\lambda}\right)^2\right]\tanh \frac{2\pi h}{\lambda} \qquad (6.5a)$$

If $\sigma$ is negligibly small, Eq. (6.3a) is recovered. This result shows that the effect of surface tension is to increase the propagation speed of the wave.

For deep liquids, the parameter $2\pi h/\lambda$ is large, so that Eq. (6.5a) becomes

$$\frac{c^2}{gh} = \frac{\lambda}{2\pi h}\left[1 + \frac{\sigma}{\rho g}\left(\frac{2\pi}{\lambda}\right)^2\right]$$

If, in addition, the parameter inside the brackets is sufficiently large that

$$\frac{\sigma}{\rho g}\left(\frac{2\pi}{\lambda}\right)^2 \gg 1$$

then the expression for the propagation speed will reduce to the following:

$$\frac{c^2}{gh} = \frac{2\pi\sigma}{\rho g\lambda h} \qquad (6.5b)$$

Waves which satisfy the foregoing conditions and so travel at the speed defined

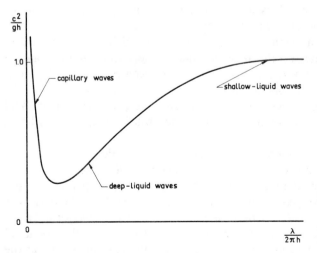

**FIGURE 6.5**
Propagation speed for sinusoidal waves including the effects of surface tension.

by Eq. (6.5b) are called *capillary waves*. It will be noted that the propagation speed of capillary waves depends upon the wavelength $\lambda$, so that an arbitrarily shaped wave will disperse because of the different propagation speeds of its Fourier components.

The propagation speed of sinusoidal waves, as predicted by Eq. (6.5a), is shown in Fig. 6.5 as a function of the parameter $\lambda/(2\pi h)$. It will be seen that the effect of surface tension modifies our previous result only in the deep-liquid end of the spectrum. This is because the condition

$$\frac{\sigma}{\rho g}\left(\frac{2\pi}{\lambda}\right)^2 \geqslant 1$$

is realized only for small values of $\lambda$, which in turn, corresponds to deep-liquid waves.

## 6.5  SHALLOW-LIQUID WAVES OF ARBITRARY FORM

It was deduced from the results of the previous two sections that waves of arbitrary form will disperse unless the liquid is shallow. That is, owing to the different propagation speeds of its Fourier components, an arbitrarily shaped wave will decompose unless the liquid depth $h$ is small compared with the shortest wavelength $\lambda$ of the various Fourier components which constitute the wave. The deduction that shallow-liquid waves of small amplitude will not decompose may be verified by carrying out a detailed study of such waves.

The starting point of such a study is the equations governing the dependent variables. These equations may be obtained from the continuity and Euler equations by integrating across the fluid depth and employing a one-dimensional approximation. However, it is no more difficult and considerably more instructive to derive the equations from first principles using a one-dimensional approach. The latter procedure will be followed here.

Figure 6.6a shows a portion of a liquid layer in which a surface wave of arbitrary form exists. The waveform is assumed to be such that the smallest wavelength of its various Fourier components is large compared with the mean depth $h$. Then a one-dimensional approximation may be employed. That is, the $x$ component of the velocity vector will be assumed to be constant over the fluid depth, and the $y$ component of the velocity vector will be neglected as being small.

Figure 6.6b shows an element of length $\Delta x$ of the fluid which extends from the bottom to the free surface. The mass-flow rates into the element and out of it are also indicated in Fig. 6.6b. The mass-flow rate per unit depth entering the element through the surface at $x$ is $\rho u(h + \eta)$. Then the mass flow leaving the element at $x + \Delta x$ is indicated by the first two terms of a Taylor series about the station at $x$, the remaining terms, not indicated, then being of order $(\Delta x)^2$ or smaller. A mass flux is shown leaving the control volume at the

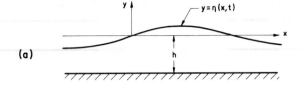

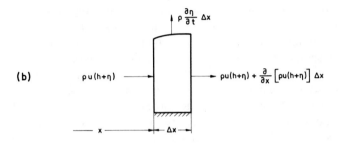

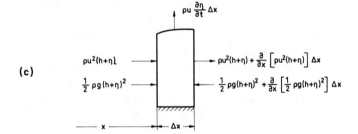

**FIGURE 6.6**
(a) Arbitrary waveform on a shallow liquid layer, (b) mass-flow-rate balance for an element, and
(c) momentum and force balance in the x direction.

top. This is due to the fact that $\eta$ depends on both $x$ and $t$ and the quantity
$\partial\eta/\partial t$ represents the vertical velocity of the free surface. Then, multiplying this
velocity by the density and the length $\Delta x$ gives a mass efflux per unit time per
unit depth. Using the expressions for these various components of mass-flow
rate, the equation of mass conservation becomes

$$\left\{\rho u(h + \eta) + \frac{\partial}{\partial x}[\rho u(h + \eta)]\,\Delta x\right\} + \rho\frac{\partial\eta}{\partial t}\,\Delta x - \rho u(h + \eta) = 0$$

The first and last terms cancel each other, and the remaining terms may be
divided by $\rho\,\Delta x$ to give

$$\frac{\partial\eta}{\partial t} + \frac{\partial}{\partial x}[u(h + \eta)] = 0$$

The limit $\Delta x \to 0$ may now be taken so that the terms which were not included
in the Taylor expansion now become identically zero. In the resulting equation,

the product $u\eta$ will be of second order and so may be neglected. Hence the linearized form of the continuity equation is

$$\frac{\partial \eta}{\partial t} + h\frac{\partial u}{\partial x} = 0 \tag{6.6a}$$

Figure 6.6c shows the same element of fluid considered above but on which the components of the $x$ momentum and the external forces are indicated. The components of the $x$ momentum are obtained by multiplying the mass-flow components obtained above by the $x$ component of the velocity vector $u$. In so doing it should be noted that the mass-flow component which leaves the control volume by way of the free surface will, in general, have an $x$ component of velocity, so that an efflux of $x$ momentum will be involved although the mass flow is essentially vertical. The forces which act on the fluid element in the $x$ direction are due to the pressure in the fluid. This pressure, in turn, will be hydrostatic in our linear approximation. Then at the reference station at $x$ the pressure will vary from atmospheric at the free surface to atmospheric plus $\rho g(h + \eta)$ at the bottom. This linear variation in pressure gives rise to a force in the positive $x$ direction of $\rho g(h + \eta)^2/2$, in which gauge pressures have been used since absolute values have no consequence here. At the location $x + \Delta x$ the first two terms in a Taylor series of this quantity are indicated in Fig. 6.6c. Then, from Newton's second law, the rate of increase in the $x$ momentum of the fluid as it passes through the control volume is equal to the net external force acting in the $x$ direction on the fluid. Thus

$$\frac{\partial}{\partial t}\left[\rho u(h + \eta)\,\Delta x\right] + \frac{\partial}{\partial x}\left[\rho u^2(h + \eta)\right]\Delta x + \rho u\frac{\partial \eta}{\partial t}\,\Delta x$$

$$= -\frac{\partial}{\partial x}\left[\frac{1}{2}\rho g(h + \eta)^2\right]\Delta x$$

The first term in this equation represents the time rate of increase of the momentum of the element of fluid, while the second and third terms represent the net convective increase associated with the various mass-flow components. The term on the right-hand side of the equation represents the net external force which comes from the hydrostatic pressures.

Dividing this equation by $\rho\,\Delta x$ gives the following form of the equation of momentum conservation:

$$\frac{\partial}{\partial t}\left[u(h + \eta)\right] + \frac{\partial}{\partial x}\left[u^2(h + \eta)\right] + u\frac{\partial \eta}{\partial t} = -g(h + \eta)\frac{\partial \eta}{\partial x}$$

Here the differentiation on the right-hand side has been carried out, and $\Delta x$ may now be permitted to tend to zero, so that the unwritten terms in the Taylor expansion vanish. This equation will now be linearized in the small quantities $u$ and $\eta$. Thus, in the first term, the product $u\eta$ is of second order and hence may be neglected. The entire second term is of second order or smaller owing to the presence of $u^2$. Likewise, the third term is quadratically small in $u$ and $\eta$. In the term on the right-hand side the product $\eta\, \partial\eta/\partial x$ is quadratically small and so may be neglected. Thus the linearized form of the equation of momentum conservation is

$$\frac{\partial u}{\partial t} + g\frac{\partial \eta}{\partial x} = 0 \tag{6.6b}$$

Equations (6.6$a$) and (6.6$b$) are, respectively, the equations of mass and momentum conservation. They represent two equations in the two unknowns $u$ and $\eta$. By forming the cross derivatives $\partial^2 u/\partial x\, \partial t$ and $\partial^2 \eta/\partial x\, \partial t$, first $\eta$ and then $u$ may be eliminated between Eqs. (6.6$a$) and (6.6$b$). This shows that the equations to be satisfied by $u$ and $\eta$ are

$$\frac{\partial^2 u}{\partial t^2} - gh\frac{\partial^2 u}{\partial x^2} = 0$$

$$\frac{\partial^2 \eta}{\partial t^2} - gh\frac{\partial^2 \eta}{\partial x^2} = 0$$

That is, both $u$ and $\eta$ must satisfy the one-dimensional wave equation. Hence $u$ and $\eta$ must be of the general form

$$u(x,t) = f_1\left(x - \sqrt{gh}\ t\right) + g_1\left(x + \sqrt{gh}\ t\right) \tag{6.6c}$$

$$\eta(x,t) = f_2\left(x - \sqrt{gh}\ t\right) + g_2\left(x + \sqrt{gh}\ t\right) \tag{6.6d}$$

where $f_1, g_1$ and $f_2, g_2$ are many differentiable functions. The first solution in each of these equations represents a wave traveling in the positive $x$ direction with velocity $\sqrt{gh}$. The second solution in both cases represents a wave traveling in the negative $x$ direction with velocity $\sqrt{gh}$. That is, if an arbitrary wave is traveling along the surface of a shallow-liquid layer, it will continue to travel with velocity $\sqrt{gh}$. This confirms the propagation speed derived earlier for a sinusoidal wave [Eq. (6.3$c$)] and shows that the shape of the wave does not change as it moves along the surface. Thus if the shape of the wave is known as a function of $x$ at some time, it will be known for all values of $x$ and $t$.

## 6.6 COMPLEX POTENTIAL FOR TRAVELING WAVES

Consider, again, the case of a small-amplitude surface wave in a fluid of arbitrary depth. For a sinusoidal wave of the form

$$\eta(x,t) = \varepsilon \sin\frac{2\pi}{\lambda}(x - ct)$$

it was shown in Sec. 6.3 that the velocity potential was

$$\phi(x, y, t) = C_2 \cos \frac{2\pi}{\lambda}(x - ct)\left(\tanh \frac{2\pi h}{\lambda} \sinh \frac{2\pi y}{\lambda} + \cosh \frac{2\pi y}{\lambda}\right)$$

The constant $C_2$ may be evaluated by completely imposing the kinematic boundary condition on the free surface. This boundary condition was used only to establish the functional form of the solution, but it was not strictly imposed. Then, as required by Eq. (6.2b), the condition $\partial\phi/\partial y(x, 0, t) = \partial\eta/\partial t(x, t)$ gives

$$C_2 \frac{2\pi}{\lambda} \cos \frac{2\pi}{\lambda}(x - ct) \tanh \frac{2\pi h}{\lambda} = -\varepsilon \frac{2\pi c}{\lambda} \cos \frac{2\pi}{\lambda}(x - ct)$$

which is satisfied by setting

$$C_2 = -\frac{c\varepsilon}{\tanh(2\pi h/\lambda)}$$

Then the velocity potential for a traveling sinusoidal wave is

$$\phi(x, y, t) = -c\varepsilon \cos \frac{2\pi}{\lambda}(x - ct)\left(\sinh \frac{2\pi y}{\lambda} + \coth \frac{2\pi h}{\lambda} \cosh \frac{2\pi y}{\lambda}\right)$$

$$(6.7a)$$

where the propagation speed $c$ must satisfy Eq. (6.3a). From Eq. (6.7a) the stream function for a traveling wave may be deduced, and so the corresponding complex potential may be obtained. This, in turn, will be used to establish the particle paths for traveling waves.

Since $u = \partial\psi/\partial y = \partial\phi/\partial x$, it follows from Eq. (6.7a) that

$$\frac{\partial\psi}{\partial y} = \frac{2\pi c}{\lambda} \varepsilon \sin \frac{2\pi}{\lambda}(x - ct)\left(\sinh \frac{2\pi y}{\lambda} + \coth \frac{2\pi h}{\lambda} \cosh \frac{2\pi y}{\lambda}\right)$$

Integrating this expression shows that $\psi(x, y, t)$ is of the form

$$\psi(x, y, t) = c\varepsilon \sin \frac{2\pi}{\lambda}(x - ct)\left(\cosh \frac{2\pi y}{\lambda} + \coth \frac{2\pi h}{\lambda} \sinh \frac{2\pi y}{\lambda}\right) + F(x)$$

where $F(x)$ is any function of $x$ which may be added through the integration. In principle a function of time could also be added, but it is known that, for a traveling wave, the time dependence above is correct. Another expression for $\psi(x, y, t)$ may be obtained from the fact that $v = -\partial\psi/\partial x = \partial\phi/\partial y$. This gives

$$\frac{\partial\psi}{\partial x} = \frac{2\pi c}{\lambda} \varepsilon \cos \frac{2\pi}{\lambda}(x - ct)\left(\cosh \frac{2\pi y}{\lambda} + \coth \frac{2\pi h}{\lambda} \sinh \frac{2\pi y}{\lambda}\right)$$

so that

$$\psi(x, y, t) = c\varepsilon \sin \frac{2\pi}{\lambda}(x - ct)\left(\cosh \frac{2\pi y}{\lambda} + \coth \frac{2\pi h}{\lambda} \sinh \frac{2\pi y}{\lambda}\right) + G(y)$$

where $G(y)$ is any function of $y$. Comparing this result with the previous expression for $\psi(x, y, t)$ shows that $F(x) = G(y) = 0$, so that

$$\psi(x, y, t) = c\varepsilon \sin \frac{2\pi}{\lambda}(x - ct)\left(\cosh \frac{2\pi y}{\lambda} + \coth \frac{2\pi h}{\lambda} \sinh \frac{2\pi y}{\lambda}\right) \quad (6.7b)$$

Equations (6.7a) and (6.7b) define, respectively, the velocity potential $\phi(x, y, t)$ and the stream function $\psi(x, y, t)$ for a traveling sinusoidal wave. Then the corresponding complex potential $F = \phi + i\psi$ may be established as follows:

$$F(z, t) = - \frac{c\varepsilon}{\sinh(2\pi h/\lambda)}$$

$$\times \left\{ \cos \frac{2\pi}{\lambda}(x - ct)\left[ \sinh \frac{2\pi h}{\lambda} \sinh \frac{2\pi y}{\lambda} + \cosh \frac{2\pi h}{\lambda} \cosh \frac{2\pi y}{\lambda} \right] \right.$$

$$- i \sin \frac{2\pi}{\lambda}(x - ct)\left[ \sinh \frac{2\pi h}{\lambda} \cosh \frac{2\pi y}{\lambda} \right.$$

$$\left. \left. + \cosh \frac{2\pi h}{\lambda} \sinh \frac{2\pi y}{\lambda} \right] \right\}$$

$$= - \frac{c\varepsilon}{\sinh(2\pi h/\lambda)}$$

$$\times \left\{ \cosh \frac{2\pi h}{\lambda}\left[ \cos \frac{2\pi}{\lambda}(x - ct)\cosh \frac{2\pi y}{\lambda} \right. \right.$$

$$\left. - i \sin \frac{2\pi}{\lambda}(x - ct)\sinh \frac{2\pi y}{\lambda} \right]$$

$$+ \sinh \frac{2\pi h}{\lambda}\left[ \cos \frac{2\pi}{\lambda}(x - ct)\sinh \frac{2\pi y}{\lambda} \right.$$

$$\left. \left. - i \sin \frac{2\pi}{\lambda}(x - ct)\cosh \frac{2\pi y}{\lambda} \right] \right\}$$

The hyperbolic functions which are inside the brackets will now be transformed into trigonometric functions having imaginary arguments using the identities $\sin i\alpha = i \sinh \alpha$ and $\cos i\alpha = \cosh \alpha$. Thus the complex potential may be written in the form

$$F(z,t) = -\frac{c\varepsilon}{\sinh(2\pi h/\lambda)}$$

$$\times \left\{ \cosh\frac{2\pi h}{\lambda} \left[ \cos\frac{2\pi}{\lambda}(x-ct)\cos\left(i\frac{2\pi y}{\lambda}\right) \right.\right.$$

$$\left. - \sin\frac{2\pi}{\lambda}(x-ct)\sin\left(i\frac{2\pi y}{\lambda}\right) \right]$$

$$+ \sinh\frac{2\pi h}{\lambda} \left[ -i\cos\frac{2\pi}{\lambda}(x-ct)\sin\left(i\frac{2\pi y}{\lambda}\right) \right.$$

$$\left.\left. -i\sin\frac{2\pi}{\lambda}(x-ct)\cos\left(i\frac{2\pi y}{\lambda}\right) \right]\right\}$$

$$= -\frac{c\varepsilon}{\sinh(2\pi h/\lambda)} \left[ \cosh\frac{2\pi h}{\lambda}\cos\frac{2\pi}{\lambda}(z-ct) \right.$$

$$\left. -i\sinh\frac{2\pi h}{\lambda}\sin\frac{2\pi}{\lambda}(z-ct) \right]$$

Here the quantities inside the brackets have been observed to be the expansions of single trigonometric functions involving $x - ct + iy = z - ct$. Again converting the hyperbolic functions inside the brackets to trigonometric functions gives

$$F(z,t) = -\frac{c\varepsilon}{\sinh(2\pi h/\lambda)}$$

$$\times \left[ \cos\left(i\frac{2\pi h}{\lambda}\right)\cos\frac{2\pi}{\lambda}(z-ct) - \sin\left(i\frac{2\pi h}{\lambda}\right)\sin\frac{2\pi}{\lambda}(z-ct) \right]$$

$$F(z,t) = -\frac{c\varepsilon}{\sinh(2\pi h/\lambda)}\cos\frac{2\pi}{\lambda}(z-ct+ih) \tag{6.7c}$$

where it has been observed that the brackets contain the expansion of a single trigonometric function. Equation (6.7c) gives the complex potential for the traveling sinusoidal wave $\eta(x,t) = \varepsilon\sin 2\pi(x-ct)/\lambda$.

## 6.7  PARTICLE PATHS FOR TRAVELING WAVES

As a wave train travels across the surface of an otherwise quiescent liquid, the individual particles of the liquid undergo small cyclical motions. The precise trajectory followed by the fluid particles may be established with the aid of the results of the previous section.

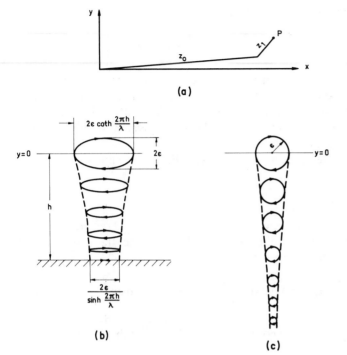

**FIGURE 6.7**

(a) Coordinate system for establishing particle paths. (b) Particle trajectories due to a sinusoidal wave, and (c) the trajectories in deep liquids.

Consider a specific particle of fluid such as the one indicated by the point $P$ in Fig. 6.7a. The instantaneous position of this particle of fluid will be indicated by a fixed-position vector $z_0$ and an additional vector $z_1$ which varies with time. That is, the length and orientation of $z_0$ remains fixed while both the length and inclination of $z_1$ will vary with time. Then, considering the complex conjugate of the variable-position vector, that is, considering $\bar{z}_1 = x_1 - iy_1$, it follows that

$$\frac{d\bar{z}_1}{dt} = \frac{dx_1}{dt} - i\frac{dy_1}{dt}$$

$$= u - iv$$

$$= W$$

$$= \frac{dF}{dz}$$

Then, using Eq. (6.7c),

$$\frac{d\bar{z}_1}{dt} = \frac{(2\pi/\lambda)c\varepsilon}{\sinh(2\pi h/\lambda)}\sin\frac{2\pi}{\lambda}(z - ct + ih)$$

Integrating this equation with respect to time gives

$$\bar{z}_1 = \frac{\varepsilon}{\sinh\left(2\pi h/\lambda\right)} \cos \frac{2\pi}{\lambda}(z - ct + ih)$$

Here the constant of integration has been taken to be zero without loss of generality. Such a constant would not affect the time dependence of $z_1$, and so it would not affect the trajectory of the fluid particle. Rather, it would only change the length of the $z_1$ position vector, which is equivalent to adjusting the choice of the constant $z_0$ position vector.

Comparing the above expression for $\bar{z}_1$ with Eq. (6.7c) shows that

$$\bar{z}_1 = -\frac{F(z,t)}{c}$$

Then it follows that $x_1 = -\phi(x, y, t)/c$ and $y_1 = \psi(x, y, t)/c$. Hence, from Eqs. (6.7a) and (6.7b), the coordinates $x_1$ and $y_1$ of the trajectory of our reference fluid particle will be given by

$$x_1 = \varepsilon \cos \frac{2\pi}{\lambda}(x - ct)\left(\sinh \frac{2\pi y}{\lambda} + \coth \frac{2\pi h}{\lambda} \cosh \frac{2\pi y}{\lambda}\right)$$

$$y_1 = \varepsilon \sin \frac{2\pi}{\lambda}(x - ct)\left(\cosh \frac{2\pi y}{\lambda} + \coth \frac{2\pi h}{\lambda} \sinh \frac{2\pi y}{\lambda}\right)$$

That is, the instantaneous coordinates of the trajectory of a fluid particle depend on both the $x$ and $y$ coordinates of the fluid particle and on the time. The time $t$ may be eliminated between these two equations to yield the trajectory of the fluid particle in the following way:

$$\sin^2 \frac{2\pi}{\lambda}(x - ct) + \cos^2 \frac{2\pi}{\lambda}(x - ct) = 1$$

Substituting from the above equations for $x_1$ and $y_1$ into this identity gives

$$\frac{x_1^2}{\varepsilon^2\left[\sinh\left(2\pi y/\lambda\right) + \coth\left(2\pi h/\lambda\right)\cosh\left(2\pi y/\lambda\right)\right]^2}$$

$$+ \frac{y_1^2}{\varepsilon^2\left[\cosh\left(2\pi y/\lambda\right) + \coth\left(2\pi h/\lambda\right)\sinh\left(2\pi y/\lambda\right)\right]^2} = 1 \quad (6.8)$$

Equation (6.8) shows that the trajectory of a fluid particle depends only on its depth of submergence. Eliminating the time also eliminated the $x$ coordinate. This might have been expected, since each particle of fluid experiences the same waves passing above it, irrespective of its $x$ coordinate. Thus the motion experienced by two particles which are separated in the $x$ direction only will be the same, only the phasing will be different. Since Eq. (6.8) is that of an ellipse, the trajectories of the fluid particles will be ellipses whose dimensions are determined by the value of $y$ for the various particles. For particles which lie on the free surface, $y = 0$, so that Eq. (6.8) becomes

$$\frac{x_i^2}{[\varepsilon \coth (2\pi h/\lambda)]^2} + \frac{y_1^2}{\varepsilon^2} = 1$$

This shows that the trajectory of particles on the free surface is that of an ellipse whose semiaxes are $\varepsilon$ in the $y$ direction and $\varepsilon \coth (2\pi h/\lambda)$ in the $x$ direction. This result is shown in Fig. 6.7b. For particles which are on the bottom, $y = -h$, so that the semiaxis in the $y$ direction becomes zero and the semiaxis in the $x$ direction becomes $\varepsilon/\sinh (2\pi h/\lambda)$. That is, the ellipse degenerates to the line $-\varepsilon/\sinh (2\pi h/\lambda) \le x_1 \le \varepsilon/\sinh (2\pi h/\lambda)$. For values of $y$ which are intermediate to $y = 0$ and $y = -h$, the particle trajectories will be ellipses as described by Eq. (6.8) and as shown in Fig. 6.7b.

For shallow liquids the ellipses shown in Fig. 6.7b merely become elongated in the $x$ direction. However, for deep liquids the ellipses become circles. This may be shown by observing that for deep liquids the parameter $2\pi h/\lambda$ will be very large, so that $\coth (2\pi h/\lambda)$ will be unity. Then Eq. (6.8) becomes

$$\frac{x_1^2}{\varepsilon^2 [\sinh (2\pi y/\lambda) + \cosh (2\pi y/\lambda)]^2}$$

$$+ \frac{y_1^2}{\varepsilon^2 [\sinh (2\pi y/\lambda) + \cosh (2\pi y/\lambda)]^2} = 1$$

This is the equation of a circle of radius $\varepsilon|\sinh (2\pi y/\lambda) + \cosh (2\pi y/\lambda)|$. That is, the radius will be $\varepsilon$ at the free surface and will decrease as $y$ becomes more and more negative. The particle trajectories for deep liquids are shown in Fig. 6.7c.

## 6.8 STANDING WAVES

Up to this point we have been dealing with traveling waves, that is, waves which move along the surface of the liquid. We now consider standing waves, which are waves that remain stationary—the surface moves vertically only. An interesting way of obtaining the equation of a standing wave is to superimpose two identical traveling waves which are moving in opposite directions. Thus consider two traveling waves $\eta_1$ and $\eta_2$ as follows:

$$\eta_1(x, t) = \tfrac{1}{2}\varepsilon \sin \frac{2\pi}{\lambda}(x - ct)$$

$$\eta_2(x, t) = \tfrac{1}{2}\varepsilon \sin \frac{2\pi}{\lambda}(x + ct)$$

Let $\eta(x, t)$ represent the free-surface profile which results from superimposing these two traveling waves. Then

$$\eta(x,t) = \tfrac{1}{2}\varepsilon\left[\sin\frac{2\pi}{\lambda}(x - ct) + \sin\frac{2\pi}{\lambda}(x + ct)\right]$$

$$= \tfrac{1}{2}\varepsilon\left(\sin\frac{2\pi x}{\lambda}\cos\frac{2\pi ct}{\lambda} - \cos\frac{2\pi x}{\lambda}\sin\frac{2\pi ct}{\lambda}\right.$$

$$\left. + \sin\frac{2\pi x}{\lambda}\cos\frac{2\pi ct}{\lambda} + \cos\frac{2\pi x}{\lambda}\sin\frac{2\pi ct}{\lambda}\right)$$

$$= \varepsilon\sin\frac{2\pi x}{\lambda}\cos\frac{2\pi ct}{\lambda}$$

That is, the superposition of two identical traveling waves results in a wave which, at any time, is a sine function in $x$ and which, for any value of $x$, oscillates vertically in time. Such a wave, in which the entire surface oscillates in time, is called a standing wave.

The complex potential for a sinusoidal-shaped standing wave may be obtained by superimposing the complex potentials for two traveling waves moving in opposition to each other. Thus, Eq. (6.7c) will be used to obtain the complex potential for two waves, each of amplitude $\varepsilon/2$ and wavelength $\lambda$, one of which is traveling in the positive $x$ direction with velocity $c$ and the other of which is traveling in the opposite direction with the same velocity. Hence

$$F(z,t) = \frac{c\varepsilon/2}{\sinh(2\pi h/\lambda)}\left[-\cos\frac{2\pi}{\lambda}(z - ct + ih) + \cos\frac{2\pi}{\lambda}(z + ct + ih)\right]$$

The cosine functions will now be expanded, taking $z + ih$ as one element and $ct$ as the other. This gives

$$F(z,t) = \frac{c\varepsilon/2}{\sinh(2\pi h/\lambda)}$$

$$\times\left[-\cos\frac{2\pi}{\lambda}(z + ih)\cos\frac{2\pi ct}{\lambda} - \sin\frac{2\pi}{\lambda}(z + ih)\sin\frac{2\pi ct}{\lambda}\right.$$

$$\left. + \cos\frac{2\pi}{\lambda}(z + ih)\cos\frac{2\pi ct}{\lambda} - \sin\frac{2\pi}{\lambda}(z + ih)\sin\frac{2\pi ct}{\lambda}\right] \tag{6.9}$$

$$F(z,t) = -\frac{c\varepsilon}{\sinh(2\pi h/\lambda)}\sin\frac{2\pi}{\lambda}(z + ih)\sin\frac{2\pi ct}{\lambda}$$

Equation (6.9) gives the complex potential for a standing sinusoidal wave of wavelength $\lambda$ which is oscillating in time with frequency $2\pi c/\lambda$.

## 6.9 PARTICLE PATHS FOR STANDING WAVES

Following the procedure employed in Sec. 6.7 for traveling waves, the particle paths for standing waves may be established from the complex potential. Using the same coordinate system as was used in Sec. 6.7, it follows as before that

$$\frac{d\bar{z}_1}{dt} = \frac{dF}{dz}$$

Then, using Eq. (6.9),

$$\frac{d\bar{z}_1}{dt} = -\frac{(2\pi/\lambda)c\varepsilon}{\sinh(2\pi h/\lambda)}\cos\frac{2\pi}{\lambda}(z + ih)\sin\frac{2\pi ct}{\lambda}$$

Integrating with respect to time and neglecting the constant of integration as before gives the following expression for $\bar{z}_1$:

$$\bar{z}_1 = \frac{\varepsilon}{\sinh(2\pi h/\lambda)}\cos\frac{2\pi}{\lambda}(z + ih)\cos\frac{2\pi ct}{\lambda}$$

Writing $z + ih = x + i(y + h)$ and expanding the trigonometric function of this argument gives

$$\bar{z}_1 = \frac{\varepsilon}{\sinh(2\pi h/\lambda)}$$

$$\times \cos\frac{2\pi ct}{\lambda}\left[\cos\frac{2\pi x}{\lambda}\cosh\frac{2\pi}{\lambda}(y + h) - i\sin\frac{2\pi x}{\lambda}\sinh\frac{2\pi}{\lambda}(y + h)\right]$$

in which the trigonometric terms having imaginary arguments have been converted to hyperbolic terms. The quantity $\bar{z}_1$ is complex and so may be written in the polar form

$$\bar{z}_1 = r_1 e^{-i\theta_1}$$

where $r_1$ and $\theta_1$ are defined by

$$r_1 = \frac{\varepsilon}{\sinh(2\pi h/\lambda)}$$

$$\times \cos\frac{2\pi ct}{\lambda}\sqrt{\cos^2\frac{2\pi x}{\lambda}\cosh^2\frac{2\pi}{\lambda}(y + h) + \sin^2\frac{2\pi x}{\lambda}\sinh^2\frac{2\pi}{\lambda}(y + h)} \tag{6.10a}$$

$$\theta_1 = \tan^{-1}\left[\tan\frac{2\pi x}{\lambda}\tanh\frac{2\pi}{\lambda}(y + h)\right] \tag{6.10b}$$

Equations (6.10a) and (6.10b) show that, for given values of $x$ and $y$, the polar angle $\theta_1$ of the particle trajectory is constant whereas the radius $r_1$ oscillates in time. Thus the particle trajectories will be straight lines whose inclination will depend upon the location of the particle under consideration. In

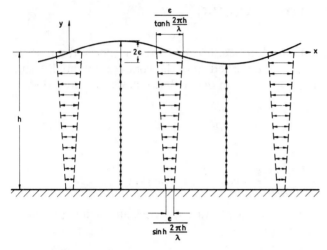

**FIGURE 6.8**
Particle trajectories induced by a sinusoidal standing wave of amplitude $\varepsilon$ and wavelength $\lambda$.

particular, when $x = n\lambda/2$. Eqs. (6.10a) and (6.10b) reduce to

$$r_1 = \varepsilon \cos \frac{2\pi ct}{\lambda} \frac{\cosh (2\pi/\lambda)(y + h)}{\sinh (2\pi h/\lambda)}$$

$$\theta_1 = 0 \quad \text{or} \quad \pi$$

This describes a family of horizontal lines whose length $r_1$ decreases with the depth of submergence. The location $x = n\lambda/2$ corresponds to the nodes of the free surface, that is, the points of the free surface which have no vertical motion. The horizontal motion of these points, which is shown in Fig. 6.8, is necessary to satisfy the continuity equation as the maximum amplitude of the wave shifts from one side of the node to the other as the surface oscillations take place.

Midway between the nodes, that is, at $x = (2n + 1)\lambda/4$, Eqs. (6.10a) and (6.10b) show that

$$r_1 = \varepsilon \cos \frac{2\pi ct}{\lambda} \frac{\sinh (2\pi/\lambda)(y + h)}{\sinh (2\pi h/\lambda)}$$

$$\theta_1 = \frac{\pi}{2} \quad \text{or} \quad \frac{3\pi}{2}$$

This defines a family of vertical lines whose length $r_1$ decreases as the submergence increases and reaches zero on the bottom, $y = -h$. This motion is also shown in Fig. 6.8. As the boundary condition requires, the vertical motion vanishes on $y = -h$.

## 6.10   WAVES IN RECTANGULAR VESSELS

The fact that standing waves may exist on the surface of an infinite expanse of liquid raises the question of whether standing waves may exist on the surface of a liquid which is contained in a vessel of finite extent. In this section rectangular vessels will be considered, and it will be shown, as might be expected, that only standing waves whose wavelengths coincide with a discrete spectrum of values may exist on such liquid surfaces.

Figure 6.9a shows a two-dimensional rectangular container of width $2l$ which contains a liquid of average depth $h$. For this configuration, we ask the following question: What type of steady-state waves, if any, may exist on the surface of the liquid? Any waves which may exist will have to satisfy the following partial differential equation and boundary conditions:

$$\frac{\partial^2 \phi}{\partial x^2} + \frac{\partial^2 \phi}{\partial y^2} = 0 \tag{6.11a}$$

$$\frac{\partial^2 \phi}{\partial t^2}(x,h,t) + g\frac{\partial \phi}{\partial y}(x,h,t) = 0 \tag{6.11b}$$

$$\frac{\partial \phi}{\partial y}(x,0,t) = 0 \tag{6.11c}$$

$$\frac{\partial \phi}{\partial x}(\pm l, y, t) = 0 \tag{6.11d}$$

The first boundary condition is the pressure condition at the free surface in which the kinematic condition has been employed, and the other boundary conditions prevent normal velocity components on the bottom and side surfaces of the container. Since the free-surface profile is not being specified a priori here, the kinematic condition at the free surface should not be imposed separately.

Since a steady-state wave solution is being sought, the velocity potential should have a trigonometric time dependence. It may be observed that the existence of the sidewalls at $x = \pm l$ eliminates the possibility of traveling waves, since the particle paths for traveling waves are ellipses, so that the wall boundary conditions could not be satisfied. The time variation will therefore be of the standing-wave type and will be chosen to be $\sin(2\pi ct/\lambda)$. There is no loss in generality with this choice, since any phase change merely corresponds to a shifting of the time origin, which is of no consequence here. Thus the appropriate separable solution to Eq. (6.11a) will be of the form

$$\phi(x,y,t) = \left(A_1 \sin\frac{2\pi x}{\lambda} + A_2 \cos\frac{2\pi x}{\lambda}\right)$$

$$\times \left(B_1 \sinh\frac{2\pi y}{\lambda} + B_2 \cosh\frac{2\pi y}{\lambda}\right)\sin\frac{2\pi ct}{\lambda}$$

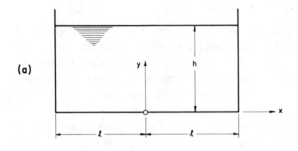

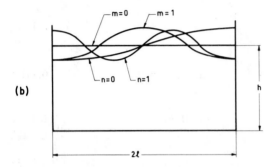

**FIGURE 6.9**
(*a*) Geometry for liquid in a rectangular container and (*b*) the first four fundamental modes of surface oscillation.

The $x$ dependence has been chosen to be trigonometric in view of the homogeneous boundary conditions at $x = \pm l$. Then, in order to satisfy Laplace's equation, the $y$ dependence must be exponential or hyperbolic. In view of the finite extent of the domain and the homogeneous boundary condition at $y = 0$, the hyperbolic form has been employed. The boundary condition (6.11*c*) requires that $B_1 = 0$, so that the velocity potential becomes of the form

$$\phi(x, y, t) = \left( D_1 \sin \frac{2\pi x}{\lambda} + D_2 \cos \frac{2\pi x}{\lambda} \right) \cosh \frac{2\pi y}{\lambda} \sin \frac{2\pi ct}{\lambda}$$

The pressure condition on the free surface [Eq. (6.11*b*)] then requires that

$$\left[ -\left( \frac{2\pi c}{\lambda} \right)^2 \cosh \frac{2\pi h}{\lambda} + g \frac{2\pi}{\lambda} \sinh \frac{2\pi h}{\lambda} \right]$$

$$\times \left( D_1 \sin \frac{2\pi x}{\lambda} + D_2 \cos \frac{2\pi x}{\lambda} \right) \sin \frac{2\pi ct}{\lambda} = 0$$

Since this equation is to be satisfied for all values of $x$ and all values of $t$, it follows that the quantity inside the brackets must be zero. This gives

$$\frac{c^2}{gh} = \frac{\lambda}{2\pi h} \tanh \frac{2\pi h}{\lambda}$$

That is, the pressure condition on the free surface establishes the frequencies of the wave motion. It will be seen that this result agrees with Eq. (6.3$a$) and that each Fourier component of the waveform has a different frequency of motion.

The final boundary condition which is to be satisfied is that of no horizontal velocity component at the vertical walls of the container. Thus Eq. (6.11$d$) requires that

$$\frac{2\pi}{\lambda}\left(D_1 \cos \frac{2\pi l}{\lambda} \mp D_2 \sin \frac{2\pi l}{\lambda}\right) \cosh \frac{2\pi y}{\lambda} \sin \frac{2\pi ct}{\lambda} = 0$$

which will be satisfied for all values of $y$ and $t$ if

$$D_1 \cos \frac{2\pi l}{\lambda} = \pm D_2 \sin \frac{2\pi l}{\lambda}$$

This condition may be satisfied by setting $D_1 = D_2 = 0$, but then $\phi = 0$, which is the trivial solution. For a nontrivial solution either $D_1$ or $D_2$ at least, must be different from zero.

Suppose, first, that $D_1$ is different from zero and $D_2 = 0$. Then

$$\cos \frac{2\pi l}{\lambda_n} = 0$$

$$\therefore \quad \lambda_n = \frac{4l}{2n + 1}$$

where the subscript $n$ has been associated with the quantity $\lambda$ in anticipation of the fact that the foregoing transcendental equation may be satisfied in an infinite number of ways. That is, one way of satisfying the side boundary conditions is to choose the above values of $\lambda_n$ so that the corresponding velocity potentials will be of the form

$$\phi_n(x, y, t) = D_{1n} \sin \frac{(2n + 1)\pi x}{2l} \cosh \frac{(2n + 1)\pi y}{2l} \sin \frac{(2n + 1)\pi c_n t}{2l}$$

where $c_n$ is related to $\lambda_n$ through the identity which resulted from imposing the pressure condition on the free surface.

Next, suppose $D_1 = 0$ and $D_2$ is different from zero. Then

$$\sin \frac{2\pi l}{\lambda_m} = 0$$

$$\therefore \quad \lambda_m = \frac{2l}{m}$$

Thus another way of satisfying the side boundary conditions is to adopt the above value for $\lambda_m$ so that the corresponding velocity potentials will be

$$\phi_m(x, y, t) = D_{2m} \cos \frac{m\pi x}{l} \cosh \frac{m\pi y}{l} \sin \frac{m\pi c_m t}{l}$$

where $c_m$ is related to $\lambda_m$. The first two surface modes corresponding to $\phi_n$ and $\phi_m$ are shown in Fig. 6.9b.

It will be seen that, out of the continuous spectrum of wavelengths which may exist, only those waves whose particle paths are vertical at $x = \pm l$ are permissible solutions. This gives rise to an even spectrum of modes (corresponding to $D_1 = 0$), and an odd spectrum of modes (corresponding to $D_2 = 0$). That is, there is a discrete spectrum of wavelengths whose particle paths are vertical at $x = \pm l$ and which may therefore satisfy the boundary conditions at the sidewalls.

The individual solutions given by $\phi_n$ and $\phi_m$ may be superimposed to describe more general waveforms. Thus a more general solution will be obtained by superimposing all the $\phi_n$ solutions and all the $\phi_m$ solutions. This gives

$$\phi(x, y, t) = \sum_{n=0}^{\infty} D_{1n} \sin \frac{(2n + 1)\pi x}{2l} \cosh \frac{(2n + 1)\pi y}{2l} \sin \frac{(2n + 1)\pi c_n t}{2l}$$

$$+ \sum_{m=0}^{\infty} D_{2m} \cos \frac{m\pi x}{l} \cosh \frac{m\pi y}{l} \sin \frac{m\pi c_m t}{l} \tag{6.12a}$$

where

$$\frac{c_n^2}{gh} = \frac{2l}{(2n + 1)\pi h} \tanh \frac{(2n + 1)\pi h}{2l} \tag{6.12b}$$

and

$$\frac{c_m^2}{gh} = \frac{1}{m\pi h} \tanh \frac{m\pi h}{l} \tag{6.12c}$$

The coefficients $D_{1n}$ and $D_{2m}$ which appear in Eq. (6.12a) are undetermined at this point. If the initial shape and velocity of the free surface are specified, these constants may be evaluated. An example of how this may be utilized is to establish the response of a body of water to an earthquake. The body of water may be an artificial reservoir or a lake whose shape may be approximated by a rectangular container. Seismographic records for the area would indicate the magnitude and frequency of the expected accelerations. These data may be Fourier-analyzed and used to establish a surface waveform and oscillation frequency at the end of the earthquake, which would be the beginning of the standing-wave oscillations. The constants $D_{1n}$ and $D_{2m}$ may be used to fit these data, and then Eq. (6.12a) will describe the subsequent motion.

## 6.11   WAVES IN CYLINDRICAL VESSELS

An analysis similar to that which was presented in the previous section may be carried out for cylindrical containers. Figure 6.10a shows a cylindrical container of radius $a$ which contains a liquid whose average depth is $h$. Then, in terms of the cylindrical coordinates $R$, $\theta$, and $z$ and the time $t$, the problem to be solved

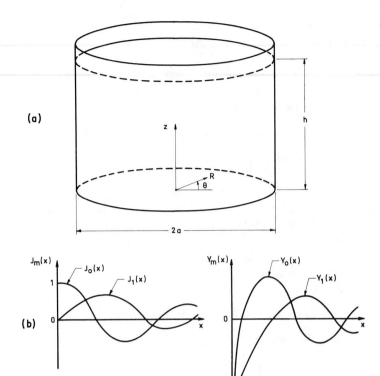

**FIGURE 6.10**
(a) Geometry for liquid in a cylindrical container and (b) Bessel functions of the first and second kind.

for the velocity potential $\phi(R, \theta, z, t)$ is

$$\frac{1}{R}\frac{\partial}{\partial R}\left(R\frac{\partial \phi}{\partial R}\right) + \frac{1}{R^2}\frac{\partial^2 \phi}{\partial \theta^2} + \frac{\partial^2 \phi}{\partial z^2} = 0 \qquad (6.13a)$$

$$\frac{\partial^2 \phi}{\partial t^2}(R, \theta, h, t) + g\frac{\partial \phi}{\partial z}(R, \theta, h, t) = 0 \qquad (6.13b)$$

$$\frac{\partial \phi}{\partial z}(R, \theta, 0, t) = 0 \qquad (6.13c)$$

$$\frac{\partial \phi}{\partial R}(a, \theta, z, t) = 0 \qquad (6.13d)$$

The solutions to this problem will describe the possible waveforms which may exist on the surface of the liquid which is in the container.

The solution to the foregoing problem may be obtained by the method of separation of variables. Thus a solution is sought in the form

$$\phi(R, \theta, z, t) = \mathscr{R}(R)T(\theta)Z(z) \sin \omega t$$

Here the time dependence has again been taken to be sinusoidal, corresponding to standing waves. Substituting this expression for $\phi$ into Eq. (6.13a) and multiplying by $R^2/\phi$ gives

$$\frac{R}{\mathscr{R}} \frac{d}{dR}\left(R \frac{d\mathscr{R}}{dR}\right) + \frac{1}{T}\frac{d^2T}{d\theta^2} + \frac{R^2}{Z}\frac{d^2Z}{dz^2} = 0$$

Following the usual argument of separation of variables, it is observed that the second term in this equation contains all the $\theta$ dependence and that it is a function of $\theta$ only. Then this term must equal a constant. This constant will be chosen to be $-m^2$, where $m$ is an integer. The significance of the minus sign is that trigonometric rather than exponential $\theta$ dependence will result and the significance of $m$'s being an integer is that $\phi(\theta) = \phi(\theta + 2\pi)$ will be satisfied, as is required. The solution for $T(\theta)$ is then

$$T(\theta) = A_1 \sin m\theta + A_2 \cos m\theta$$

The remaining differential equation is, after dividing by $R^2$,

$$\frac{1}{R\mathscr{R}} \frac{d}{dR}\left(R \frac{d\mathscr{R}}{dR}\right) + \frac{m^2}{R^2} + \frac{1}{Z}\frac{d^2Z}{dz^2} = 0$$

The separation-of-variables argument now requires that the last term be equal to a constant. Since trigonometric $z$ dependence does not fit the physical circumstances, this constant will be chosen as $k^2$. Then the solution for $Z(z)$ will be

$$Z(z) = B_1 \sinh kz + B_2 \cosh kz$$

Here the hyperbolic form has been used in preference to the exponential form in view of the finite extent of the $z$ domain.

The remaining differential equation, after multiplying by $R^2\mathscr{R}$, becomes

$$R \frac{d}{dR}\left(R \frac{d\mathscr{R}}{dR}\right) + (k^2R^2 - m^2)\mathscr{R} = 0$$

But this is Bessel's equation of order $m$ whose solution is

$$\mathscr{R}(R) = D_{1m}J_m(kR) + D_{2m}Y_m(kR)$$

where $J_m$ is Bessel's function of the first kind and $Y_m$ is Bessel's function of the second kind. The first two Bessel functions of each kind are shown schematically in Fig. 6.10b.

Since the Bessel functions of the second kind, $Y_m(x)$, diverge for $x = 0$ for all values of $m$, the coefficients $D_{2m}$ must be zero. Thus the radial dependence of the velocity potential will be proportional to $J_m(kR)$. Then, for any integer

$m$, the solution by separation of variables is

$$\phi_m(R,\theta,z,t) = (A_{1m}\sin m\theta + A_{2m}\cos m\theta)$$
$$\times(B_{1m}\sinh kz + B_{2m}\cosh kz)J_m(kR)\sin \omega t$$

The boundary condition (6.13c) requires that $B_{1m}$ be zero, while the pressure condition at the free surface [Eq. (6.13b)] determines the oscillation frequency to be

$$\omega^2 = gk \tanh kh$$

Thus the velocity potential will be of the form

$$\phi_m(R,\theta,z,t) = (K_{1m}\sin m\theta + K_{2m}\cos m\theta)\cosh kzJ_m(kR)\sin \omega t$$

The remaining boundary condition [Eq. (6.13d)] then requires, for a nontrivial solution,

$$J'_m(ka) = 0$$

where the prime denotes differentiation. This transcendental equation may be satisfied by any of an infinite number of discrete values of $k$. These values will be distinguished by employing a double subscript on $k$. Thus $k_{mn}$ will denote the $n$th root of the $J_m$ Bessel function in the equation

$$J'_m(k_{mn}a) = 0$$

Values of $k_{mn}a$ which satisfy this equation may be found in tables of functions. For example, the first few roots of $J'_0(k_{0n}a) = 0$ are given below.

| $n$ | 0 | 1 | 2 | 3 |
|---|---|---|---|---|
| $k_{0n}a$ | 3.832 | 7.016 | 10.174 | 13.324 |

From the foregoing analysis, one solution to the problem posed for the velocity potential is

$$\phi_{mn}(R,\theta,z,t) = (K_{1mn}\sin m\theta + K_{2mn}\cos m\theta)\cosh k_{mn}zJ_m(k_{mn}R)\sin \omega_{mn}t$$

Here a double subscript has been associated with the oscillation frequency $\omega$, since this quantity is related to the separation constant $k$. The above expression for $\phi_{mn}$ represents a valid solution to our problem for any integer $m$ and any integer $n$. Then a more general solution may be obtained by superimposing all such solutions to give

$$\phi(R,\theta,z,t) = \sum_{m=0}^{\infty}\sum_{n=0}^{\infty}(K_{1mn}\sin m\theta + K_{2mn}\cos m\theta)$$
$$\times\cosh k_{mn}zJ_m(k_{mn}R)\sin \omega_{mn}t \qquad (6.14a)$$

where
$$\omega_{mn}^2 = gk_{mn}\tanh (k_{mn}h) \qquad (6.14b)$$

and
$$J'_m(k_{mn}a) = 0 \qquad (6.14c)$$

As was the case in the previous section, the remaining arbitrary constants may be defined by specifying the nature of the free surface at some value of the time.

A simple illustration of the validity of the above result may be obtained by use of a cup of coffee or some other liquid. If such a cup is jarred by striking it squarely on a flat surface, it may be induced to vibrate in a purely radial mode. That is, the fundamental mode in which the surface $R = a$ vibrates in and out may be induced. This motion causes surface waves which will also have no $\theta$ dependence. Then, putting $m = 0$ in Eq. (6.14$a$) shows that the velocity potential will be proportional to $J_0(k_{0n}R)$. Thus the surface will adopt the shape of the $J_0$ Bessel function, which is shown in Fig. 6.10$b$. This shape may be actually observed, and under certain conditions the peak at the axis of the vessel may become very pronounced. Of course, the analysis is no longer valid under these conditions, since large amplitudes lead to large slopes, which violates the assumptions which were made in the linearization.

## 6.12   PROPAGATION OF WAVES AT AN INTERFACE

As a final example of surface waves, the behavior of propagating waves at the separation of two dissimilar fluids will be investigated. Figure 6.11 shows a wavy surface $y = \eta(x, t)$ below which a fluid of density $\rho_1$ flows with mean velocity $U_1$ in the $x$ direction. Above the interface is a fluid of density $\rho_2$ whose mean velocity is $U_2$ in the $x$ direction.

For the foregoing configuration we specify a sinusoidal waveform at the interface and ask what is the propagation speed of the wave? That is, we specify the equation of the interface to be

$$\eta(x, t) = \varepsilon e^{i(2\pi/\lambda)(x - \sigma t)}$$

This represents a sinusoidal wave of amplitude $\varepsilon$ and wavelength $\lambda$. If $\sigma$ is real, the wave is traveling in the $x$ direction with velocity $\sigma$, whereas if $\sigma$ is imaginary, the wave is decaying (if $\sigma/i$ is negative) or is growing (if $\sigma/i$ is positive). The last situation represents an unstable interface.

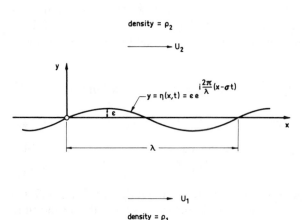

**FIGURE 6.11**
Wave-shaped interface separating two different fluids traveling at different average speeds.

Since the two fluids now have nonzero mean velocities, the linearization of the boundary conditions must be reexamined here. Letting the subscript $i$ be 1 or 2 for the lower and upper fluids, respectively, the velocity $\mathbf{u}_i$ may be written as follows:

$$\mathbf{u}_i = U_i \mathbf{e}_x + \nabla \phi_i$$

where $\phi_i$ is the velocity potential for the perturbation to the uniform flow caused by the waves at the interface. Then the material derivative becomes

$$\frac{D}{Dt} = \frac{\partial}{\partial t} + \mathbf{u}_i \cdot \nabla$$

$$= \frac{\partial}{\partial t} + U_i \frac{\partial}{\partial x} + \nabla \phi_i \cdot \nabla$$

The second term on the right of this identity is new and is of first order in this case. The third term on the right involves velocity components which are derived from the perturbation velocity potential, and hence these velocity components will be small.

Using this result for the material derivative, the kinematic condition on the free surface, $D(y - \eta)/Dt = 0$, becomes

$$-\frac{\partial \eta}{\partial t} - \cdot U_i \frac{\partial \eta}{\partial x} + \frac{\partial \phi_i}{\partial y} - \nabla \phi_i \cdot \nabla \eta = 0$$

The last term on the left-hand side of this equation is quadratically small, for small-amplitude waves, and so may be neglected. Thus the revised kinematic boundary condition on the free surface is

$$\frac{\partial \phi_i}{\partial y}(x, 0, t) = \frac{\partial \eta}{\partial t}(x, t) + U_i \frac{\partial \eta}{\partial x}(x, t) \qquad (6.15a)$$

Comparison of this result with Eq. (6.2b) shows that the last term in the above equation is new and that this term vanishes for $U_i = 0$.

From Eq. (6.1c) the Bernoulli equation for a constant-pressure surface in which $F(t)$ is absorbed into the velocity potential, as before, is

$$\rho_i \frac{\partial \phi_i}{\partial t} + \tfrac{1}{2}\rho_i \mathbf{u}_i \cdot \mathbf{u}_i + \rho_i g \eta = \text{constant}$$

Substituting our expansion for $\mathbf{u}_i$ into this equation and neglecting quadratic terms in the perturbation velocity gives

$$\rho_i \frac{\partial \phi_i}{\partial t}(x, 0, t) + \rho_i U_i \frac{\partial \phi_i}{\partial x}(x, 0, t) + \rho_i g \eta(x, t) = \text{constant} \quad (6.15b)$$

Here the term $\rho_i U_i^2/2$ has been absorbed into the constant on the right-hand side of this equation.

Using Eqs. (6.15a) and (6.15b), we may now define the problem to be satisfied by the velocity potentials $\phi_1$ and $\phi_2$. In the region $y < 0$, the velocity potential $\phi_1$ must satisfy

$$\frac{\partial^2 \phi_1}{\partial x^2} + \frac{\partial^2 \phi_1}{\partial y^2} = 0 \tag{6.16a}$$

where the velocities derived from $\phi_1$ should be finite. That is,

$$|\nabla \phi_1| = \text{finite} \tag{6.16b}$$

Similarly, in the region $y > 0$, we have

$$\frac{\partial^2 \phi_2}{\partial x^2} + \frac{\partial^2 \phi_2}{\partial y^2} = 0 \tag{6.16c}$$

$$|\nabla \phi_2| = \text{finite} \tag{6.16d}$$

At the interface, which is linearized to $y = 0$, the kinematic condition must be satisfied by $\phi_1$ and by $\phi_2$ separately. Thus from Eq. (6.15a),

$$\frac{\partial \phi_1}{\partial y}(x,0,t) = \frac{\partial \eta}{\partial t}(x,t) + U_1 \frac{\partial \eta}{\partial x}(x,t) \tag{6.16e}$$

$$\frac{\partial \phi_2}{\partial y}(x,0,t) = \frac{\partial \eta}{\partial t}(x,t) + U_2 \frac{\partial \eta}{\partial x}(x,t) \tag{6.16f}$$

Finally, the pressure condition at the interface must be satisfied, which in the present case, amounts to equating the pressure in the two fluids at the interface. Thus, from Eq. (6.15b), since the Bernoulli constant will be the same for $i = 1$ and $i = 2$, the pressure condition becomes

$$\rho_1 \frac{\partial \phi_1}{\partial t}(x,0,t) + \rho_1 U_1 \frac{\partial \phi_1}{\partial x}(x,0,t) + \rho_1 g \eta(x,t)$$

$$= \rho_2 \frac{\partial \phi_2}{\partial t}(x,0,t) + \rho_2 U_2 \frac{\partial \phi_2}{\partial x}(x,0,t) + \rho_2 g \eta(x,t) \tag{6.16g}$$

Equations (6.16a) to (6.16g) represent the problem to be satisfied by any perturbation to the uniform flows. In particular, we decided to study the effect of a sinusoidal wave at the interface, so that the equation of the interface was chosen to be

$$\eta(x,t) = \varepsilon e^{i(2\pi/\lambda)(x - \sigma t)} \tag{6.16h}$$

The solution to Eqs. (6.16a) and (6.16c) may be obtained by separation of variables. In view of the shape of the interface, as defined by Eq. (6.16h), the solutions should be trigonometric in $x$. Then the $y$ dependence will be exponential. In view of the conditions (6.16b) and (6.16d), the negative exponential should be rejected for $\phi_1$ and the positive exponential should be rejected for $\phi_2$. Thus the solutions to (6.16a) and (6.16c) which satisfy (6.16b) and (6.16d) are

$$\phi_1(x, y, t) = A_1 e^{(2\pi/\lambda)y} e^{i(2\pi/\lambda)(x-\sigma t)}$$

$$\phi_2(x, y, t) = A_2 e^{-(2\pi/\lambda)y} e^{i(2\pi/\lambda)(x-\sigma t)}$$

Imposing the kinematic surface conditions (6.16e) and (6.16f) on these solutions shows that

$$A_1 = i\varepsilon(-\sigma + U_1)$$

$$A_2 = -i\varepsilon(-\sigma + U_2)$$

Thus the velocity potentials in the lower and upper regions are, respectively,

$$\phi_1(x, y, t) = -i\varepsilon(\sigma - U_1)e^{(2\pi/\lambda)y} e^{i(2\pi/\lambda)(x-\sigma t)}$$

$$\phi_2(x, y, t) = i\varepsilon(\sigma - U_2)e^{-(2\pi/\lambda)y} e^{i(2\pi/\lambda)(x-\sigma t)}$$

These solutions satisfy all the required conditions except the pressure condition at the interface. Thus Eq. (6.16g) requires

$$\rho_1 i(\sigma - U_1)\left(i\frac{2\pi}{\lambda}\sigma\right) - \rho_1 U_1 i(\sigma - U_1)\left(i\frac{2\pi}{\lambda}\right) + \rho_1 g$$

$$= \rho_2 i(\sigma - U_2)\left(-i\frac{2\pi}{\lambda}\sigma\right) + \rho_2 U_2 i(\sigma - U_2)\left(-i\frac{2\pi}{\lambda}\right) + \rho_2 g$$

The quantity $\eta$, as defined by Eq. (6.16h), has been canceled throughout this equation as a nonzero common factor. Combining the first and second terms on each side of this equation reduces it to the form

$$-\rho_1 \frac{2\pi}{\lambda}(\sigma - U_1)^2 + \rho_1 g = \rho_2 \frac{2\pi}{\lambda}(\sigma - U_2)^2 + \rho_2 g$$

Everything in this algebraic equation is known as a priori except the quantity $\sigma$. Then, the above equation should be looked upon as a quadratic equation for $\sigma$. Solving this quadratic equation gives

$$\sigma = \frac{\rho_2 U_2 + \rho_1 U_1}{\rho_2 + \rho_1} \pm \sqrt{\left(\frac{\rho_2 U_2 + \rho_1 U_1}{\rho_2 + \rho_1}\right)^2 - \left(\frac{\rho_2 U_2^2 + \rho_1 U_1^2}{\rho_2 + \rho_1}\right) - \left(\frac{\rho_2 - \rho_1}{\rho_2 + \rho_1}\right)\frac{g\lambda}{2\pi}}$$

The first two quantities inside the square root may be combined to give the following simplified expression for $\sigma$:

$$\sigma = \frac{\rho_2 U_2 + \rho_1 U_1}{\rho_2 + \rho_1} \pm \sqrt{\left(\frac{\rho_1 - \rho_2}{\rho_1 + \rho_2}\right)\frac{g\lambda}{2\pi} - \frac{\rho_1 \rho_2}{(\rho_2 + \rho_1)^2}(U_2 - U_1)^2} \quad (6.17a)$$

Equation (6.17a) shows that $\sigma$ may be real, imaginary, or complex, depending upon the nature of the free parameters. Several special cases will be investigated.

Consider, first, the special case where $U_1 = U_2 = 0$ and $\rho_2 = 0$. This would correspond to two stationary fluids in which the density of the upper fluid is very small compared with that of the lower fluid. Such a condition would closely approximate a stationary liquid over which a stationary gas exists, for example, air over water. Then Eq. (6.17a) shows that $\sigma$ will be real, having the values

$$\sigma = \pm \sqrt{\frac{g\lambda}{2\pi}} \qquad (6.17b)$$

This agrees with Eq. (6.3b), which gives the propagation speed for surface waves in deep liquids. The minus sign in Eq. (6.17b) corresponds to a wave traveling in the negative $x$ direction. Since it turned out that $\sigma$ is real, the waves at the interface will propagate, so that the surface of separation will remain intact. That is, the interface is stable.

Next, consider the case where $\rho_2 = 0$ and the other parameters are nonzero. Physically, this would approximate the case of a gas blowing over a liquid surface. Under these conditions Eq. (6.17a) reduces to

$$\sigma = U_1 \pm \sqrt{\frac{g\lambda}{2\pi}} \qquad (6.17c)$$

But this is just Eq. (6.17b), in which a galilean transformation of magnitude $U_1$ has been applied. That is, the waves move along the surface of the liquid at the speed of the liquid plus or minus the speed of the waves on a quiescent body of the liquid. Again the interface will remain in contact and so is stable.

Consider now the case in which $\rho_2 = \rho_1$. Physically, the situation is a discontinuity in the velocity (i.e., a shear layer) in a homogeneous fluid. Then Eq. (6.17a) becomes

$$\sigma = \frac{U_2 + U_1}{2} \pm i \frac{U_2 - U_i}{2} \qquad (6.17d)$$

This result shows that unless $U_2 = U_1$ (in which case there is no shear layer), the quantity $\sigma$ will have an imaginary part which will result in the interfacial wave's growing exponentially with time. That is, the interface at the shear layer is unstable. This form of instability is known as *Helmholtz instability* or *Rayleigh instability*.

Finally, consider both fluids to be quiescent so that $U_1 = U_2 = 0$, but let their densities differ. Then Eq. (6.17a) reduces to

$$\sigma = \pm \sqrt{\frac{g\lambda}{2\pi} \left( \frac{\rho_1 - \rho_2}{\rho_1 + \rho_2} \right)} \qquad (6.17e)$$

For $\rho_1 > \rho_2$, that is, for the heavier fluid on the bottom, $\sigma$ will be real, so that the interface will be stable. However, for $\rho_2 > \rho_1$, that is, for the heavier fluid on top, $\sigma$ will be imaginary, so that the interface will be unstable. This form of instability is known as *Taylor instability*.

## PROBLEMS

**6.1.** Use Eq. (6.7c) for the complex potential for a traveling wave on a quiescent liquid surface to deduce that the complex potential for a stationary wave on the surface of a liquid layer whose mean velocity is $c$ in the negative $x$ direction is

$$F(z) = -cz - \frac{c\varepsilon}{\sinh(2\pi h/\lambda)} \cos \frac{2\pi}{\lambda}(z + ih)$$

Show that, in very deep liquids, the above result becomes

$$F(z) = -cz - c\varepsilon e^{-i(2\pi/\lambda)z} \qquad (6.18a)$$

Use this last result to determine the stream function $\psi(x, y)$ for a stationary wave on the surface of a deep-liquid layer whose mean velocity is $c$. Hence show that the streamline $\psi(x, \eta) = 0$ gives the equation of the free surface as

$$\eta = \varepsilon e^{(2\pi/\lambda)\eta} \sin \frac{2\pi x}{\lambda} \qquad (6.18b)$$

**6.2.** The Bernoulli equation for the situation depicted in Prob. 6.1 is

$$\frac{p}{\rho} + \tfrac{1}{2}\mathbf{u} \cdot \mathbf{u} + gy = \frac{P}{\rho} + \tfrac{1}{2}c^2\left[1 + \varepsilon^2\left(\frac{2\pi}{\lambda}\right)^2\right]$$

where $P$ is the pressure of the free surface. Use this equation, together with Eqs. (4.5) and (6.18a), to show that setting $p = P$ when $y = \eta$ requires that the quantity $c$, the velocity difference between the mean velocity and the wave-train velocity, must satisfy the equation

$$c^2 = \frac{2g\eta}{\varepsilon^2(2\pi/\lambda)^2 + (4\pi/\lambda)\eta - \varepsilon^2(2\pi/\lambda)^2 e^{(4\pi/\lambda)\eta}}$$

Linearize this expression for small values of $\varepsilon/\lambda$ and hence confirm Eq. (6.3b). Show that by neglecting only those terms which are of fourth order or smaller, the expression for $c$ becomes

$$c^2 = \frac{g\lambda/(2\pi)}{1 - (2\pi/\lambda)^2\varepsilon^2}$$

That is, the effect of finite wave amplitude is to increase the wave speed on the liquid surface.

**6.3.** By applying a galilean transformation to the coordinate system which was used in arriving at Eq. (6.7c), show that the stream function for the stationary wave shown

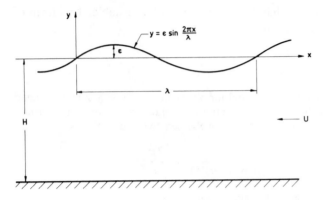

**FIGURE 6.12**
Stationary sinusoidal wave on the surface of a liquid layer which is moving with a uniform velocity.

in Fig. 6.12 is

$$\psi(x, y) = -Uy + \frac{U\varepsilon}{\sinh\left(2\pi H/\lambda\right)} \sin\frac{2\pi x}{\lambda} \sinh\frac{2\pi}{\lambda}(y + H) \quad (6.19a)$$

where

$$\frac{U^2}{gH} = \frac{\lambda}{2\pi H} \tanh\frac{2\pi H}{\lambda} \quad (6.19b)$$

Here $H$ is the mean depth of the liquid while $\varepsilon$ and $\lambda$ are, respectively, the amplitude and wavelength of the stationary wave on the surface.

**6.4.** The result of Prob. 6.3 may be used to obtain a solution to the problem of steady flow over a wave-shaped wall. The configuration for which the solution is sought is shown in Fig. 6.13. To obtain the solution from the results of Prob. 6.3, observe that the solution obtained there will, at some depth $h < H$, have a wave-shaped streamline of amplitude $\varepsilon_0$ and wavelength $\lambda$. This streamline may be considered to be a wall, so that if the free surface is taken to be $\psi$, the boundary defined by $\psi = Uh$ will correspond to $y = -h + \eta_0$ where $\eta_0 = \varepsilon_0 \sin(2\pi x/\lambda)$. In this way, show from the linear theory that the ratio of the wave amplitude to the wall

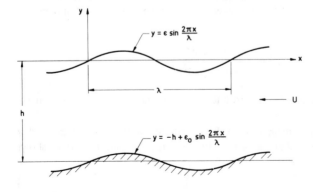

**FIGURE 6.13**
Steady, uniform flow over a sinusoidal bed producing a stationary wave train on the surface.

amplitude is

$$\frac{\varepsilon}{\varepsilon_0} = \frac{1}{\cosh\left(2\pi h/\lambda\right) - \left(g\lambda/2\pi U^2\right)\sinh\left(2\pi h/\lambda\right)}$$

where the various parameters are defined in Fig. 6.13.

**6.5.** Consider two traveling waves defined by

$$\eta_1(x,t) = \tfrac{1}{2}\varepsilon \sin \frac{2\pi}{\lambda_1}(x - c_1 t)$$

$$\eta_2(x,t) = \tfrac{1}{2}\varepsilon \sin \frac{2\pi}{\lambda_2}(x - c_2 t)$$

Show that the equation of the surface which results from superimposing these two waves will be defined by

$$\eta(x,t) = \varepsilon \cos \pi\left[\left(\frac{1}{\lambda_1} - \frac{1}{\lambda_2}\right)x - \left(\frac{c_1}{\lambda_1} - \frac{c_2}{\lambda_2}\right)t\right]$$

$$\times \sin \pi\left[\left(\frac{1}{\lambda_1} + \frac{1}{\lambda_2}\right)x - \left(\frac{c_1}{\lambda_1} + \frac{c_2}{\lambda_2}\right)t\right]$$

Hence show that if $\lambda_1$ and $\lambda_2$ differ by only a small amount and that if $c_1$ and $c_2$ differ by only a small amount, this surface profile may be considered to be of the form

$$\eta(x,t) = A(x,t) \sin \pi\left[\left(\frac{1}{\lambda_1} + \frac{1}{\lambda_2}\right)x - \left(\frac{c_1}{\lambda_1} + \frac{c_2}{\lambda_2}\right)t\right]$$

which represents a traveling wave whose amplitude is changing slowly with time compared with the frequency of the oscillations. This situation is shown in Fig. 6.14. The phenomenon depicted in this figure is referred to as *beating*, and it occurs in situations where two similar waves or signals are superimposed.

**FIGURE 6.14**
Beating phenomenon which results from superimposing two waves of almost equal frequencies.

**6.6.** (*a*) The potential energy per wavelength of a wave train is given by

$$V = \int_0^\lambda \tfrac{1}{2}\rho g \eta^2 \, dx$$

Use this expression to show that the potential energy per wavelength of the wave $\eta = \varepsilon \sin 2\pi(x - ct)/\lambda$ is

$$T = \tfrac{1}{4}\rho g \varepsilon^2 \lambda$$

(b) The kinetic energy per wavelength of a wave train is given by

$$T = \tfrac{1}{2}\rho \int_0^\lambda \phi \frac{\partial \phi}{\partial y}\Big|_{y=0} dx$$

Use this expression and Eq. (6.7a) to show that the kinetic energy per wavelength of the same sinusoidal wave is

$$T = \tfrac{1}{4}\rho g \varepsilon^2 \lambda$$

**6.7.** The work done on a vertical plane in a liquid layer is given by

$$WD = \int_{-h}^0 p \frac{\partial \phi}{\partial x} dy$$

Use the linearized form of the Bernoulli equation and Eq. (6.7a) to show that the work done across a vertical plane of liquid which has a traveling wave defined by the expression $\eta = \varepsilon \sin 2\pi(x - ct)/\lambda$ on its surface is

$$WD = \tfrac{1}{2}\rho g c \varepsilon^2 \sin^2 \frac{2\pi}{\lambda}(x - ct)\left[ 1 + \frac{2\pi h/\lambda}{\sinh(2\pi h/\lambda)\cosh(2\pi h/\lambda)} \right]$$

Hence show that for deep liquids the time average of the work done is one-half of the sum of the kinetic energy per wavelength and the potential energy per wavelength. That is

$$(WD)_{\text{ave}} = \tfrac{1}{2}(T + V)$$

**6.8.** The distribution of vorticity in a lake is assumed to be represented by the following relationship

$$\Omega(x, y, t) = \omega(x, y, t) + \beta y$$

In this equation $\Omega$ is the total vorticity, $\omega$ is the intrinsic vorticity and $\beta$ is a constant. $x$ and $y$ are coordinates which lie in the plane of the surface of the lake.

A traveling wave is initiated on the surface of the lake for which the stream function may be taken to be

$$\psi(x, y, t) = A e^{i(kx + ly + \sigma t)}$$

where $A$ is a constant, $k$ and $l$ represent the wavelengths in the $x$ and $y$ directions, respectively, and $\sigma$ represents the speed at which the wave is traveling. Use the definition of vorticity and the definition of the stream function to show that $\omega$ is proportional to $\psi$, and find the constant of proportionality.

Using linear theory, show that the material derivative of the total vorticity $\Omega$ is proportional to the stream function $\psi$, and find the constant of proportionality. Also find the value of the speed $\sigma$ which makes the material derivative of $\Omega$ zero.

**6.9.** One very simple way of representing a boundary layer is to consider it to be a layer of zero fluid velocity. Fig. 6.15 shows such a model of a boundary layer for uniform flow over a flat plate. For this configuration, carry out a stability analysis of the interface, similar to that which was carried out in Sec. 6.12, by imposing a wave of the form

$$\eta(x, t) = \varepsilon e^{i(2\pi/\lambda)(x - \sigma t)}$$

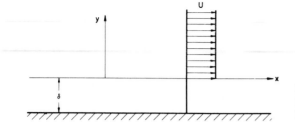

**FIGURE 6.15**
Stagnant layer of fluid of thickness $\delta$ below a uniform stream of magnitude $U$.

on the interface. Determine whether the interface is unstable, and if it is unstable, determine the fastest growing wavelength of the instability.

**6.10.** A fluid of density $\rho_1$ occupies the space $-h < y < 0$ while a different fluid of density $\rho_2$ occupies the space $0 < y < h$. Solid boundaries exist at both $y = -h$ and $y = +h$, and both fluids are originally at rest. A small-amplitude traveling wave of wavelength $\lambda$ is introduced along the interface separating the two fluids. Using a linear theory, determine the speed at which the wave will travel along the interface and discuss the conditions under which the amplitude of the wave will decay with time, or grow with time.

## FURTHER READING—PART II

The topic of ideal-fluid flow is probably the most studied branch of fluid mechanics, and it well represented in the literature. Most texts on fluid mechanics have at least one chapter on the subject, and some books are entirely to devoted to it. Listed below are some books which, collectively, cover the subject in some depth. The book by Sir Horace Lamb was first published in 1879, and it does not utilize vector analysis or tensor analysis. However, this book has been a standard reference for many years and it continues to be a valuable source of information.

Lamb, Sir Horace: "Hydrodynamics," 6th ed., Dover Publications, Inc., New York, N.Y., 1932.

Lighthill, James: "Waves in Fluids," Cambridge University Press, London, 1978.

Milne-Thompson, L. M.: "Theoretical Hydrodynamics," 4th ed., The Macmillan Company, New York, N.Y., 1962.

Panton, Ronald L.: "Incompressible Flow," John Wiley & Sons, New York, N.Y. 1984.

Robertson, James M.: "Hydrodynamics in Theory and Application," Prentice-Hall, Inc., Englewood Cliffs, N.J., 1965.

# PART
# III

# VISCOUS
# FLOWS OF
# INCOMPRESSIBLE
# FLUIDS

In this section problems will be solved and phenomena will be established in which the viscosity of the fluid is intrinsically important. The treatment is divided into four chapters. Chapter 7 covers the exact solutions to the Navier-Stokes equations. Although these solutions are relatively few in number, they are cherished. They are used as the basis for perturbation schemes to solve problems which are close to the exact solution configurations, they are used to test the accuracy of numerical techniques, and they are used to calibrate instruments.

Chapter 8 deals with approximate solutions to the Navier-Stokes equations which are valid for small Reynolds numbers. This is achieved by reducing the

Navier-Stokes equations through the so-called "Stokes approximation" and by studying the solutions to the resulting equations. Such solutions are valuable in their own right, and they have physical counterparts. In addition, they form the basis of approximate solutions to other problems.

Chapter 9 deals with large-Reynolds-number flows. Specifically, the Prandtl boundary-layer approximation to the Navier-Stokes equations is examined. Some exact solutions to these equations are first obtained through similarity methods. The Kármán-Pohlhausen method is then covered as an example of an approximate solution to the boundary-layer equations. The stability of boundary layers is also introduced.

The final chapter in this part of the book deals with buoyancy-driven flows. The Boussinesq approximation to the Navier-Stokes and thermal energy equations is introduced in Chapter 10. Solutions to the resulting equations are presented for vertical isothermal surfaces, a line source of heat, and a point source of heat. The stability of horizontal fluid layers is also discussed with a view to establishing the condition for the onset of thermal convection.

The governing equations for this part of the book are the continuity equation and the Navier-Stokes equations. Thus from Eqs. (1.3c) and (1.9b) the vector form of the governing equations is

$$\nabla \cdot \mathbf{u} = 0 \tag{III.1}$$

$$\frac{\partial \mathbf{u}}{\partial t} + (\mathbf{u} \cdot \nabla)\mathbf{u} = -\frac{1}{\rho}\nabla p + \nu \nabla^2 \mathbf{u} + \mathbf{f} \tag{III.2}$$

As was the case in the previous parts of the book, these equations form a complete set for the unknown quantities $p$ and $\mathbf{u}$. This is due to the assumption of incompressibility, which has the mathematical consequence of uncoupling the equations of dynamics from those of thermodynamics.

The boundary condition which is to be imposed on the velocity vector $\mathbf{u}$ is the no-slip boundary condition which is given by Eq. (1.14). This boundary condition is

$$\mathbf{u} = \mathbf{U} \quad \text{on solid boundaries} \tag{III.3}$$

where $\mathbf{u}$ is the fluid velocity and $\mathbf{U}$ is the velocity of the solid which forms the boundary with the fluid. This condition states that the fluid which is adjacent to a solid boundary adheres to that boundary and does not slip. This boundary condition is much stronger than that which was used in the study of ideal fluids. The essential difference is that the inclusion of the viscous terms in Eq. (III.2) has raised the order of the governing partial differential equation by one. Thus the true physical boundary condition may be accommodated in this part, whereas it could not be satisfied completely in the previous part of the book.

It was observed in Chap. 3 that the equations of mass conservation, the Navier-Stokes equations in this instance, may be alternatively phrased in terms of the vorticity $\boldsymbol{\omega}$. Although we will not solve problems here from the vorticity formulation, it is sometimes of interest to examine solutions from the point of view of the distribution of vorticity. For such cases Eq. (3.4$a$) shows that the vorticity equation is

$$\frac{\partial \omega}{\partial t} + (\mathbf{u} \cdot \nabla)\omega = (\omega \cdot \nabla)\mathbf{u} + \nu \nabla^2 \omega + \nabla \times \mathbf{f} \qquad \text{(III.4)}$$

in which the possibility of nonconservative body forces has been included for generality.

The solutions to the foregoing equations of viscous flow which will be established in this part of the book will all correspond to laminar flow. Viscous flows may be divided into two principal categories, laminar flows and turbulent flows. The phenomena and treatment of turbulent flows is somewhat different from the other fundamental aspects of fluid flow, and it is usually treated separately in specialized books. This procedure will be adopted here, so that only laminar flows will be considered.

# CHAPTER

# 7

# EXACT
# SOLUTIONS

In this chapter some exact solutions to the equations governing the motion of an incompressible, viscous liquid will be established. It is perhaps because so few exact solutions have been found that they are so important. The basic difficulty in obtaining exact solutions to viscous-flow problems lies in the existence of the nonlinear convection terms in Eq. (III.2). Furthermore, these nonlinear terms cannot be circumvented in this instance in the manner which was adopted in the study of ideal fluids. This, in turn, is due to the inapplicability of Kelvin's theorem due to viscosity, so that viscous flows are not potential. In addition, the Bernoulli equations do not apply.

The exact solutions may be divided into two broad categories. In one of these categories, the nonlinear term $(\mathbf{u} \cdot \nabla)\mathbf{u}$ is identically zero owing to the simple nature of the flow field. Examples of this situation which are covered in this chapter are Couette flow, Poiseuille flow, the flow between rotating cylinders, Stokes' problems, and pulsating flow between parallel surfaces.

The second broad category of exact solutions is that for which the nonlinear convective terms are not identically zero. Examples which are presented here include stagnation-point flow, the flow in convergent and divergent channels, and the flow over a porous wall.

## 7.1 COUETTE FLOW

One of the simplest viscous-flow fields is that for flow between two parallel surfaces. Figure 7.1$a$ shows two parallel surfaces whose size in the $z$ direction is supposed to be very large compared with their separation distance $h$. The flow

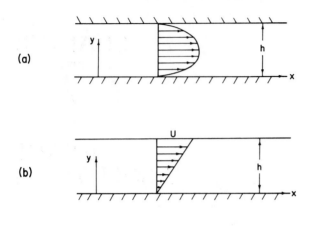

(a)

(b)

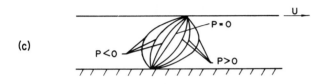

(c)

**FIGURE 7.1**
(a) Flow between parallel surfaces, (b) plane Couette flow, (c) general Couette flow.

between these plates is taken to be in the $x$ direction, and since there is no flow in the $y$ direction, the pressure will be a function of $x$ only. That is, since there are no inertia, viscous, or external forces in the $y$ direction, there can be no pressure gradient in that direction. Using the fact that $u = u(y)$ only and $v = w = 0$ together with the fact that $p = p(x)$ only, Eq. (III.2) becomes, for a force-free fluid field

$$0 = -\frac{dp}{dx} + \mu \frac{d^2u}{dy^2}$$

Here, the continuity equation is identically satisfied and the nonlinear convection terms are identically zero by virtue of the simplicity of the flow field. The Navier-Stokes equations reduce to the above ordinary differential equation, which states that there is a balance between the pressure force in the fluid and the viscous-shear force at all points in the fluid. Since $dp/dx$ is a function of $x$ only, this equation may be integrated twice with respect to $y$ to give

$$u(y) = \frac{1}{\mu}\frac{dp}{dx}\left(\frac{y^2}{2} + Ay + B\right)$$

where $A$ and $B$ are constants of integration. The boundary condition $u(0) = 0$ requires that $B = 0$ while the condition $u(h) = 0$ requires that $A = -h/2$. Thus the velocity profile will be given by the equation

$$u(y) = -\frac{1}{2\mu}\frac{dp}{dx}y(h - y)$$

It is usual to introduce a dimensionless pressure parameter, which is defined as follows:

$$P = -\frac{h^2}{2\mu U}\frac{dp}{dx}$$

Here $U$ is any characteristic velocity such as the mean-flow velocity. In terms of this pressure parameter the expression for the velocity profile between the parallel plates becomes

$$\frac{u(y)}{U} = P\frac{y}{h}\left(1 - \frac{y}{h}\right) \tag{7.1a}$$

Equation (7.1a) shows that the fluid flows in the direciton of the negative pressure gradient and that the velocity profile across the flow field is parabolic. The maximum velocity therefore occurs at the centerline between the two plates (that is, at $y = h/2$), and the magnitude of the maximum velocity is $PU/4$. It should be noted that for this type of flow the pressure gradient is the driving mechanism, so that if there is no external pressure gradient there will be no flow.

Another way of inducing a flow between two parallel surfaces, apart from applying a pressure gradient, is to move one of the two surfaces. Figure 7.1b depicts such a situation, which is referred to as *plane Couette flow*. The surface $y = 0$ is held fixed while the surface $y = h$ is moved in the $x$ direction with constant velocity $U$. As before, the only nonzero velocity component will be $u$, and it will be a function of $y$ only. Also, there will be no pressure gradient in the $y$ direction, as before, and here it is assumed that there is no external pressure gradient in the $x$ direction. Then the governing equations reduce to the same equation as before but without the pressure term. That is, the velocity must satisfy the equation

$$0 = \mu\frac{d^2u}{dy^2}$$

Integrating this equation gives

$$u(y) = Ay + B$$

where $A$ and $B$ are constants of integration. The boundary condition $u(0) = 0$ requires that $B = 0$, while the condition $u(h) = U$ requires that $A = U/h$. Thus the velocity profile for plane Couette flow is

$$\frac{u(y)}{U} = \frac{y}{h} \tag{7.1b}$$

This result shows that the velocity profile induced in a fluid by moving one of the boundaries at constant velocity is linear across the gap between the two boundaries.

A more general situation is one in which either of the two surfaces is moving at constant velocity and there is also an external pressure gradient. Such a situation is referred to as *general Couette flow*. The velocity profile for general Couette flow may be obtained by superimposing Eqs. (7.1*a*) and (7.1*b*), since the governing equations which led to these results are linear. Thus it follows that

$$\frac{u(y)}{U} = \frac{y}{h} + P\frac{y}{h}\left(1 - \frac{y}{h}\right)$$   (7.1*c*)

The velocity profiles corresponding to this equation are shown in Fig. 7.1*c*. It will be seen from Fig. 7.1 that for $P = 0$ plane Couette flow is recovered, while for $P \neq 0$ the pressure gradient will either assist or resist the viscous shear motion. For $P > 0$ (that is, for $dp/dx < 0$) the pressure gradient will assist the viscously induced motion to overcome the shear force at the lower surface. For $P < 0$ (that is, for $dp/dx > 0$) the pressure gradient will resist the motion which is induced by the motion of the upper surface. In this case a region of reverse flow may occur near the lower surface as shown in Fig. 7.1*c*.

## 7.2   POISEUILLE FLOW

The steady flow of a viscous fluid in a conduit of arbitrary but constant cross section is referred to as Poiseuille flow. Figure 7.2*a* shows an arbitrary cross section in the *yz* plane with a steady flow in the *x* direction. Here again, the transverse velocity components $v$ and $w$ will be zero, while $u$ will be a funciton of $y$ and $z$ only. The pressure cannot vary in the transverse directions, since there is no motion or forces in these directions; hence $p$ will be a function of $x$ only. With these conditions the governing equations (III.1) and (III.2) reduce to

$$0 = -\frac{dp}{dx} + \mu\left(\frac{\partial^2 u}{\partial y^2} + \frac{\partial^2 u}{\partial z^2}\right)$$

Again, owing to the simple geometry of the flow field, the nonlinear term $(\mathbf{u} \cdot \nabla)\mathbf{u}$ is identically zero and the continuity equation is identically satisfied for any velocity distribution $u(y, z)$. The remaining equation which has to be satisfied is a Poisson type of equation. In standard form this equation is

$$\frac{\partial^2 u}{\partial y^2} + \frac{\partial^2 u}{\partial z^2} = \frac{1}{\mu}\frac{dp}{dx}$$   (7.2*a*)

where the nonhomogeneous term must be a constant at most. There is no general solution to Eq. (7.2*a*) for arbitrary cross sections, but solutions for a few specific cross sections do exist.

Consider, first, the special case in which the cross section in the *yz* plane is circular with radius $a$, as shown in Fig. 7.2*b*. With this geometry the preferred coordinate system is cylindrical coordinates. Then, let the cross section of the conduit be represented by the cylindrical coordinates $R$ and $\theta$ rather than the cartesian coordinates $y$ and $z$, so that the independent coordinates are now $R$, $\theta$, and $x$. In this coordinate system the axial velocity $u$ will be independent of $\theta$

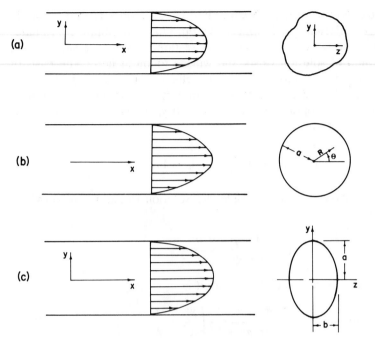

**FIGURE 7.2**
Viscous flow along conduits of various cross sections, (*a*) arbitrary, (*b*) circular, (*c*) elliptic.

and $x$, so that Eq. (7.2*a*) will become

$$\frac{1}{R}\frac{d}{dR}\left(R\frac{du}{dR}\right) = \frac{1}{\mu}\frac{dp}{dx}$$

Since the pressure gradient is independent of $R$, this equation may be integrated twice with respect to $R$ to give

$$u(R) = \frac{1}{\mu}\frac{dp}{dx}\frac{R^2}{4} + A\log R + B$$

where $A$ and $B$ are constants of integration. The condition $u(0) = $ finite requires that $A = 0$, while the condition $u(a) = 0$ requires that $B = -(dp/dx)a^2/(4\mu)$. Thus the velocity profile in the conduit will be of the form

$$u(R) = -\frac{1}{4\mu}\frac{dp}{dx}(a^2 - R^2) \qquad (7.2b)$$

This result is similar in nature to that for the flow between two parallel surfaces. The flow depends upon the external pressure for its existence, and the resultant velocity profile is parabolic.

For elliptic cross sections, as shown in Fig. 7.2c, a proper procedure would be to express the laplacian which appears in Eq. (7.2a) in elliptic coordinates and proceed as above with the circular cross section. However, a simpler and more direct method of solution exists and will be followed here. The basis of this method is the observation that, for the ellipse shown in Fig. 7.2c, the quantity $y^2/a^2 + z^2/b^2 - 1$ is zero on the boundary. This motivates us to look for a solution which is proportional to this quantity, so that a solution to Eq. (7.2a) is sought in the form

$$u(y, z) = \alpha\left(\frac{y^2}{a^2} + \frac{z^2}{b^2} - 1\right)$$

Direct substitution shows that this is indeed a solution to Eq. (7.2a) provided the value of $\alpha$ is

$$\alpha = \frac{1}{2\mu}\frac{dp}{dx}\frac{a^2b^2}{a^2 + b^2}$$

Thus the velocity profile for an elliptic conduit is

$$u(y, z) = \frac{1}{2\mu}\frac{dp}{dx}\frac{a^2b^2}{a^2 + b^2}\left(\frac{y^2}{a^2} + \frac{z^2}{b^2} - 1\right) \tag{7.2c}$$

## 7.3 FLOW BETWEEN ROTATING CYLINDERS

An exact solution to the Navier-Stokes equations exists for the case of a fluid which is contained between two concentric circular cylinders either or both of which is rotating at constant speed about its axis. The cylinders are assumed to be long compared with their diameter, so that the flow field will be two-dimensional. Figure 7.3 shows the geometry under consideration. The outer cylinder has a radius $R_0$ and it is rotating in the clockwise direction with angular, velocity $\omega_0$, while the radius of the inner cylinder is $R_i$ and its angular velocity is $\omega_i$.

Cylindrical coordinates are preferred for the geometry shown, and the only nonzero velocity component in this coordinate system will be the tangential velocity $u_\theta$. Furthermore, this velocity component will depend upon $R$ only. For this type of velocity field and in the absence of any external body forces, the pressure can depend upon $R$ only. Using these observations the governing equations [Eqs. (III.1) and (III.2)] become

$$-\frac{u_\theta^2}{R} = -\frac{1}{\rho}\frac{dp}{dR}$$

$$0 = \frac{d^2u_\theta}{dR^2} + \frac{d}{dR}\left(\frac{u_\theta}{R}\right)$$

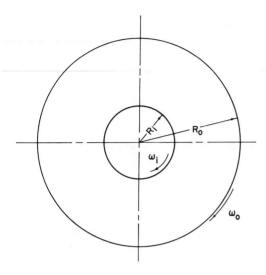

**FIGURE 7.3**
Geometry for flow between concentric rotating circular cylinders.

Because of the simple geometry of the flow field the continuity equation is identically satisfied. The first of the above equations shows that there is a balance between the centrifugal force which acts on an element of fluid and a force which is produced by the induced pressure field. The second equation above states that there is a balance between the viscous stresses in the fluid.

The foregoing equations may be readily integrated by establishing $u_\theta$ from the second equation and then determining the pressure $p$ from the first equation. Integrating the second equation twice with respect to $R$ gives

$$u_\theta(R) = A\frac{R}{2} + \frac{B}{R}$$

where $A$ and $B$ are constants of integration. The boundary conditions $u_\theta(R_0) = \omega_0 R_0$ and $u_\theta(R_i) = \omega_i R_i$ give

$$A = \frac{2(\omega_0 R_0^2 - \omega_i R_i^2)}{R_0^2 - R_i^2}$$

$$B = -R_i^2 R_0^2 \frac{\omega_0 - \omega_i}{R_0^2 - R_i^2}$$

Thus the velocity distribuiton in the fluid between the two cylinders will be

$$u_\theta(R) = \frac{1}{R_0^2 - R_i^2}\left[(\omega_0 R_0^2 - \omega_i R_i^2)R - (\omega_0 - \omega_i)\frac{R_i^2 R_0^2}{R}\right] \qquad (7.3a)$$

Using Eq. (7.3a) and the remaining equation which is to be satisfied, it follows that the pressure $p$ must satisfy the equation

$$\frac{dp}{dR} = \frac{\rho}{\left(R_0^2 - R_i^2\right)^2}$$

$$\left[\left(\omega_0 R_0^2 - \omega_i R_i^2\right)^2 R - 2(\omega_0 - \omega_i)(\omega_0 R_0^2 - \omega_i R_i^2)\frac{R_i^2 R_0^2}{R} + (\omega_0 - \omega_i)^2 \frac{R_i^4 R_0^4}{R^3}\right]$$

Integrating this equation shows that the pressure distribution will be

$$p(R) = \frac{\rho}{\left(R_0^2 - R_i^2\right)^2}\left[\left(\omega_0 R_0^2 - \omega_i R_i^2\right)^2 \frac{R^2}{2} - 2(\omega_0 - \omega_i)\right.$$

$$\left. \times \left(\omega_0 R_0^2 - \omega_i R_i^2\right)R_i^2 R_0^2 \log R - (\omega_0 - \omega_i)^2 \frac{R_i^4 R_0^4}{2R^2}\right] + C$$

$$(7.3b)$$

where $C$ is a constant of integration which may be evaluated in any particular problem by specifying the value of the pressure on $R = R_0$ or on $R = R_i$.

As special cases Eqs. (7.3) describe the flow field due to a single cylinder which is rotating in a fluid of infinite extent and a cylinder filled with fluid which is rotating. Some aspects of these special cases will be investigated in the problems at the end of the chapter.

## 7.4 STOKES' FIRST PROBLEM

The fluid-mechanics problem which is referred to as Stokes' first problem has counterparts in many branches of engineering and physics. In the fluid-mechanics context the situation which is being considered is shown in Fig. 7.4a. The $x$ axis coincides with an infinitely long flat plate above which a fluid exists. Initially, both the plate and the fluid are at rest. Suddenly, the plate is jerked into motion in its own plate with a constant velocity. Under these conditions, what will be the response of the fluid to this motion of the boundary?

To answer this question, we appropriately reduce the equations of motion and obtain a solution to them. Since the motion of the boundary is in the $x$ direction, it may be reasonably assumed that the motion of the fluid will also be in that direction. Thus the only nonzero velocity component will be $u$, and this velocity component will be a function of $y$ and $t$ only. Then the pressure will be independent of $y$, and since $u$ is independent of $x$, so will $p$ be independent of $x$. That is, the pressure will be constant everywhere in the fluid. Using these properties of the flow field, the governing equations reduce to the following linear partial differential equation:

$$\frac{\partial u}{\partial t} = \nu \frac{\partial^2 u}{\partial y^2}$$

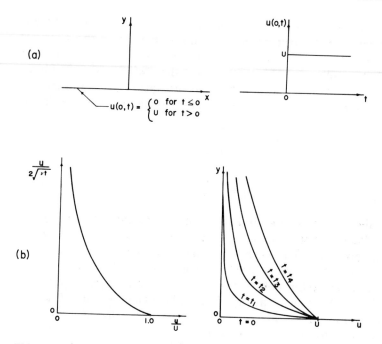

**FIGURE 7.4**

(*a*) Definition sketch for Stokes' first problem and (*b*) the solution curves in terms of the similarity variable and in terms of the dimensional variables.

The boundary conditions are

$$u(0, t) = \begin{cases} 0 & \text{for } t \leq 0 \\ U & \text{for } t > 0 \end{cases}$$

$$u(y, t) = \text{finite}$$

This problem lends itself to solution by Laplace transforms and by similarity methods. Since similarity solutions are the only ones which exist for some nonlinear problems which arise in boundary-layer theory and other situations, this method of solution will be employed here to establish a base for future considerations.

Similarity solutions are a special class of solutions which exist for problems which are governed by parabolic partial differential equations in two independent variables where there is no geometric length scale in the problem. Stokes' first problem meets these requirements. It may be observed that had there been a second plate at some plane $y = h$, the geometric length scale $h$ would exist, and so the conditions for a similarity solution would no longer exist. In the absence of such a length scale, however, it may be anticipated that the fluid velocity $u$ will reach some specified value, say $0.4U$, at different values of $y$ which will depend upon the value of the time $t$. That is, anticipating the results to be of the form indicated in Fig. 7.4b, it may be observed that at some time $t_1$ the velocity will have a value of $0.4U$ at some distance $y_1$ from the plate. At some later time $t_2$ the same velocity magnitude of $0.4U$ will exist at some different distance $y_2$, and so on. This suggests that there will be some combination of $y$ and $t$, such as $y/t^n$, such that when this quantity is constant, the velocity will be constant. That is, it is expected that a solution will exist in the form

$$\frac{u(y,t)}{U} = f(\eta)$$

where

$$\eta = \alpha \frac{y}{t^n}$$

Here $\eta(y,t)$ is called the *similarity variable* and $\alpha$ is a constant of proportionality which will be defined later to render $\eta$ dimensionless. This assumed form of solution has the property that when $\eta$ is constant (which corresponds to $y \sim t^n$), $u = $ constant. If indeed a similarity solution exists to our problem, substitution of our assumed form of solution into the governing partial differential equation will result in an ordinary differential equation with $f$ as the dependent variable and $\eta$ as the independent variable. That is, it will be possible to eliminate $y$ and $t$ in terms of $\eta$ only.

From the assumed form of solution the following expressions for the derivatives are obtained:

$$\frac{\partial u}{\partial t} = -Un\frac{\alpha y}{t^{n+1}}f' = -Un\frac{\eta}{t}f'$$

$$\frac{\partial u}{\partial y} = U\frac{\alpha}{t^n}f'$$

$$\frac{\partial^2 u}{\partial y^2} = U\frac{\alpha^2}{t^{2n}}f''$$

Here the primes denote differentiation of $f$ with respect to $\eta$. Substitution of these expressions into the governing equation yields the identity

$$-Un\frac{\eta}{t}f' = \nu U\frac{\alpha^2}{t^{2n}}f''$$

This is not an ordinary differential equation for arbitrary values of $n$, but for $n = \frac{1}{2}$ the explicit time dependence will be eliminated, yielding an ordinary differential equation. That is, for $n = \frac{1}{2}$ a similarity solution is obtained. For this value of $n$ the differential equation for $f$ and the definition of the similarity variable are as follows:

$$f'' + \frac{\eta}{2\nu\alpha^2}f' = 0$$

$$\eta = \alpha\frac{y}{t^{1/2}}$$

The quantity $\alpha$ may now be determined in terms of the parameter $\nu$ (and $U$ if necessary) to render $\eta$ dimensionless. The dimensions of $y/t^{1/2}$ are a length divided by the square root of time. Since the dimensions of $\nu$ are a length squared divided by time, it is sufficient to take $\alpha$ equal to $1/\sqrt{\nu}$. For convenience in the solution of the differential equation a factor of 2 is also included, so that the similarity variable becomes

$$\eta = \frac{y}{2\sqrt{\nu t}}$$

and the differential equation to be solved becomes

$$f'' + 2\eta f' = 0$$

This equation may be integrated successively by rewriting it as follows:

$$\frac{d}{d\eta}(\log f') = -2\eta$$

hence

$$\log f' = -\eta^2 + \log A$$

where the constant of integration has been chosen as $\log A$. Then, combining the two logarithmic quantities and taking exponentials of both sides of the resulting equation gives

$$f' = Ae^{-\eta^2}$$

$$\therefore \quad f(\eta) = A\int_0^\eta e^{-\xi^2}\,d\xi + B$$

where $B$ is another constant of integration and $\xi$ is a dummy variable of integration. The boundary condition $u(0, t) = U$ for $t > 0$ requires that $f(0) = 1$. This, in turn, requires that $B = 1$. The initial condition $u(y, 0) = 0$ for $y \geq 0$ requires that $f(\eta) \to 0$ when $\eta \to \infty$. This gives

$$0 = U\left(A\int_0^\infty e^{-\xi^2}\,d\xi + 1\right) = U\left(A\frac{\sqrt{\pi}}{2} + 1\right)$$

so that the value of the constant $A$ is $-2/\sqrt{\pi}$. Then, using the definition of the similarity variable, the expression for the velocity becomes

$$\frac{u(y, t)}{U} = 1 - \frac{2}{\sqrt{\pi}}\int_0^{y/2\sqrt{\nu t}} e^{-\xi^2}\,d\xi$$

But the second term on the right-hand side of this equation is the error function whose argument is the upper limit of integration. Thus the solution to Stokes' first problem may be written as

$$\frac{u(y, t)}{U} = 1 - \operatorname{erf}\left(\frac{y}{2\sqrt{\nu t}}\right) \qquad (7.4)$$

Values of the error function are presented in many tables of functions. Figure

7.4*b* shows the functional form of the error function and the dimensional velocity profiles which are generated by this single similarity curve.

As one might expect intuitively, the disturbance caused by the impulsive motion of the boundary diffuses into the fluid as the time from the initiation of the motion progresses. An estimate of the depth of fluid which is affected by the movement of the boundary may be obtained by observing from detailed plots of the error function that $u/U$ is reduced to about 0.04 when $\eta = \frac{3}{2}$. That is, for values of $\eta$ greater than $\frac{3}{2}$ the motion of the fluid is small, and the fluid may be considered to be unaffected by the moving boundary. Then, denoting the value of $y$ by $\delta$ at which $u/U$ is 0.04 shows that

$$\eta = \tfrac{3}{2} = \frac{\delta}{2\sqrt{\nu t}}$$

$$\therefore \quad \delta = 3\sqrt{\nu t}$$

That is, the thickness of the fluid layer which is affected by the motion of the boundary is proportional to the square root of the time and to the square root of the kinematic viscosity of the fluid. This result shows the role played by the kinematic viscosity in the diffusion of momentum through fluids.

## 7.5   STOKES' SECOND PROBLEM

Another problem to which an exact solution exists is geometrically identical to the one treated in the previous section, but the principal boundary condition is different. Stokes' second problem differs from Stokes' first problem only in the condition that the boundary $y = 0$ is oscillating in time rather than impulsively starting into motion. The geometry of the flow and the nature of the boundary condition are indicated in Fig. 7.5*a*. Since the geometry is the same as that of the previous section and since the motion is again in the plane of the boundary itself, the differential equation to be satisfied by $u(y, t)$ will be the same. That is, the problem to be solved becomes

$$\frac{\partial u}{\partial t} = \nu \frac{\partial^2 u}{\partial y^2}$$

$$u(0, t) = U \cos nt$$

$$u(y, t) = \text{finite}$$

Since the boundary $y = 0$ is oscillating in time, it is to be expected that the fluid will also oscillate in the $x$ direction in time with the same frequency. However, it is to be expected that the amplitude of the motion and the phase shift relative to the motion of the boundary will depend upon $y$. Thus a steady-state solution is sought of the form

$$u(y, t) = \text{Re}\left[w(y)e^{int}\right]$$

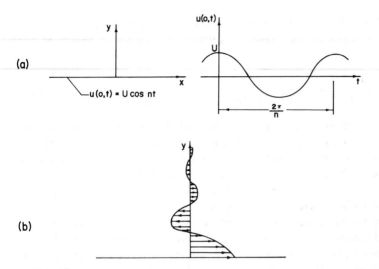

**FIGURE 7.5**
(a) Definition sketch for Stokes' second problem and (b) the nature of the velocity profiles.

where the symbol Re signifies the real part of the quantity which is inside the brackets. Substituting this assumed form of solution into the partial differential equation for $u(y, t)$ gives

$$\text{Re}\left[inw(y)e^{int}\right] = \nu\,\text{Re}\left[\frac{d^2w}{dy^2}e^{int}\right]$$

$$\therefore \quad \frac{d^2w}{dy^2} - i\frac{n}{\nu}w = 0$$

Noting that $\sqrt{i} = \pm(1 + i)/\sqrt{2}$, the solution to this differential equation is

$$w(y) = A\exp\left[-(1 + i)\sqrt{\frac{n}{2\nu}}\,y\right] + B\exp\left[(1 + i)\sqrt{\frac{n}{2\nu}}\,y\right]$$

The condition that the velocity be finite requires that the constant $B$ should be zero. Thus the quantity $w(y)$ will be of the form

$$w(y) = A\exp\left(-\sqrt{\frac{n}{2\nu}}\,y\right)\exp\left(-i\sqrt{\frac{n}{2\nu}}\,y\right)$$

The expression for the velocity then becomes

$$u(y, t) = \text{Re}\left[A\exp\left(-\sqrt{\frac{n}{2\nu}}\,y\right)\exp i\left(nt - \sqrt{\frac{n}{2\nu}}\,y\right)\right]$$

$$= A\exp\left(-\sqrt{\frac{n}{2\nu}}\,y\right)\cos\left(nt - \sqrt{\frac{n}{2\nu}}\,y\right)$$

The constant $A$ may be evaluated by imposing the boundary condition $u(0, t) = U \cos nt$, which requires $A = U$. Thus the velocity distribution in Stokes' second problem will be given by

$$\frac{u(y, t)}{U} = \exp\left(-\sqrt{\frac{n}{2\nu}}\, y\right) \cos\left(nt - \sqrt{\frac{n}{2\nu}}\, y\right) \tag{7.5}$$

Equation (7.5) describes a velocity which is oscillating in time with the same frequency as the boundary $y = 0$. The amplitude has its maximum value at $y = 0$ and decreases exponentially as $y$ increases. Also, Eq. (7.5) shows that there is a phase shift in the motion of the fluid and that this phase shift is proportional to $y$ and to the square root of $n$. The type of velocity profile which Eq. (7.5) represents is illustrated in Fig. 7.5b.

A measure of the distance away from the moving boundary within which the fluid is influenced by the motion of the boundary may be obtained as follows. The amplitude of the motion at any plane $y = $ constant may be obtained by letting the trigonometric term in Eq. (7.5) assume its maximum value of unity. Then, if the value of $y$ at which the amplitude of the motion is $1/e^2$ of its maximum value $U$ is denoted by $\delta$, it follows from Eq. (7.5) that

$$\frac{1}{e^2} = \exp\left(-\sqrt{\frac{n}{2\nu}}\, \delta\right)$$

hence

$$\delta = 2\sqrt{\frac{2\nu}{n}}$$

The quantity $\delta$ is a distance such that for $y > \delta$ the fluid may be considered to be essentially unaffected by the motion of the boundary. Again it is seen that viscous effects extend over a distance which is proportional to $\sqrt{\nu}$. It is also observed that $\delta$ varies inversely as the square root of the frequency of the motion. That is, the faster the motion the smaller will be the distance over which the adjacent fluid will be influenced.

## 7.6   PULSATING FLOW BETWEEN PARALLEL SURFACES

Another type of unsteady-flow situation for which an exact solution exists is that of an oscillating pressure in a fluid layer which is bounded by two parallel planes. We consider the two parallel surfaces to be located at $y = \pm a$ and consider the pressure gradient in the $x$ direction to oscillate in time. Then the velocity will be in the $x$ direction only and will also oscillate in time. That is, the only nonzero velocity component will be $u(y, t)$. Using these features of the flow, the governing equations reduce to the following single equation:

$$\frac{\partial u}{\partial t} = -\frac{1}{\rho}\frac{\partial p}{\partial x} + \nu\frac{\partial^2 u}{\partial y^2}$$

where $u(a, t) = u(-a, t) = 0$. The pressure gradient is assumed to oscillate in time so that $\partial p/\partial x$ will be taken to be of the form

$$\frac{\partial p}{\partial x} = P_x \cos nt$$

where $P_x$ is a constant which represents the magnitude of the pressure-gradient oscillations.

This problem may be treated in the same manner as that of the previous section. That is, by virtue of the oscillatory nature of the pressure gradient it may be expected that the velocity of the fluid will also oscillate in time, and with the same frequency, but possibly with a phase lag relative to the oscillations in the pressure. Thus the pressure gradient and the velocity may be represented as follows:

$$\frac{\partial p}{\partial x} = \text{Re}\left(P_x e^{int}\right)$$

$$u(y, t) = \text{Re}\left[w(y)e^{int}\right]$$

Substituting these expressions into the governing equation gives

$$\text{Re}\left(inwe^{int}\right) = -\frac{1}{\rho}\,\text{Re}\left(P_x e^{int}\right) + \nu\,\text{Re}\left(\frac{d^2w}{dy^2}e^{int}\right)$$

Thus the quantity $w(y)$ must satisfy the following nonhomogeneous ordinary differential equation:

$$\frac{d^2w}{dy^2} - \frac{in}{\nu}w = \frac{P_x}{\rho\nu}$$

The general solution to this differential equation consists of a constant particular integral plus the general solution to the homogeneous equation. This gives

$$w(y) = i\frac{P_x}{\rho n} + A\cosh\left[(1+i)\sqrt{\frac{n}{2\nu}}\,y\right] + B\sinh\left[(1+i)\sqrt{\frac{n}{2\nu}}\,y\right]$$

where the quantity $(1+i)/\sqrt{2}$ has been used for $\sqrt{i}$ and the hyperbolic form of solution has been chosen rather than the exponential form due to the finite extent of the flow field in the $y$ direction. The boundary conditions $u(a, t) = 0$ and $u(-a, t) = 0$ give, respectively,

$$0 = i\frac{P_x}{\rho n} + A\cosh\left[(1+i)\sqrt{\frac{n}{2\nu}}\,a\right] + B\sinh\left[(1+i)\sqrt{\frac{n}{2\nu}}\,a\right]$$

$$0 = i\frac{P_x}{\rho n} + A\cosh\left[(1+i)\sqrt{\frac{n}{2\nu}}\,a\right] - B\sinh\left[(1+i)\sqrt{\frac{n}{2\nu}}\,a\right]$$

The solution to this pair of algebraic equations for the undetermined constants

*A* and *B* is

$$A = - \frac{iP_x}{\rho n \cosh \left[ (1 + i)\sqrt{(n/2\nu)}\, a \right]}$$

$$B = 0$$

Thus the solution for $w(y)$ is

$$w(y) = i\frac{P_x}{\rho n}\left\{ 1 - \frac{\cosh\left[(1+i)\sqrt{(n/2\nu)}\, y\right]}{\cosh\left[(1+i)\sqrt{(n/2\nu)}\, a\right]} \right\}$$

Then the expression for the velocity in the fluid becomes

$$u(y,t) = \mathrm{Re}\left( i\frac{P_x}{\rho n}\left\{ 1 - \frac{\cosh\left[(1+i)\sqrt{(n/2\nu)}\, y\right]}{\cosh\left[(1+i)\sqrt{(n/2\nu)}\, a\right]} \right\} e^{int} \right) \qquad (7.6)$$

This expression may be decomposed to yield the real part explicitly. Although the concepts are straightforward, the details are cumbersome; hence the compact form of Eq. (7.6) will be considered to be the final expression. It is evident from the result that the velocity oscillates with the same frequency as the pressure gradient but that a phase lag, which depends upon $y$, exists. Thus the motion of the fluid which is adjacent to the boundaries will have a timewise phase shift relative to the motion near the centerline of the boundaries. The amplitude of the motion near the boundaries will also be different from that near the centerline, and in order to satisfy the boundary conditions, this amplitude will approach zero as the boundaries are approached.

## 7.7   STAGNATION-POINT FLOW

In all the foregoing flow situations the geometry of the flow field was such that the nonlinear inertia terms $(\mathbf{u} \cdot \nabla)\mathbf{u}$ were identically zero. The flow in the vicinity of a plane stagnation point is an example of a flow field in which these inertia terms are not zero yet one for which an exact solution exists.

Figure 7.6*a* shows the situation under consideration. A fluid stream whose velocity vector coincides with the $y$ axis impinges on a plane boundary which coincides with the $x$ axis. The boundary may be considered to be curved, such as the surface of a circular cylinder, provided the region under consideration is small in extent compared with the radius of curvature of the surface. This problem was investigated by Hiemenz, and the flow field is frequently referred to as *Hiemenz flow*. The basis of the solution is to modify the potential-flow solution in such a way that the Navier-Stokes equations are still satisfied and such that the no-slip boundary condition may be satisfied.

potential-flow solution for the situation under consideration was established in Chap. 4, and the complex potential for the flow in a sector of angle $\pi/n$ is given by Eq. (4.10). Using this result and the value $n = 2$, the

(a)

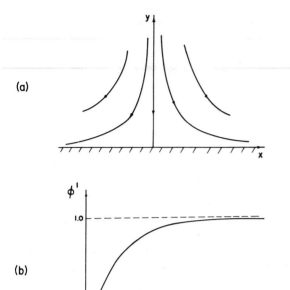

(b)

**FIGURE 7.6**
(*a*) Flow configuration for a plane stagnation point and (*b*) the functional form of the solution.

velocity components for the potential flow are

$$u = 2Ux$$
$$v = -2Uy$$

Then, from the Bernoulli equation, the pressure distribution will be

$$p = p_0 - 2\rho U^2(x^2 + y^2)$$

where $p_0$ is the Bernoulli constant which corresponds to the pressure at the stagnation point.

The foregoing velocity and pressure distributions satisfy the potential-flow problem exactly, and like all potential flows, they also satisfy the equations of motion for a viscous, incompressible fluid exactly. This may be readily shown by observing that the difference between the potential-flow equations and the equations governing the flow of a viscous, incompressible fluid is the presence of the viscous term $\nu \nabla^2 \mathbf{u}$ in the latter. But for potential flows $\mathbf{u} = \nabla\phi$, so that

$$\nabla^2 \mathbf{u} = \nabla^2(\nabla\phi) = \nabla(\nabla^2\phi) = 0$$

That is, the viscous-shear terms in the Navier-Stokes equations are identically zero for potential-flow fields.

Although potential-flow fields satisfy the equations of motion for a viscous, incompressible fluid, they do not satisfy the no-slip boundary condition. Thus, for the case of stagnation-point flow, Hiemenz attempted to modify the potential-flow field in such a way that meeting this boundary condition would be possible. Thus the $x$ component of velocity is taken to be

$$u = 2Uxf'(y)$$

where the prime denotes differentiation with respect to $y$. Then, the continuity equation requires that

$$\frac{\partial v}{\partial y} = -\frac{\partial u}{\partial x} = -2Uf'(y)$$

so that the vertical component of the velocity will be of the form

$$v = -2Uf(y)$$

Defining the velocity field in this way satisfies the continuity equation for all functions $f(y)$, and if we stipulate that $f(y) \to y$ as $y \to \infty$, the potential-flow solution will be recovered far from the boundary.

The equations of motion have yet to be satisfied, and this will impose further restrictions on the function $f$. The equations to be satisfied are

$$u\frac{\partial u}{\partial x} + v\frac{\partial u}{\partial y} = -\frac{1}{\rho}\frac{\partial p}{\partial x} + v\left(\frac{\partial^2 u}{\partial x^2} + \frac{\partial^2 u}{\partial y^2}\right)$$

$$u\frac{\partial v}{\partial x} + v\frac{\partial v}{\partial y} = -\frac{1}{\rho}\frac{\partial p}{\partial y} + v\left(\frac{\partial^2 v}{\partial x^2} + \frac{\partial^2 v}{\partial y^2}\right)$$

Substituting the expressions obtained above for $u$ and $v$ into these equations shows that the following pair of equations must be satisfied:

$$4U^2x(f')^2 - 4U^2xff'' = -\frac{1}{\rho}\frac{\partial p}{\partial x} + 2Uvxf'''$$

$$4U^2ff' = -\frac{1}{\rho}\frac{\partial p}{\partial y} - 2Uvf''$$

The second of these equations will be used to establish the pressure distribuiton, and this result will be used to eliminate the pressure from the first equation. The result will be a nonlinear ordinary differential equation which the function $f(y)$ must satisfy.

Integrating the last equation with respect to $y$ gives the following expression for the pressure:

$$p(x, y) = -2\rho U^2(f)^2 - 2\rho Uvf' + g(x)$$

where $g(x)$ is some function of $x$ which may be determined by comparison with the potential-flow pressure distribution which should be recovered for large values of $y$. Recalling that $f(y) \to y$ for large values of $y$ shows that, for large values of $y$,

$$p(x, y) \to -2\rho U^2 y^2 - 2\rho Uv + g(x)$$

which, by comparison with the potential-flow pressure, requires that

$$g(x) = p_0 - 2\rho U^2 x^2 + 2\rho U \nu$$

Then the pressure distribution in the viscous fluid will be

$$p(x, y) = p_0 - 2\rho U^2 (f)^2 + 2\rho U \nu (1 - f') - 2\rho U^2 x^2$$

So far we have satisfied the continuity equation and the equation of $y$ momentum. From the above result it follows that $\partial p / \partial x = -4\rho U^2 x$, so that the equation of $x$ momentum becomes

$$4U^2 x (f')^2 - 4U^2 x f f'' = 4U^2 x + 2U \nu x f'''$$

In standard form, with the highest derivative to the left, this equation becomes

$$\frac{\nu}{2U} f''' + f f'' - (f')^2 + 1 = 0$$

The boundary condition $u(x, 0) = 0$ requires that $f'(0) = 0$ while the condition $v(x, 0) = 0$ requires that $f(0) = 0$. In addition, the condition that the potential-flow solution be recovered as $y \to \infty$ requires that $f(y) \to y$, or that $f'(y) \to 1$, as $y \to \infty$. Thus the boundary conditions which accompany the foregoing ordinary differential equation are

$$f(0) = f'(0) = 0$$

$$f'(y) \to 1 \qquad \text{as } y \to \infty$$

That is, the potential-flow solution may be modified to satisfy not only the governing equations but also the viscous boundary condition provided the modifying function $f(y)$ satisfies the foregoing conditions. Clearly, it would be preferable to solve a problem which is free of the parameter $\nu/(2U)$, for then the result will be valid for all kinematic viscosities and all flow velocities. It is possible to render the foregoing problem free from parameters by making the following change of variables. Let

$$\phi(\eta) = \sqrt{\frac{2U}{\nu}} f(y)$$

and

$$\eta = \sqrt{\frac{2U}{\nu}} y$$

Then, in terms of $\phi(\eta)$, the problem to be solved in order to satisfy all the requirements is

$$\phi''' + \phi \phi'' - (\phi')^2 + 1 = 0 \qquad\qquad (7.7a)$$

$$\phi(0) = \phi'(0) = 0 \qquad\qquad (7.7b)$$

$$\phi'(\eta) \to 1 \qquad \text{as } \eta \to \infty \qquad\qquad (7.7c)$$

where the primes denote differentiation with respect to $\eta$. This nonlinear problem must be solved numerically, but this is a much easier task than solving the original system of partial differential equations numerically. For this reason the solution is usually considered to be exact.

To summarize, the velocity and pressure fields in stagnation-point flow are given by

$$u(x, y) = 2Ux\phi' \tag{7.8a}$$

$$v(x, y) = -\sqrt{2U\nu}\,\phi \tag{7.8b}$$

$$p(x, y) = p_0 - \rho U\nu\phi^2 + 2\rho U\nu(1 - \phi') - 2\rho U^2 x^2 \tag{7.8c}$$

where $\phi(\eta)$ is the solution to Eqs. (7.7) and $\eta = \sqrt{2U}\,y/\sqrt{\nu}$. The nature of this solution is shown in Fig. 7.6b in the form of a curve of $\phi'$ as a function of $\eta$. From quantitative plots of this type it is found that the value of $\eta$ for which $\phi' = 0.99$ is about 2.4.

From this qualitative figure and the supplementary quantitative data, it is evident that $\phi$ may be considered to be unity (and hence the potential-flow solution is recovered) when $\eta = 2.4$. Then if $\delta$ denotes the value of $y$ at this edge of the viscous layer, it follows that

$$\sqrt{\frac{2U}{\nu}}\,\delta = 2.4$$

hence
$$\delta = 2.4\sqrt{\frac{\nu}{2U}}$$

That is, viscous effects are confined to a layer adjacent to the boundary, whose thickness varies as the square root of the kinematic viscosity of the fluid and inversely as the square root of the velocity-magnitude parameter.

## 7.8 FLOW IN CONVERGENT AND DIVERGENT CHANNELS

Another flow field in which the continuity equation is not identically satisfied and in which the nonlinear inertia terms are not identically zero is that of flow in a convergent or divergent channel. For such flow fields an exact solution to the governing equations exists in the sense of the previous section—that is, the system of partial differential equations may be reduced to a simple numerical problem.

Figure 7.7a shows the flow configurations for flow in a converging channel and flow in a diverging channel. The preferred coordinate system for such configurations is cylindrical coordinates $R$, $\theta$, and $z$. Then, of the three velocity components only the radial component $u_R$ will be different from zero, and this velocity component will depend upon $R$ and $\theta$ only. Thus the continuity

(a)

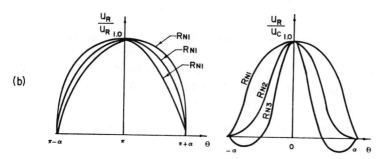

(b)

**FIGURE 7.7**

(*a*) Flow configuration and (*b*) velocity profiles for flow in convergent and divergent channels.

equation and the Navier-Stokes equations become

$$\frac{1}{R}\frac{\partial}{\partial R}(Ru_R) = 0$$

$$u_R\frac{\partial u_R}{\partial R} = -\frac{1}{\rho}\frac{\partial p}{\partial R} + \nu\left[\frac{1}{R}\frac{\partial}{\partial R}\left(R\frac{\partial u_R}{\partial R}\right) - \frac{u_R}{R^2} + \frac{1}{R^2}\frac{\partial^2 u_R}{\partial \theta^2}\right]$$

$$0 = -\frac{1}{\rho R}\frac{\partial p}{\partial \theta} + \nu\left(\frac{2}{R^2}\frac{\partial u_R}{\partial \theta}\right)$$

A separable form of solution will be sought to these equations. That is, a solution for the velocity will be sought in the form

$$u_R(R,\theta) = f(R)F(\theta)$$

Then the continuity equation shows that $Ru_R$ must be a constant, so that $u_R$ must be proportional to $R^{-1}$. Thus the velocity distribution will be of the form

$$u_R(R,\theta) = \frac{\nu}{R}F(\theta) \tag{7.9a}$$

where the kinematic viscosity has been used as a proportionality factor in order to render the function $F(\theta)$ dimensionless.

   Having satisfied the continuity equation, the two components of the Navier-Stokes equations must next be satisfied. This will impose some restrictions on the function $F(\theta)$. Substitution of Eq. (7.9a) into the reduced form of the Navier-Stokes equations shows that the following pair of equations must be satisfied:

$$-\frac{\nu^2}{R^3}(F)^2 = -\frac{1}{\rho}\frac{\partial p}{\partial R} + \frac{\nu^2}{R^3}F''$$

$$0 = -\frac{1}{\rho R}\frac{\partial p}{\partial \theta} + 2\frac{\nu^2}{R^3}F'$$

where the primes denote differentiation with respect to $\theta$. This pair of equations may be reduced to a single equation by forming the second cross derivative of $p$, namely, $\partial^2 p/\partial R\,\partial\theta$, and thus eliminating the pressure between the two equations. The resulting ordinary differential equation for $F(\theta)$ is

$$F''' + 4F' + 2FF' = 0$$

This equation may be immediately integrated once with respect to $\theta$ to give

$$F'' + 4F + (F)^2 = K$$

where $K$ is a constant of integration. In order to further reduce this equation, new dependent and independent variables are introduced. Thus let $G(F) = F'$ be the new dependent variable and $F$ be the new independent variable. Then

$$\frac{dG}{dF} = \frac{d}{dF}(F') = \frac{d\theta}{dF}\frac{d}{d\theta}(F') = \frac{F''}{F'}$$

$$= \frac{F''}{G}$$

Using this result to eliminate $F''$, the differential equation to be satisfied becomes

$$G\frac{dG}{dF} + 4F + (F)^2 = K$$

That is, in terms of $G(F)$ the differential equation is reduced to first order. But this equation may be integrated directly to yield $G$ as follows. Rewriting the above equation in the form

$$\frac{d}{dF}\left(\frac{1}{2}G^2\right) = K - 4F - F^2$$

and integrating with respect to $F$ gives

$$\tfrac{1}{2}G^2 = A + KF - 2F^2 - \tfrac{1}{3}F^3$$

where $A$ is a constant of integration. Solving this equation for $G(F)$ gives

$$G(F) = \frac{dF}{d\theta} = \sqrt{2(A + KF - 2F^2 - \tfrac{1}{3}F^3)}$$

Although this equation cannot be solved to give an explicit expression for $F$ in terms of $\theta$, the result may be put in the form of an integral expression for $\theta$ as a function of $F$. The expression is the following elliptic integral:

$$\theta = \int_0^F \frac{d\xi}{\sqrt{2\left(A + K\xi - 2\xi^2 - \frac{1}{3}\xi^3\right)}} + B \qquad (7.9b)$$

where $\xi$ is a dummy variable of integration and $B$ is a constant of integration. Equation (7.9b) represents a fairly simple numerical problem whose solution, when coupled with Eq. (7.9a), defines the velocity distribution.

The physical boundary conditions which have to be satisfied are

$$u_R(\alpha) = u_R(-\alpha) = 0 \qquad \text{(divergent)}$$

$$u_R(\pi + \alpha) = u_R(\pi - \alpha) = 0 \qquad \text{(convergent)}$$

These boundary conditions represent the no-slip condition at the walls of the channel. In addition, since the velocity profiles will be symmetrical about the reference axis, it follows that

$$\frac{\partial u_R}{\partial \theta}(R, 0) = 0 \qquad \text{(divergent)}$$

$$\frac{\partial u_R}{\partial \theta}(R, \pi) = 0 \qquad \text{(convergent)}$$

Then, from Eq. (7.9a), the conditions which the function $F(\theta)$ must satisfy are

$$F(\alpha) = F(-\alpha) = F'(0) = 0 \qquad \text{(divergent)} \qquad (7.9c)$$

$$F(\pi + \alpha) = F(\pi - \alpha) = F'(\pi) = 0 \qquad \text{(convergent)} \qquad (7.9d)$$

These boundary conditions are sufficient to determine the constants $A$, $B$, and $K$ which appear in Eq. (7.9b). Equations (7.9) describe velocity profiles which have the form indicated in Fig. 7.7b.

In Fig. 7.7b the various curves are identified by the Reynolds number where $R_{N1} > R_{N2} > R_{N3}$. Here the Reynolds number is defined as

$$R_N = \frac{u_c R}{\nu}$$

where $u_c$ is the velocity of the fluid along the centerline of the channel. It may be seen that the nature of the velocity profile in a convergent channel may be quite different from that in a divergent channel, particularly at low Reynolds numbers. The adverse pressure gradient which exists in a divergent channel may overcome the inertia of the fluid near the wall (where viscous effects have reduced the velocity), resulting in a reversed-flow configuration. This separation of the flow is well established experimentally, particularly at large values of the angle $\alpha$.

## 7.9   FLOW OVER A POROUS WALL

The foregoing exact solutions to the equations of viscous flow of an incompressible fluid either were sufficiently simple that the nonlinear inertia terms dropped out or these terms were nonzero and a reduction to a nonlinear ordinary differential equation was possible. The flow field which will be studied here is an example of a case where the nonlinear inertia terms become linearized and a closed form of solution becomes possible.

Figure 7.8 shows a flat surface over which a steady uniform flow exists. Rather than being impervious, the surface is porous and fluid is being drawn off into the porous surface such that the normal component of velocity at the surface is $V$. Porous surfaces of this type with suction beneath them are sometimes used to prevent boundary layers from separating (a topic which will be discussed in Chap. 9). However, it may be stated now that boundary-layer separation on airfoil surfaces can lead to a stalled configuration which destroys the lift generated by the airfoil. Thus it is natural that one of the applications to which boundary-layer suction has been applied is aeronautics.

A solution to the foregoing problem will be sought in which $p = $ constant and $u$ depends upon $y$ only. That is, a solution to the governing equations is being sought in which the magnitude of the suction is adjusted in such a way that the tangential velocity component is independent of $x$. For this situation the continuity and Navier-Stokes equations become

$$\frac{\partial v}{\partial y} = 0$$

$$v\frac{du}{dy} = \nu\frac{d^2u}{dy^2}$$

$$u\frac{\partial v}{\partial x} + v\frac{\partial v}{\partial y} = \nu\left(\frac{\partial^2 v}{\partial x^2} + \frac{\partial^2 v}{\partial y^2}\right)$$

with the boundary conditions

$$u(0) = 0$$

$$v(x,0) = -V$$

$$u(y) \to U \qquad \text{as } y \to \infty$$

Now the continuity equation may be integrated to show that $v(x, y)$ is actually a

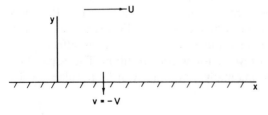

**FIGURE 7.8**
Uniform flow over a plane boundary with suction.

constant, and the boundary condition at $y = 0$ shows that this constant must be $-V$. That is,

$$v = -V$$

With this information the momentum equations reduce further to the following single equation:

$$-V\frac{du}{dy} = \nu\frac{d^2u}{dy^2}$$

It may now be seen that the inertia terms are not retained in a comprehensive form yet they are not zero. Rather, the intermediate case of a linearized form exists in which the convection velocity is $V$ rather than a variable, which $v$ would be in more general cases.

The foregoing ordinary differential equation may be integrated once with respect to $y$ to give

$$\frac{du}{dy} + \frac{V}{\nu}u = A\frac{V}{\nu}$$

where the constant of integration has been chosen as $AV/\nu$. The particular solution to the remaining equation is then $u = A$, so that the complete solution is

$$u(y) = A + Be^{-(V/\nu)y}$$

where $B$ is a constant. The boundary condition $u(0) = 0$ requires that $B = -A$, and the condition $u \rightarrow U$ as $y \rightarrow \infty$ then gives $A = U$. Hence the velocity distribution will be

$$u(y) = U(1 - e^{-(V/\nu)y}) \tag{7.10}$$

Some idea of the thickness of the layer which is affected by viscosity may be obtained by considering the value of $y$ at which $u = U(1 - 1/e^2)$ to be $\delta$. Then, from Eq. (7.10), the value of $\delta$ will be

$$\delta = 2\frac{\nu}{V}$$

That is, the distance away from the surface at which the uniform flow is essentially recovered is proportional to the kinematic viscosity of the fluid and inversely proportional to the suction velocity.

It will be observed that the solution [Eq. (7.10)] diverges for negative values of $V$ (that is, for blowing instead of suction). Some insight into the reason for this may be obtained by studying the vorticity distribution. The equation governing the vorticity is Eq. (III.4). Here we are dealing with steady flow without external body forces. In addition, the flow is two-dimensional, so that the vorticity vector will be $\boldsymbol{\omega} = (0, 0, \xi)$. Then, using the fact that $v = -V$ and

$u = u(y)$, the vorticity equation becomes

$$-V\frac{d\xi}{dy} = \nu\frac{d^2\xi}{dy^2}$$

This equation may be integrated once with respect to $y$ to give

$$-V\xi = \nu\frac{d\xi}{dy}$$

Interpreted physically, the term on the left-hand side of this equation represents the convection of vorticity, which is toward the boundary (negative $y$ direction), and the convection velocity is $V$. The term on the right-hand side represents the diffusion of vorticity in which the diffusion coefficient is the kinematic viscosity of the fluid. Thus the equation states that the convection rate of vorticity toward the wall due to the suction is just balancing the diffusion of vorticity away from the wall. It is this balance which makes possible a solution of the form assumed. However, if blowing instead of suction exists, the convection and the diffusion will both take place in the same direction, so that a solution of the form $u = u(y)$ only no longer exists.

## PROBLEMS

**7.1.** To establish the manner in which Couette flow is established, find the velocity in a fluid which is bounded by two parallel surfaces in which everything is quiescent for $t < 0$ and for which the upper surface is impulsively set into motion along its own plane with constant velocity $U$ at time $t = 0$. (This is readily done by writing the solution in its asymptotic form, corresponding to $t \to \infty$, plus a separable solution.)

**7.2.** For Poiseuille flow through an elliptic pipe of semiaxes $a$ and $b$, find the ratio $b/a$ which gives the maximum flow rate for a given flow area and a given pressure gradient.

For a given pressure gradient, find the ratio of the discharge for an elliptic pipe to that of a circular pipe having the same flow area. Evaluate this ratio for $a/b = \frac{8}{7}$.

**7.3.** Fig. 7.9 shows a conduit whose cross section is the shape of an equilateral triangle. For the coordinate system shown in the figure, the equations of the three sides are

$$z + \frac{b}{2\sqrt{3}} = 0$$

$$z + \sqrt{3}\,y - \frac{b}{\sqrt{3}} = 0$$

$$z - \sqrt{3}\,y - \frac{b}{\sqrt{3}} = 0$$

Following the procedure used in Sec. 7.2, look for a solution for the velocity distribution in such a conduit of the following form

$$u(y, z) = \alpha\left(z + \frac{b}{2\sqrt{3}}\right)\left(z + \sqrt{3}\,y - \frac{b}{\sqrt{3}}\right)\left(z - \sqrt{3}\,y - \frac{b}{\sqrt{3}}\right)$$

Determine the value of the constant $\alpha$ which makes this assumed form of solution exact, the value of the constant being expressed in terms of the applied pressure gradient.

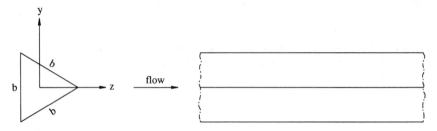

**FIGURE 7.9**
Conduit with cross-section in shape of an equilateral triangle.

**7.4.** Figure 7.10 shows two parallel, vertical surfaces and a horizontal surface. The space defined by these surfaces, $0 < y < b$ for $0 < z$, is filled with a viscous, incompressible fluid. The horizontal surface, $z = 0$, is moving in the positive $x$ direction with constant velocity $U$. The other surfaces are stationary, as is the fluid —other than the motion which is induced by the moving surface.

Derive an expression for the velocity distribution in the $yz$ plane if the flow is steady and if there are no body forces. Also obtain an expression for the volumetric flow rate of the fluid which is induced to flow in the $x$ direction by the moving surface.

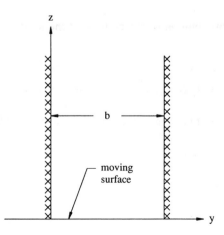

**FIGURE 7.10**
Fluid enclosed by vertical stationary and horizontal moving surfaces.

**7.5.** Two concentric circular cylinders enclose a viscous fluid. If the inner cylinder is at rest and the outer cylinder rotates at a constant angular velocity, calculate the torque required to rotate the outer cylinder and that required to hold the inner cylinder at rest.

**7.6.** Using the solution for flow between concentric rotating cylinders, deduce the velocity distribution created by a circular cylinder which is rotating in an infinite fluid which is otherwise at rest. Compare this result with that for a line vortex of strength $\Gamma = 2\pi R_i^2 \omega_i$ in an inviscid fluid which is at rest at infinity.

**7.7.** Obtain the velocity distribution for the modified Stokes second problem which consists of a fluid which is contained between two infinite parallel surfaces which are separated by a distance $h$. The upper surface is held fixed while the lower surface oscillates with velocity $U \cos nt$.

**7.8.** Equation (7.6) gives the velocity profile between two parallel surfaces due to an oscillating pressure gradient. A Reynolds number for such a flow may be defined by the quantity

$$R_N = \frac{a^2 n}{2\nu}$$

For this situation, consider the two asymptotic limits which are described below.

(a) For $R_N \ll 1$ it might be expected that viscous effects will dominate. Expand the expression for the velocity for this case and obtain an explicit expression for the leading term in this expansion. Interpret the result physically.

(b) For $R_N \gg 1$ it might be expected that viscous effects will be small everywhere except in the vicinity of the walls. Expand the expression for the velocity for this case, and interpret the result which is obtained.

**7.9.** For potential flow due to a line vortex the vorticity is concentrated along the axis of the vortex. Thus the problem to be solved for the decay of a line vortex due to the viscosity of the fluid is the following:

$$\frac{\partial \omega}{\partial t} = \nu \nabla^2 \omega$$

$$\omega(R,0) = \begin{cases} 0 & \text{for } R > 0 \\ \Gamma & \text{for } R = 0 \end{cases}$$

Here $\omega(R, t)$ is the vorticity, and the maximum circulation around the vortex for any time $t \geq 0$ is $\Gamma$. Look for a similarity solution to this problem of the form

$$\omega(R,t) = \frac{\Gamma}{2\pi\nu t} f\left(\frac{R}{\sqrt{\nu t}}\right)$$

Hence obtain expressions for the velocity $u_\theta(R, t)$ and the pressure $p(R, t)$ in the fluid.

**7.10.** The following flow field satisfies the continuity equation everywhere except at $R = 0$, where a singularity exists:

$$u_R = -aR$$

$$u_\theta = \frac{K}{R}$$

$$u_z = 2az$$

Show that this flow field also satisfies the Navier-Stokes equations everywhere except at $R = 0$, and find the pressure distribution in the flow field.

Modify the foregoing expressions to the following:

$$u_R = -aR$$

$$u_\theta = \frac{K}{R}f(R)$$

$$u_z = 2az$$

Determine the function $f(R)$ such that the modified expression satisfies the governing equations for a viscous, incompressible fluid and such that the original flow field is recovered as $R \to \infty$.

# CHAPTER
# 8

# LOW-REYNOLDS-NUMBER SOLUTIONS

For flow problems in which an exact solution is not known, it may be possible to obtain an approximate solution. By an approximate solution we mean an analytic expression which approximately satisfies the governing equations rather than a numerical approximation to these equations. In this chapter the full governing equations will be approximated for flows involving low Reynolds numbers, and some exact solutions to the resulting equations will be established.

The fundamental low-Reynolds-number approximation is the Stokes approximation, and this is the first topic in the chapter. Some fundamental solutions are used to establish more practical solutions. In this way the flow in the vicinity of a rotating sphere is obtained, as is the solution for uniform flow past a sphere. The case of uniform flow past a circular cylinder is examined to illustrate the consequences of the Stokes approximation. Finally, an alternative low-Reynolds-number approximation, the Oseen approximation, is briefly discussed. A detailed study of the Oseen equations is not made, but the nature and utility of the approximation is discussed.

## 8.1 THE STOKES APPROXIMATION

The Stokes equations are a special case of the Navier-Stokes equations corresponding to very slow motion of a viscous fluid. Under these conditions the inertia of the fluid may be neglected in comparison with the other forces which will act on it. Since the Reynolds number may be considered to be the ratio of the inertia forces of the fluid to the viscous forces, the condition of negligible inertia forces amounts to very small Reynolds numbers. The essential feature of the Stokes approximation is that all the convective inertia components are assumed to be small compared with the viscous forces. Then, from Eqs. (III.1) and (III.2), the equations governing the Stokes approximation for a flow field without body forces are

$$\nabla \cdot \mathbf{u} = 0 \tag{8.1a}$$

$$\frac{\partial \mathbf{u}}{\partial t} = -\frac{1}{\rho} \nabla p + \nu \nabla^2 \mathbf{u} \tag{8.1b}$$

Equations (8.1b), which represent three scalar equations, are usually referred to as the *Stokes equations*. These equations, together with the continuity equation [Eq. (8.1a)], represent four scalar equations in the four unknowns $\mathbf{u}$ and $p$. The great simplification in this approximation is that the governing equations are now linear.

The foregoing equations may be extracted from the Navier-Stokes equations in a more formal manner. Since this alternative approach must be used when employing higher-order corrections, it will be outlined here. The first step is to nondimensionalize the dependent and the independent variables as follows. Let

$$\mathbf{u} = U\mathbf{u}^*$$

$$p = \frac{\rho \nu U}{l} p^*$$

$$x_i = l x_i^*$$

$$t = \frac{l^2}{\nu} t^*$$

Here $U$ is a characteristic velocity of the fluid (such as the free-stream velocity) and $l$ is a characteristic length scale (such as a body dimension). The starred quantities are the dimensionless variables where the kinematic viscosity $\nu$ has been used to nondimensionalize the pressure and the time. Experience gained in the previous chapter indicates that the time $l^2/\nu$ corresponds to the time required for viscous diffusion to traverse the distance $l$.

The variables which appear above are now substituted into the Navier-Stokes equations to yield the equations governing the dimensionless variables. The resulting vector equation is

$$\frac{\nu U}{l^2} \frac{\partial \mathbf{u}^*}{\partial t^*} + \frac{U^2}{l} (\mathbf{u}^* \cdot \nabla^*) \mathbf{u}^* = -\frac{\nu U}{l^2} \nabla^* p^* + \frac{\nu U}{l^2} \nabla^{*2} \mathbf{u}^*$$

where the gradient and the laplacian operators are now expressed in terms of the dimensionless space coordinates. Multiplying this equation by $l^2/(\nu U)$ and introducing the Reynolds number $R_N = Ul/\nu$ gives

$$\frac{\partial \mathbf{u}^*}{\partial t^*} + R_N(\mathbf{u}^* \cdot \nabla^*)\mathbf{u}^* = -\nabla^* p^* + \nabla^{*2}\mathbf{u}^*$$

In this form it is evident that the Stokes equations [Eqs. (8.1b)] may be obtained from the Navier-Stokes equations by taking the limit $R_N \to 0$ while holding the coordinates fixed. This suggests that higher-order approximations to the Stokes solution for any given problem could be obtained by expanding the dependent variables in ascending powers of the Reynolds number. The sequence of differential equations which would have to be solved could then be obtained from the above form of the Navier-Stokes equations by a limiting procedure. Thus the Stokes equations may be considered to be an asymptotic limit of the Navier-Stokes equations corresponding to zero Reynolds number, while the space coordinates remain of order unity.

An alternative form of Eqs. (8.1a) and (8.1b) exists which is frequently useful. In this alternative form the velocity and the pressure equations are separated so that the velocity and the pressure fields may be established separately. To obtain the equation governing the velocity field, the curl of the curl of Eq. (8.1b) is taken. Having done this, the identities $\nabla \times (\nabla \times \mathbf{u}) = \nabla(\nabla \cdot \mathbf{u}) - \nabla^2 \mathbf{u}$ and $\nabla \times \nabla p = 0$ are employed. The resulting equation is

$$\frac{\partial}{\partial t}\left[\nabla(\nabla \cdot \mathbf{u}) - \nabla^2 \mathbf{u}\right] = \nu \nabla^2 \left[\nabla(\nabla \cdot \mathbf{u}) - \nabla^2 \mathbf{u}\right]$$

Finally, using the continuity equation, the pressure-free form of the momentum equation becomes

$$\nabla^2 \frac{\partial \mathbf{u}}{\partial t} = \nu \nabla^4 \mathbf{u} \tag{8.2a}$$

To obtain the equation governing the pressure, the divergence of Eq. (8.1b) is taken. This gives

$$\nabla^2 p = 0 \tag{8.2b}$$

The advantage in the above formulation is that the pressure field has been separated, mathematically, from the velocity field. However, the price we pay is that the highest differentials are now fourth-order instead of second-order.

Solutions to the Stokes equations may be obtained in either of two basic ways. Using the governing equations and the appropriate boundary conditions, the boundary-value problem for each geometry of interest may be solved. Alternatively, basic solutions to the governing equations may be established and superimposed to obtain other solutions. This is the procedure which was used in Chaps. 4 and 5, and it will be used again here. The principal value of this approach is that it leads to a clear understanding of which elements of a mathematical solution are responsible for producing forces and torques.

## 8.2  UNIFORM FLOW

The simplest solution to the Stokes equations is that for a uniform flow. It may be simply observed that for a constant velocity vector and a constant pressure, Eqs. (8.1a) and (8.1b) are identically satisfied. That is, for any constant $U$, the following velocity and pressure fields satisfy the Stokes approximation to the Navier-Stokes equations:

$$\mathbf{u} = U\mathbf{e}_x \qquad\qquad (8.3a)$$

$$p = \text{constant} \qquad\qquad (8.3b)$$

where $\mathbf{e}_x$ is the unit vector in the $x$ direction, which is the reference direction. Clearly this velocity and pressure distribution does not create a force or turning moment on the system. Some of the other fundamental solutions which will be considered later correspond to point singularities, and some of these singularities correspond to point forces or turning moments acting on the fluid.

## 8.3  DOUBLET

It was pointed out in Sec. 7.7 that any potential flow is an exact solution of the full Navier-Stokes equations, since the viscous term is identically zero for potential flows. Then, for any steady potential flow, the Stokes equations will be satisfied provided the pressure term is zero. That is, for steady flow, all the inertia terms are zero to the Stokes approximation, and for potential flows the viscous term will be zero. Hence, potential flows are also solutions to the Stokes equations provided $\nabla p = 0$ or $p = \text{constant}$.

In the study of the flow of ideal fluids it could be shown, through Kelvin's theorem, that an initially irrotational flow would remain irrotational irrespective of the shape of any bodies it may flow around. Here, we consider the nature of the velocity field assumed by an irrotational motion if it exists. The solution to any real flow problem may contain such a component in its solution in addition to other fundamental solutions, some of which may correspond to rotational motion. It should be noted that the velocity field corresponding to viscous irrotational motion is not related to the pressure field in the manner which existed for ideal-fluid flows. The conditions which were stipulated in deriving the Bernoulli equation are violated for viscous flows, so that the pressure and the velocity are no longer connected by such a relationship. Indeed, it was shown above that any velocity distribution which was irrotational had to be accompanied by a constant pressure in order to satisfy the Stokes equations.

If an irrotational flow field exists, the velocity will be derivable from a velocity potential, and from the continuity equation, the velocity potential must satisfy Laplace's equation. Then, for three-dimensional potential flows the mathematical problem is the same as that of Chap. 5. That is, if we are interested in axisymmetric flow fields, we may use the coordinate system defined in Fig. 5.1 and the solutions to Eq. (5.1) which have already been established.

But the solution which was identified as a doublet [Eq. (5.7a)] was of the functional form.

$$\phi(r, \theta) = A\frac{\cos \theta}{r^2}$$

$$= A\frac{x}{r^3}$$

where the fact that $x = r \cos \theta$ has been used. Then, since $\mathbf{u} = \nabla\phi$, the velocity vector will have a component along the $x$ axis and a radial component, giving

$$\mathbf{u} = A\left(\frac{\mathbf{e}_x}{r^3} - \frac{3x\mathbf{e}_r}{r^4}\right) \tag{8.4a}$$

where $\mathbf{e}_x$ and $\mathbf{e}_r$ are unit vectors in the $x$ direction and in the radial direction, respectively. This formulation of expressing variables in terms of streamwise and radial components will be found to be useful in the present application.

The velocity field described by Eq. (8.4a) cannot be proved to be valid from upstream irrotational conditions but is presented here only as a possible form for a viscous fluid. Then, in order for this flow field to satisfy the present form of the momentum equations, the pressure distribution must be

$$p = \text{constant} \tag{8.4b}$$

Although the solution defined by Eqs. (8.4a) and (8.4b) represents the flow field generated by a singularity which is located at the origin, this singularity does not exert a force or a moment on the surrounding fluid. This may be simply argued from the fact that the pressure is constant and the flow configuration is such that there is no net momentum flux acting on the fluid.

## 8.4 ROTLET

In this section a solution to the Stokes equations will be sought in which the vorticity is different from zero and the pressure is constant. That is, a rotational-flow solution will be sought. The resulting solution will involve a singularity at the origin which is known as a rotlet.

Consider steady flow fields of the form

$$\mathbf{u} = \mathbf{r} \times \nabla\chi$$

where $\mathbf{r}$ is the position vector. In tensor notation, this expression becomes

$$u_i = \varepsilon_{ijk} x_j \frac{\partial\chi}{\partial x_k}$$

Then, the divergence of this velocity vector will be given by

$$\frac{\partial u_i}{\partial x_i} = \varepsilon_{ijk}\left(\frac{\partial x_j}{\partial x_i}\frac{\partial\chi}{\partial x_k} + x_j\frac{\partial^2\chi}{\partial x_i \partial x_k}\right)$$

The first term on the right-hand side of this equation is zero, since $\partial x_j/\partial x_i$ is zero unless $i = j$ and $\varepsilon_{ijk}$ is zero when $i = j$. That is, the product of the pseudoscalar $\varepsilon_{ijk}$ and the symmetric tensor $\partial x_j / \partial x_i$ is zero. Likewise, the second term inside the parentheses in the above equation is a symmetric tensor, so that the product of this quantity with the pseudoscalar $\varepsilon_{ijk}$ will be zero. That is, the continuity equation will be satisfied identically for all forms of the scalar $\chi$.

Since the flow is assumed to be steady and since the pressure has been taken to be constant, the Stokes equations reduce to

$$\nabla^2 \mathbf{u} = 0$$

But, for the form of the velocity vector introduced above it follows that

$$\nabla^2 u_i = \frac{\partial^2 u_i}{\partial x_l \partial x_l}$$

$$= \varepsilon_{ijk} \left( \frac{\partial^2 x_j}{\partial x_l \partial x_l} \frac{\partial \chi}{\partial x_k} + x_j \frac{\partial}{\partial x_k} \frac{\partial^2 \chi}{\partial x_l \partial x_l} \right)$$

The first term on the right-hand side of this equation is zero, since $\partial^2 x_j / (\partial x_l \partial x_l) = 0$ for all $j$ and $l$. Also, the second term on the right-hand side will be zero if $\nabla^2 \chi = 0$. That is, the velocity distribution $\mathbf{u} = \mathbf{r} \times \nabla \chi$ will be a valid solution to the Stokes equations for a constant pressure distribution provided $\nabla^2 \chi = 0$. The problem, therefore, again reduces to that of obtaining axisymmetric solutions to the three-dimensional Laplace's equation.

From Chap. 5, the first separable solution which was obtained corresponded to a source and was of the form $\chi \sim 1/r$. This gives $\nabla \chi \sim \mathbf{e}_r$ so that $\mathbf{u} = 0$, and hence that particular solution is of no interest in this case. The next solution, corresponding to a doublet, was of the form

$$\chi = B \frac{\cos \theta}{r^2}$$

$$= B \frac{x}{r^3}$$

The velocity field corresponding to this solution will be

$$\mathbf{u} = B \mathbf{r} \times \nabla \left( \frac{x}{r^3} \right)$$

$$= B \mathbf{r} \times \left( \frac{\mathbf{e}_x}{r^3} - 3 \frac{x \mathbf{e}_r}{r^4} \right)$$

Since $\mathbf{r} = r \mathbf{e}_r$ and since $\mathbf{e}_r \times \mathbf{e}_r = 0$, this velocity distribution may be represented by the simplified expression

$$\mathbf{u} = B \frac{\mathbf{e}_r \times \mathbf{e}_x}{r^2} \tag{8.5a}$$

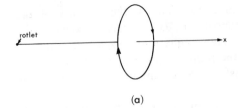

(a)

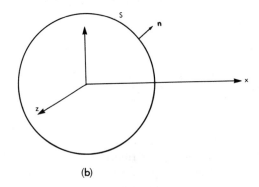

(b)

**FIGURE 8.1**
(a) Typical streamline due to rotlet and (b) spherical control surface surrounding rotlet.

while the corresponding pressure distribution is

$$p = \text{constant} \qquad (8.5b)$$

The streamlines corresponding to Eq. (8.5a) must be perpendicular to both $\mathbf{e}_r$ and $\mathbf{e}_x$. That is, the streamlines are circles whose centers lie on the $x$ axis. A typical streamline is shown in Fig. 8.1a, in which the direction is shown for $B > 0$. It is the nature of the streamlines which suggests the name rotlet for the singularity which exists in the solution (8.5a) at $r = 0$.

The rotlet does not exert a force on the fluid, but it does exert a turning moment on it. This may be verified by constructing a spherical control surface around the rotlet as shown in Fig. 8.1b. Then, if $\mathbf{F}$ is the force acting on the fluid which is contained within the control surface $S$ and if $\mathbf{n}$ is the unit outward normal to $S$, it follows that

$$F_i = -\int_S \sigma_{ij} n_j \, dS$$

where $\sigma_{ij}$ is the stress tensor. But, for an incompressible, newtonian fluid, Eq. (1.7) shows that the stress tensor may be expressed by

$$\sigma_{ij} = -p\delta_{ij} + \mu\left(\frac{\partial u_i}{\partial x_j} + \frac{\partial u_j}{\partial x_i}\right)$$

Using this result, the order of magnitude of the force $F_i$ may be evaluated as follows. Substituting for $\sigma_{ij}$ gives

$$F_i = -\int_S \left[ -p\delta_{ij} + \mu \left( \frac{\partial u_i}{\partial x_j} + \frac{\partial u_j}{\partial x_i} \right) \right] n_j \, dS$$

Since $p$ = constant here, the first component of the above integral will be zero. Also, Eq. (8.5a) shows that $u_i \sim r^{-2}$, and it is known that $dS \sim r^2$ for an element of surface of a sphere. Hence

$$F_i \sim \frac{1}{r^3} r^2 = \frac{1}{r}$$

Then by considering the control surface $S$ to be of very large radius, it is clear that $F_i = 0$. That is, there is no force acting on the fluid due to the rotlet.

The torque, or turning moment **M**, exerted on the fluid by the singularity, may be calculated as follows:

$$\mathbf{M} = \int_S \mathbf{r} \times \mathbf{P} \, dS$$

where **P** is the surface-force vector which results from the existence of the stress tensor $\sigma_{ij}$. In tensor notation, this expression becomes

$$M_i = \int_S \varepsilon_{ijk} x_j P_k \, dS$$

Using the relation $P_k = \sigma_{kl} n_l$ and the expression for $\sigma_{kl}$ which was used above, the value of the turning moment becomes

$$M_i = \int_S \varepsilon_{ijk} x_j \left[ -p\delta_{kl} + \mu \left( \frac{\partial u_k}{\partial x_l} + \frac{\partial u_l}{\partial x_k} \right) \right] n_l \, dS$$

Noting that, since $p$ = constant, the first component of this integral is zero and using the fact that $n_l = x_l/r$ for the spherical control surface of radius $r$, the expression for the turning moment on the fluid becomes

$$M_i = \frac{\mu}{r} \int_S \varepsilon_{ijk} x_j x_l \left( \frac{\partial u_k}{\partial x_l} + \frac{\partial u_l}{\partial x_k} \right) dS$$

The above expression is valid for any velocity distribution whatsoever. In this particular case, the velocity (8.5a) is a homogeneous function of degree 2. A homogeneous function of order $m$ is one which satisfies the condition

$$f\left( \frac{x}{\lambda}, \frac{y}{\lambda}, \frac{z}{\lambda} \right) = \lambda^m f(x, y, z)$$

For such functions, Euler's theorem states that

$$x \frac{\partial f}{\partial x} + y \frac{\partial f}{\partial y} + z \frac{\partial f}{\partial z} = -mf$$

Then, for the velocity distribution under consideration, which is given by Eq. (8.5a) and which is homogeneous of degree 2, it follows that

$$x_l \frac{\partial u_k}{\partial x_l} = -2u_k$$

This identifies one term which appears in the integrand of the expression for the moment $M_i$. Another term which appears in the integrand may be evaluated as follows:

$$x_l \frac{\partial u_l}{\partial x_k} = \frac{\partial}{\partial x_k}(x_l u_l) - u_l \frac{\partial x_l}{\partial x_k}$$

$$= -u_k$$

Here, it is noted that the first term on the right-hand side of this identity is zero, since $\partial(x_l u_l)/\partial x_k = \nabla(\mathbf{r} \cdot \mathbf{u})$ and the velocity vector $\mathbf{u}$ is perpendicular to the position vector $\mathbf{r}$, as may be seen from Eq. (8.5a). Also, $\partial x_l/\partial x_k = \delta_{lk}$, so that $u_l \partial x_l/\partial x_k = u_l \delta_{lk} = u_k$. Then the expression for the moment exerted on the body of fluid by the rotlet at the origin becomes

$$M_i = \frac{\mu}{r}\int_s \varepsilon_{ijk} x_j(-2u_k - u_k)\, dS$$

Or, in vector form, this expression is

$$\mathbf{M} = -3\frac{\mu}{r}\int_s \mathbf{r} \times \mathbf{u}\, dS$$

But, using Eq. (8.5a),

$$\mathbf{r} \times \mathbf{u} = \mathbf{r} \times B\frac{\mathbf{e}_r \times \mathbf{e}_x}{r^2}$$

$$= \frac{B}{r^2}(\mathbf{r} \cdot \mathbf{e}_x)\mathbf{e}_r - \frac{B}{r^2}(\mathbf{r} \cdot \mathbf{e}_r)\mathbf{e}_x$$

$$= B\frac{x}{r^2}\mathbf{e}_r - \frac{B}{r}\mathbf{e}_x$$

Thus the expression for the moment $\mathbf{M}$ becomes

$$\mathbf{M} = -3B\mu\int_s \left(\frac{x}{r}\mathbf{e}_r - \mathbf{e}_x\right)\frac{dS}{r^2}$$

This integral may now be evaluated explicitly using the following relations, which are obtained from Appendix A:

$$x = r \cos\theta$$

$$\mathbf{e}_r = \cos\theta\,\mathbf{e}_x + \sin\theta\cos\omega\,\mathbf{e}_y + \sin\theta\sin\omega\,\mathbf{e}_z$$

$$dS = r^2 \sin\theta\, d\theta\, d\omega$$

Using these results, the expression for **M** becomes

$$\mathbf{M} = -3B\mu \int_0^{2\pi} d\omega \int_0^{\pi} \left[ (\cos^2 \theta - 1)\mathbf{e}_x + \sin \theta \cos \theta \cos \omega \mathbf{e}_y \right.$$
$$\left. + \sin \theta \cos \theta \sin \omega \mathbf{e}_z \right] \sin \theta \, d\theta \quad (8.5c)$$

$$\mathbf{M} = 8\pi B\mu \mathbf{e}_x$$

That is, Eqs. (8.5a) and (8.5b) represent a valid solution to the Stokes equations, and they correspond to a singularity at the origin called a rotlet. This singularity exerts no force on the surrounding fluid, but it does exert a turning moment on it. The magnitude of this turning moment is proportional to the velocity-magnitude parameter, and it acts, according to the right-hand-screw rule, in the $x$ direction in the same algebraic sense as the velocity-magnitude parameter.

## 8.5  STOKESLET

So far, all our fundamental solutions to the Stokes equations have corresponded to a constant pressure. In general situations it is to be expected that the pressure distribution will not be constant, so that another fundamental solution will be sought, and this time the pressure will be assumed to be different from a constant value. Then the pressure must be a nontrivial solution to Eq. (8.2b), which is Laplace's equation. Having so determined the pressure, the corresponding velocity distribution will be obtained from Eq. (8.1b).

Since the pressure $p$ satisfies the three-dimensional Laplace equation, the possible fundamental solutions may be written down immediately from Sec. 5.3. The source type of solution, in which $p \sim 1/r$, turns out to be of no special interest. The next highest form of separable solution, which is the doublet type of solution, is $p \sim \cos \theta / r^2$. For reasons which will become apparent shortly, the constant of proportionality will be taken as $2c\mu$, so that the pressure field which is being considered is

$$p = 2c\mu \frac{x}{r^3}$$

where $\cos \theta = x/r$ has been used. Then, if the flow is assumed to be steady, the equation to be satisfied by the velocity is, from Eq. (8.1b),

$$\nabla^2 \mathbf{u} = \frac{1}{\mu} \nabla p$$

Of the three scalar equations represented by this vector equation, the equation for the velocity component $u$ is the most complicated, because of the nature of the pressure distribution; so this equation will be dealt with last. From the expression for $p$, the equation to be satisfied by the velocity component $v$ is

$$\nabla^2 v = -6c \frac{xy}{r^5}$$

The particular integral to this nonhomogeneous partial differential equation, which is $v = cxy/r^3$, is readily obtained if one is familiar with the properties of harmonic functions of different degrees. Alternatively, this result may be deduced from the following identities:

$$\nabla^2\left(\frac{1}{r^n}\right) = \frac{n(n-1)}{r^{n+2}}$$

$$\nabla^2(\phi\psi) = \psi\nabla^2\phi + \phi\nabla^2\psi + 2\nabla\phi \cdot \nabla\psi$$

Here, $r^2 = x^2 + y^2 + z^2$ and $\phi, \psi$ are any two functions. Then if $\phi = 1/r^3$ and $\psi = xy$, it is readily verified that

$$\nabla^2\phi = \frac{6}{r^5}$$

$$\nabla\phi = -\frac{3}{r^4}\mathbf{e}_r$$

$$\nabla^2\psi = 0$$

$$\nabla\psi = y\mathbf{e}_x + x\mathbf{e}_y$$

$$\therefore \quad \nabla\phi \cdot \nabla\psi = -6\frac{xy}{r^5}$$

where the first result follows from the first of the above identities. Then, using the second of the above identities,

$$\nabla^2(\phi\psi) = \nabla^2\left(\frac{xy}{r^3}\right) = -6\frac{xy}{r^5}$$

That is, the particular solution to the equation for the velocity component $v$ is

$$v = c\frac{xy}{r^3}$$

The equation to be satisfied by the $w$ component of the velocity vectors is

$$\nabla^2 w = -6c\frac{xz}{r^5}$$

The particular solution to this equation is obtained in exactly the same way as that for $v$ and may be deduced to be

$$w = c\frac{xz}{r^3}$$

The equation to be satisfied by the $u$ component of the velocity vector is

$$\nabla^2 u = c\left(\frac{2}{r^3} - 6\frac{x^2}{r^5}\right)$$

In view of the solutions for $v$ and $w$, it might be expected that the solution to this equation is $u = cx^2/r^3$. This is indeed the case, as may be confirmed by setting $\phi = 1/r^3$ and $\psi = x^2$ and employing the identities mentioned above. Thus

$$\nabla^2\phi = \frac{6}{r^5}$$

$$\nabla\phi = -\frac{3}{r^4}\mathbf{e}_r$$

$$\nabla^2\psi = 2$$

$$\nabla\psi = 2x\mathbf{e}_x$$

$$\therefore \quad \nabla\phi \cdot \nabla\psi = -6\frac{x^2}{r^5}$$

hence

$$\nabla^2(\phi\psi) = \nabla^2\left(\frac{x^2}{r^3}\right) = \frac{2}{r^3} - 6\frac{x^2}{r^5}$$

Thus the particular integral to the equation for $u$ is

$$u = c\frac{x^2}{r^3}$$

To each of the foregoing particular integrals may be added solutions to the homogeneous equations $\nabla^2 u = 0$, $\nabla^2 v = 0$, and $\nabla^2 w = 0$. Denoting these solutions by $u'$, $v'$, and $w'$, respectively, the complete solution for the velocity $\mathbf{u}$ corresponding to the pressure distribution $p = 2c\mu x/r^3$ is

$$\mathbf{u} = c\left(\frac{x^2}{r^3}\mathbf{e}_x + \frac{xy}{r^3}\mathbf{e}_y + \frac{xz}{r^3}\mathbf{e}_z\right) + \mathbf{u}'$$

$$= c\frac{x}{r^2}\mathbf{e}_r + \mathbf{u}'$$

where $\mathbf{u}' = (u', v', w')$. The quantity $\mathbf{u}'$, which must satisfy the equation $\nabla^2\mathbf{u}' = 0$, will now be determined such that the continuity equation is satisfied. Taking the divergence of the velocity $\mathbf{u}$ shows that

$$\nabla \cdot \mathbf{u} = c\nabla \cdot \left(\frac{x}{r^3}\mathbf{r}\right) + \nabla \cdot \mathbf{u}'$$

but

$$\nabla \cdot \left(\frac{x}{r^3}\mathbf{r}\right) = \mathbf{r} \cdot \nabla\left(\frac{x}{r^3}\right) + \frac{x}{r^3}(\nabla \cdot \mathbf{r})$$

$$= \left[x\left(\frac{1}{r^3} - \frac{3x^2}{r^5}\right) - y\frac{3xy}{r^5} - z\frac{3xz}{r^5}\right] + 3\frac{x}{r^3}$$

$$= \frac{x}{r^3}$$

where the fact that $\nabla \cdot \mathbf{r} = 3$ has been used. Hence

$$\nabla \cdot \mathbf{u} = c \frac{x}{r^3} + \nabla \cdot \mathbf{u}'$$

Thus by choosing $\mathbf{u}' = c(\mathbf{e}_x/r)$, the continuity equation will be satisfied. It will be noted that this form of $\mathbf{u}'$ also satisfies the homogeneous equation $\nabla^2 \mathbf{u}' = 0$. Thus a valid velocity distribution has been found which satisfies the Stokes equations corresponding to a doublet type of pressure field. This solution is

$$p = 2c\mu \frac{x}{r^3} \qquad (8.6a)$$

$$\mathbf{u} = c\left( \frac{x}{r^3} \mathbf{e}_r + \frac{1}{r} \mathbf{e}_x \right) \qquad (8.6b)$$

The solution represented by the above equations has a singularity at the origin, and this singularity is known as a stokeslet. Although the stokeslet does not exert a torque on the surrounding fluid, it does exert a force on it. The magnitude of this force may be established as follows: In tensor notation, the force $F_i$ acting on the fluid will be

$$F_i = -\int_S \sigma_{ij} n_j \, dS$$

where, for a newtonian fluid,

$$\sigma_{ij} = -p \, \delta_{ij} + \mu \left( \frac{\partial u_i}{\partial x_j} + \frac{\partial u_j}{\partial x_i} \right)$$

and for this particular flow field the pressure is given by Eq. (8.6a). Hence the expression for the force $F_i$ becomes

$$F_i = -\int_S \left[ -2c\mu \frac{x}{r^3} \delta_{ij} + \mu \left( \frac{\partial u_i}{\partial x_j} + \frac{\partial u_j}{\partial x_i} \right) \right] n_j \, dS$$

Now if the surface $S$ is considered to be a sphere of radius $r$, then $n_j = x_j/r$, so that the expression for $F_i$ becomes

$$F_i = -\int_S \left[ -2c\mu \frac{x}{r^3} \frac{x_i}{r} - \frac{\mu}{r} x_j \left( \frac{\partial u_i}{\partial x_j} + \frac{\partial u_j}{\partial x_i} \right) \right] dS$$

But in this particular case the velocity distribution, which is given by Eq. (8.6b), is homogeneous of degree 1. Hence, from Euler's theorem, it follows that

$$x_j \frac{\partial u_i}{\partial x_j} = -u_i$$

$$= -c\left( \frac{x}{r^3} x_i + \frac{\delta_{i1}}{r} \right)$$

A second quantity which appears in the integrand of the foregoing integral may be evaluated as follows:

$$x_j \frac{\partial u_j}{\partial x_i} = \frac{\partial}{\partial x_i}(x_j u_j) - u_j \frac{\partial x_i}{\partial x_i}$$

$$= \frac{\partial}{\partial x_i}(\mathbf{r} \cdot \mathbf{u}) - u_i$$

where the fact that $u_j \, \partial x_j / \partial x_i = u_j \, \delta_{ij} = u_i$ has been used. Hence, using Eq. (8.6b),

$$x_j \frac{\partial u_j}{\partial x_i} = \frac{\partial}{\partial x_i}\left(2c\frac{x}{r}\right) - c\left(\frac{xx_i}{r^3} + \frac{\delta_{i1}}{r}\right)$$

$$= c\left(\frac{2}{r}\delta_{i1} - 2\frac{xx_i}{r^3}\right) - c\left(\frac{xx_i}{r^3} + \frac{\delta_{i1}}{r}\right)$$

$$= c\left(\frac{\delta_{i1}}{r} - 3\frac{xx_i}{r^3}\right)$$

Using these results, the expression for the force acting on the fluid due to the stokeslet becomes

$$F_i = -\int_s \left[-2c\mu\frac{xx_i}{r^4} - c\frac{\mu}{r}\left(\frac{xx_i}{r^3} + \frac{\delta_{i1}}{r}\right) + c\frac{\mu}{r}\left(\frac{\delta_{i1}}{r} - 3\frac{xx_i}{r^3}\right)\right] dS$$

$$= 6c\mu\int_s \frac{xx_i}{r^4} \, dS$$

Alternatively, in vector notation, this expression becomes

$$\mathbf{F} = 6c\mu\int_s \frac{x}{r^3}\mathbf{e}_r \, dS$$

This integral may be evaluated explicitly using the following relations:

$$x = r\cos\theta$$
$$\mathbf{e}_r = \cos\theta\,\mathbf{e}_x + \sin\theta\cos\omega\,\mathbf{e}_y + \sin\theta\sin\omega\,\mathbf{e}_z$$
$$dS = r^2\sin\theta \, d\theta \, d\omega$$

Thus the force acting on the fluid due to the stokeslet is

$$\mathbf{F} = 6c\mu\int_0^{2\pi} d\omega \int_0^{\pi} \cos\theta(\cos\theta\,\mathbf{e}_x + \sin\theta\cos\omega\,\mathbf{e}_y + \sin\theta\sin\omega\,\mathbf{e}_z)\sin\theta \, d\theta$$

$$\mathbf{F} = 8\pi c\mu\,\mathbf{e}_x \tag{8.6c}$$

That is, the stokeslet exerts a force on the surrounding fluid whose strength is proportional to the pressure parameter $c$ and whose direction is in the positive $x$ direction for $c > 0$.

## 8.6 ROTATING SPHERE IN A FLUID

The foregoing fundamental solutions are sufficient to establish more practical solutions for low-Reynolds-number flows. One of these solutions corresponds to a solid sphere which is rotating in an otherwise quiescent fluid. Consider such a sphere to be rotating about the $x$ axis with constant angular velocity $\Omega$. The nature of the resulting flow field may be expected to be similar to that of a rotlet. Hence let the velocity distribution correspond to Eq. (8.5a) and see if the boundary conditions may be satisfied. Then

$$\mathbf{u} = B\frac{\mathbf{e}_r \times \mathbf{e}_x}{r^2}$$

This velocity distribution gives a finite velocity as $r \to \infty$ as required. The other boundary condition is, on $r = a$, $\mathbf{u} = \Omega a \mathbf{e}_r \times \mathbf{e}_x$. This condition is satisfied for $B = \Omega a^3$, so that the required velocity distribution is

$$\mathbf{u} = \Omega\frac{a^3}{r^2}\mathbf{e}_r \times \mathbf{e}_x \tag{8.7a}$$

Since the rotlet is located at $r = 0$ and since the surface of the sphere is $r = a$, there is no singularity in the fluid.

Since the rotlet was found to exert a moment on the surrounding fluid, the surrounding fluid will exert a moment on the surface $r = a$ in this case. The magnitude of this moment is given by Eq. (8.5c). Hence, using $\mathbf{B} = \Omega a^3$, the moment acting *on the sphere* will be

$$\mathbf{M} = -8\pi\mu\Omega a^3 \mathbf{e}_x \tag{8.7b}$$

This moment acts in a direction which opposes the motion of the sphere, as might be expected.

## 8.7 UNIFORM FLOW PAST A SPHERE

The solution corresponding to uniform flow past a sphere may be obtained by superimposing the solutions for a uniform flow, a doublet, and a stokeslet. Hence from Eqs. (8.3a), (8.3b), (8.4a), (8.4b), (8.6a), and (8.6b), the assumed forms of the velocity and pressure fields are

$$\mathbf{u} = U\mathbf{e}_x + A\left(\frac{\mathbf{e}_x}{r^3} - 3\frac{x\mathbf{e}_r}{r^4}\right) + c\left(\frac{x}{r^2}\mathbf{e}_r + \frac{\mathbf{e}_x}{r}\right)$$

$$p = 2c\mu\frac{x}{r^3}$$

Far from the origin this velocity field reduces to that of a uniform flow as required. The simplest way of imposing the near boundary condition is to observe that at the rear stagnation point, $\mathbf{u} = 0$. Hence substituting $x = r = a$

and setting $\mathbf{u} = 0$ in the foregoing expression for the velocity gives

$$0 = U\mathbf{e}_x + A\left(\frac{\mathbf{e}_x}{a^3} - 3\frac{\mathbf{e}_r}{a^3}\right) + c\left(\frac{\mathbf{e}_r}{a} + \frac{\mathbf{e}_x}{a}\right)$$

Setting the coefficients of $\mathbf{e}_x$ and $\mathbf{e}_r$ equal to zero separately yields the following pair of equations:

$$0 = U + \frac{A}{a^3} + \frac{c}{a}$$

$$0 = -3\frac{A}{a^3} + \frac{c}{a}$$

The solution to these algebraic equations is

$$A = -U\frac{a^3}{4}$$

$$c = -\tfrac{3}{4}Ua$$

Thus the velocity and pressure distributions are

$$\mathbf{u} = U\left[\mathbf{e}_x - \frac{1}{4}\frac{a}{r}\left(\frac{a^2}{r^2} + 3\right)\mathbf{e}_x + \frac{3}{4}\frac{ax}{r^2}\left(\frac{a^2}{r^2} - 1\right)\mathbf{e}_r\right] \qquad (8.8a)$$

$$p = -\tfrac{3}{2}uU\frac{ax}{r^3} \qquad (8.8b)$$

In this form it is readily confirmed that Eq. (8.8a) satisfies the boundary condition $\mathbf{u} = 0$ over the entire surface $r = a$.

Neither the uniform flow nor the doublet exerts a force on the fluid. However, the stokeslet exerts a force on the surrounding fluid, and the magnitude of this force is given by Eq. (8.6c). Then, since the stokeslet is inside the spherical surface $r = a$, the surrounding fluid will exert an equal but opposite force on the sphere. Thus from Eq. (8.6c) and using the fact that $c = -3Ua/4$, the magnitude of the force acting on the sphere will be

$$\mathbf{F} = 6\pi\mu Ua\mathbf{e}_x \qquad (8.8c)$$

This is the famous *Stokes' drag law* for a sphere in a uniform flow, and it is valid for low Reynolds numbers. Since the direction of this force is clearly in the direction of the uniform flow, this result is frequently quoted in terms of the dimensionless drag coefficients, which involves the scalar magnitude of the force only. This drag coefficient is defined as follows:

$$C_D = \frac{|\mathbf{F}|/A}{\tfrac{1}{2}\rho U^2}$$

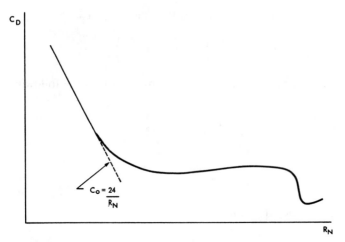

**FIGURE 8.2**
Drag coefficient as a function of Reynolds number for a sphere.

where $A = \pi a^2$ is the frontal area of the sphere. Thus, using Eq. (8.8c),

$$C_D = \frac{24}{\mathbf{R}_N} \qquad (8.8d)$$

where $\mathbf{R}_N = \rho U 2a/\mu$ is the Reynolds number of the flow. This result is shown in Fig. 8.2, which shows the form of the drag coefficient as a function of the Reynolds number for a sphere. Out of the entire range of Reynolds number, Eq. (8.8d) is the only closed-form analytic solution which exists. It is valid for low Reynolds numbers, for which the viscous forces dominate the inertia forces, but it is found experimentally that the result is valid for Reynolds numbers which are less than unity.

## 8.8  UNIFORM FLOW PAST A CIRCULAR CYLINDER

It will be shown in this section that the solution obtained above for a sphere in three dimensions has no counterpart in two dimensions. This will be shown by attempting to solve the Stokes equations for uniform flow past a circular cylinder. Since the fundamental solutions have not been established for two dimensions, the alternative approach of solving the boundary-value problem for the stream function will be adopted.

For steady flow, the Stokes equations are

$$0 = -\frac{1}{\rho}\nabla p + \nu \nabla^2 \mathbf{u}$$

Taking the curl of this equation gives the following equation for the vorticity to the Stokes approximation:

$$\nabla^2 \omega = 0$$

But in two dimensions the only nonzero component of the vorticity vector is $\zeta$, which is the vorticity in the $z$ direction. Hence

$$\nabla^2 \zeta = 0$$

but

$$\zeta = \frac{\partial v}{\partial x} - \frac{\partial u}{\partial y}$$

$$= -\left(\frac{\partial^2 \psi}{\partial x^2} + \frac{\partial^2 \psi}{\partial y^2}\right)$$

where the stream function which satisfies the continuity equation has been introduced. Thus the equation to be satisfied by the vorticity is

$$\left(\frac{\partial^2}{\partial x^2} + \frac{\partial^2}{\partial y^2}\right)^2 \psi = 0$$

This is the biharmonic equation, and in cylindrical coordinates $(R, \theta)$ it becomes

$$\left(\frac{\partial^2}{\partial R^2} + \frac{1}{R}\frac{\partial}{\partial R} + \frac{1}{R^2}\frac{\partial^2}{\partial \theta^2}\right)^2 \psi = 0$$

Noting that the stream function for a uniform flow is $\psi = Uy = UR \sin \theta$, we look for a solution to the above equation of the form

$$\psi(R, \theta) = f(R) \sin \theta$$

where $f(R) \to UR$ as $R \to \infty$. Substituting this form of solution into the biharmonic equation gives

$$\left(\frac{d^2}{dR^2} + \frac{1}{R}\frac{d}{dR} - \frac{1}{R^2}\right)^2 f = 0$$

This is an equidimensional equation whose solution is of the form

$$f(R) = AR^3 + BR \log R + CR + \frac{D}{R}$$

Thus the stream function is of the form

$$\psi(R, \theta) = \left(AR^3 + BR \log R + CR + \frac{D}{R}\right) \sin \theta$$

In order to recover a uniform flow far from the cylinder, $\psi(R, \theta)$ must tend to $UR \sin \theta$ as $R \to \infty$. Hence

$$A = B = 0$$

and

$$C = U$$

Hence the expression for the stream function reduces to

$$\psi(R, \theta) = \left( UR + \frac{D}{R} \right) \sin \theta$$

The near boundary condition requires that on the surface of the cylinder both the tangential and the radial velocity components should vanish. That is, on $R = a$ both $\partial\psi/\partial\theta$ and $\partial\psi/\partial R$ should vanish. Since $\partial\psi/\partial\theta$ should be zero for all values of $\theta$, the tangential velocity component being zero is equivalent to requiring that $\psi(a, \theta) =$ constant, where the constant may be taken to be zero. That is, the no-slip boundary condition on the surface of the cylinder requires that

$$\psi(a, \theta) = 0$$

$$\frac{\partial\psi}{\partial R}(a, \theta) = 0$$

It is now evident that there is no choice of the constant $D$ in our solution which satisfies these two boundary conditions. If we had satisfied the near boundary conditions first with the solution, it would have been found that it was impossible to satisfy the far boundary condition. Thus we conclude that there is no solution to the two-dimensional Stokes equations which can satisfy both the near and the far boundary conditions. The lack of such a solution is known as *Stokes' paradox*.

The difference between the two-dimensional Stokes equations and the three-dimensional Stokes equations is best explained by reexamining the Stokes approximation. In terms of dimensionless variables, the Navier-Stokes equations were shown to be

$$\frac{\partial\mathbf{u}^*}{\partial t^*} + R_N(\mathbf{u}^* \cdot \nabla^*)\mathbf{u}^* = -\nabla^* p^* + \nabla^{*2}\mathbf{u}^*$$

so that the Stokes equations correspond to the limit $R_N \to 0$. Thus a more accurate solution for low Reynolds numbers could be sought in the form

$$\psi = \psi_0 + R_N\psi_1 + O(R_N^2)$$

which represents an asymptotic expansion for the stream function, which is valid for low Reynolds numbers. Then, by employing a limiting procedure, the problem for $\psi_0$ may be solved, then the problem for $\psi_1$, and so on. It has been shown here and in Sec. 8.7 that a solution corresponding to $\psi_0$ exists for a sphere but not for a cylinder. However, it is found that the problem for $\psi_1$ in the case of the sphere has no solution, which is known as *Whitehead's paradox*. Thus a basic difficulty has been encountered, and this difficulty appears in the first-order problem in two dimensions and in the second-order problem in three dimensions.

In mathematical terminology, the difficulty encountered above is referred to as a *singular perturbation*. That is, the Stokes approximation is really the first-order problem arising out of a perturbation type of solution to the Navier-

Stokes equations, and the inability of this type of solution to match the required boundary conditions renders the perturbation singular. In two dimensions the difficulty associated with this singular perturbation appears immediately, whereas in three dimensions the difficulty is postponed to the second-order term in the expansion.

In more physical terms, the difficulty encountered is associated with the neglect of the convection of momentum of the fluid, an assumption which is invalid far from the body. The limit $R_N \to 0$ is equivalent to completely neglecting the convection in the fluid in comparison with the viscous diffusion in the fluid. Because of the nature of the viscous boundary condition near the body, viscous diffusion will be large near the body, whereas convection will be small because of the retardation of the velocity by the body. However, far from the body the velocity gradients will die down, so that viscous diffusion will be reduced. At the same time the fluid velocity will be close to that of the free-stream velocity. That is, the convection in the fluid will become more important while the viscous diffusion will become less important. This means that the conditions which are required to satisfy the Stokes approximation will be violated.

The nature of the failure of the approximation which was introduced by Stokes is that of a nonuniform representation in space. The approximation is valid close to the body but is invalid far from the body. Thus singular perturbations are sometimes referred to as nonuniform expansions. The mathematical difficulties encountered in singular perturbations may be overcome by matched expansions, which will be described in the next section.

## 8.9  THE OSEEN APPROXIMATION

An alternative low-Reynolds-number approximation is the Oseen approximation. Oseen recognized the discrepancy which was inherent in the Stokes approximation far from the body under consideration. He noted that the Stokes approximation corresponds to convection at zero velocity and recognized that far from the body momentum will be convected with a velocity which will be close to the free-stream velocity. Thus Oseen proposed linearizing the Navier-Stokes equations such that momentum is transported not with the local velocity (as in the exact case) or with zero velocity (as in the Stokes approximation), but with the free-stream velocity. Thus if the free stream flows in the $x$ direction with velocity $U$, the equations which represent the Oseen approximation to the Navier-Stokes equations are

$$\nabla \cdot \mathbf{u} = 0 \tag{8.9a}$$

$$\frac{\partial \mathbf{u}}{\partial t} + U\frac{\partial \mathbf{u}}{\partial x} = -\frac{1}{\rho}\nabla p + \nu \nabla^2 \mathbf{u} \tag{8.9b}$$

Solutions to the above equations may be established in a manner similar to that which was used to obtain solutions to the Stokes equations. The results so obtained will be valid far from the body but will fail close to the body. This is exactly the opposite of the solutions to the Stokes equations. Thus two independent solutions are obtained, one being valid near the body and the other being valid far from the body. By matching these two solutions, a uniformly valid expression will result which will be valid for small Reynolds numbers. The details are considered to be beyond the scope of the fundamentals which are being treated in this book, but they may be found in the book by Van Dyke which is referenced at the end of Part III. This method of overcoming the difficulties encountered owing to the singular perturbation is called *the method of matched asymptotic expansions*.

## PROBLEMS

**8.1.** Using the Stokes solution for uniform flow over a sphere, integrate the pressure around the surface of the sphere to establish the pressure drag which acts on the sphere. Hence deduce what portion of the total Stokes drag is due to the pressure distribution and what portion is due to the viscous shear on the surface of the sphere.

**8.2.** A liquid drop whose viscosity is $\mu'$ moves slowly through another liquid of viscosity $\mu$ with velocity $U$. The shape of the drop may be taken as spherical, and the motion is sufficiently slow that inertia of the fluid may be neglected. The boundary conditions at the surface of the drop are that the velocity and the tangential stresses in the two fluids are the same.

   Show that a solution to the above problem exists in which the pressure inside the drop is proportional to $x$ and that outside the drop is proportional to $x/r^3$. From the solution, calculate the drag of the drop and show that it is smaller than that for a rigid sphere of the same size, the drag ratio being

$$\frac{1 + \frac{2}{3}(\mu/\mu')}{1 + \mu/\mu'}$$

**8.3.** Show that the flow field which is represented by the equations

$$u = r\mathbf{e}_r \times \mathbf{e}_x$$
$$p = 0$$

is a solution of the Stokes equations. Using this solution, find the velocity and pressure fields for a fluid which is contained between two concentric spheres of radii $r_i$ and $r_0 > r_i$ in which the outer sphere is rotating with angular velocity $\Omega_0$ about the $x$ axis and the inner sphere is rotating with angular velocity $\Omega_i$ in the same direction. Calculate the torque which acts on each of the spheres.

   From the foregoing results, deduce the velocity and pressure fields in a fluid which is contained inside a rotating sphere and find the torque which acts on the sphere.

**8.4.** Show that the flow field which is represented by the equations

$$\mathbf{u} = \nabla\chi \times \mathbf{\Omega}$$
$$p = 0$$

where $\mathbf{\Omega}$ is a constant vector, is a solution of the Stokes equations provided that $\chi$

satisfies the equation

$$\nabla^2 \chi - \frac{1}{\nu} \frac{\partial \chi}{\partial t} = 0$$

Solve this equation for $\chi$, and hence find the velocity field generated by a sphere of radius $a$ which is rotating with a periodic angular velocity $|\Omega| e^{i\omega t}$.

Specialize the solution obtained above for the case $\omega a^2/\nu \ll 1$. Also, for $\omega a^2/\nu \gg 1$, find the thickness of the fluid layer for which the velocity is greater than $e^{-3}$ of the velocity on the surface of the sphere.

**8.5.** Obtain a solution to the Stokes equations for the stream function $\psi(R, \theta)$ in the following form

$$\psi(R, \theta) = Rf(\theta)$$

Show that this solution can be used to represent Stokes flow in a right-angled corner in which the vertical surface $x = 0$ is stationary and the horizontal surface $y = 0$ is moving in the negative $x$ direction with a constant velocity $U$. Estimate the range of validity of this solution by evaluating from the solution the order of magnitude of the inertia and viscous terms indicate below.

$$\text{inertia:} \qquad \rho u_R \frac{\partial u_R}{\partial R}$$

$$\text{viscous:} \qquad \frac{\mu}{R^2} \frac{\partial^2 u_R}{\partial \theta^2}$$

# CHAPTER
# 9

# BOUNDARY LAYERS

Boundary layers are the thin fluid layers adjacent to the surface of a body in which strong viscous effects exist. Figure 9.1 shows the nature of the flow field which would exist around an arbitrary body at a Reynolds number which is not small or of order unity. The nature of such a flow field is known from information gathered from large numbers of experiments.

A dotted line is shown in Fig. 9.1 which originates at the front stagnation point and moves downstream near the top and bottom surfaces of the body. Outside of this dotted line, relative to the body, the velocity gradients are not large, and so viscous effects are negligible. Then, if compressible effects may be ignored, the fluid may be considered to be ideal and the results of Part II of this book may be employed. Thus if the flow field far upstream is uniform, it is also irrotational there, so that Kelvin's theorem guarantees us that the flow outside the dotted line is everywhere irrotational. This potential-flow field is frequently referred to as the "outer flow."

Between the dotted line and the body there are strong viscous effects due to the large velocity gradients which exist. These large velocity gradients are necessitated by the no-slip boundary condition on the solid boundary which reduces the large velocities which exist in the outer flow to zero on the surface. This is the so-called "boundary layer" or "inner flow." Here the vorticity is not zero. Vorticity is generated along the surface of the body, and it is diffused across the boundary layer and convected along the boundary layer by the mean flow.

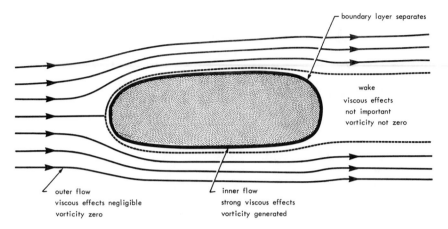

**FIGURE 9.1**
Nature of the flow field around an arbitrarily shaped body.

Toward the rear of the body, the boundary layer will encounter an adverse pressure gradient, that is, an increasing pressure. This usually causes the boundary layer to separate from the body, forming a so-called "wake region" behind the body. The velocity gradients are not large in the wake, so that viscous effects are not too important. However, all the vorticity which exists in the boundary layers is convected into the wake, so that the flow in the wake is not irrotational. If the boundary layer is still laminar at separation, a shear layer will exist of the type discussed in Sec. 6.12. Such shear layers were found to be unstable, and over a wide range of Reynolds numbers this instability manifests itself in the form of a periodic wake which is the well-known Kármán vortex street.

The coverage of boundary layers begins here with the derivation of the boundary-layer approximation to the Navier-Stokes equations. Some exact solutions to these equations are then discussed, including the Blasius solution for the boundary layer on a flat plate and the Falkner-Skan similarity solutions. The Kármán-Pohlhausen approximate method is then introduced and applied to a general boundary-layer problem. Finally, the separation of boundary layers and their stability are discussed.

## 9.1 BOUNDARY-LAYER THICKNESSES

Prior to establishing the boundary-layer equations, it is useful to establish the three types of boundary-layer thicknesses which are in common use. The most widely used of the boundary-layer thicknesses is simply referred to as the *boundary-layer thickness*, and it is denoted by $\delta$. Its usual definition is that distance $y = \delta$ from the solid boundary at which the local value of the velocity

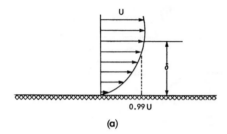

(a)

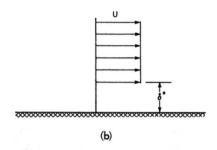

(b)

**FIGURE 9.2**
Definition sketch for (a) boundary-layer thickness and (b) displacement thickness.

reaches 0.99 of the free-stream or outer-flow value. That is,

$$y = \delta \qquad \text{when } u = 0.99U \qquad (9.1a)$$

Figure 9.2a shows the boundary-layer thickness $\delta$ for flow over a flat surface.

Another type of boundary-layer thickness which is useful under certain circumstances is the *displacement thickness*, which is denoted by $\delta^*$. This thickness is defined as the distance which the undisturbed outer flow is displaced from the boundary by a stagnant layer which removes the same mass flow from the flow field as the actual boundary layer. That is, $\delta^*$ is the thickness of a zero-velocity layer which has the same mass-flow defect as the actual boundary layer. This thickness is illustrated in Fig. 9.2b. In mathematical terms, the volume of fluid which is absent owing to the presence of the boundary-layer model is $U\delta^*$. Equating this to the volume of fluid which is absent owing to the actual boundary layer gives the equation which defines the displacement thickness. Thus

$$U\delta^* = \int_0^\infty (U - u)\, dy$$

hence

$$\delta^* = \int_0^\infty \left(1 - \frac{u}{U}\right) dy \qquad (9.1b)$$

A third type of boundary-layer thickness which is frequently used is the *momentum thickness*, denoted by $\theta$. The momentum thickness is defined as that thickness of layer which, at zero velocity, has the same momentum defect, relative to the outer flow, as the actual boundary layer. Thus the momentum thickness is a layer similar to that illustrated in Fig. 9.2b, except that momentum fluxes rather than mass flows are compared with the actual boundary layer. The mass flow which would exist through the momentum thickness at the outer velocity would be $\rho U\theta$. Hence the momentum defect due to this layer will be $\rho U^2\theta$. Equating this to the momentum defect in the actual boundary layer gives

$$\rho U^2\theta = \rho \int_0^\infty u(U - u)\, dy$$

hence

$$\theta = \int_0^\infty \frac{u}{U}\left(1 - \frac{u}{U}\right) dy \tag{9.1c}$$

The various thicknesses defined above are, to some extent, an indication of the distance over which viscous effects extend. Each of these thicknesses will be used in later sections of this chapter, but in the meantime it may be stated that the boundary-layer thickness $\delta$ is usually larger than the displacement thickness $\delta^*$ which, in turn, is usually larger than the momentum thickness $\theta$.

## 9.2  THE BOUNDARY-LAYER EQUATIONS

The boundary-layer equations may be derived from the Navier-Stokes equations by either a physically based argument or a limiting procedure as $R_N \to \infty$. The original derivation which was used by Prandtl was physical in nature and will be followed here.

Figure 9.3 shows a typical boundary-layer configuration on a plane surface or on a curved surface for which $\delta$ is small compared with the radius of curvature of the surface. The only geometric length scale in such problems is the distance $x$ from the leading edge of the surface. For all points in the boundary layer except those near the leading edge, the boundary-layer thickness will be small compared with the distance $x$. That is, except near the leading edge, $\delta/x \ll 1$. The $x$ component of velocity $u$ is of order $U$, the outer flow velocity, and $\partial/\partial x$ is of order $1/x$ in the boundary layer. Thus $\partial u/\partial x$ is of order $U/x$ and hence, from the continuity equation, $\partial v/\partial y$ is also of order $U/x$. Since $v$ will be much smaller than $u$ in the boundary layer and since $\partial/\partial y$ will be much larger than $\partial/\partial x$, this order of $\partial v/\partial y$ may be met by considering $v$ to be of order $U\delta/x$ and $\partial/\partial y$ as being of order $1/\delta$. Thus, within the boundary

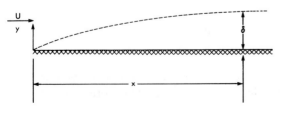

**FIGURE 9.3**
Development of a boundary layer on a plane surface.

layer, the following order of magnitudes will exist:

$$u \sim U$$

$$v \sim U\frac{\delta}{x}$$

$$\frac{\partial}{\partial x} \sim \frac{1}{x}$$

$$\frac{\partial}{\partial y} \sim \frac{1}{\delta}$$

The Navier-Stokes equations for the steady two-dimensional flow under consideration are

$$u\frac{\partial u}{\partial x} + v\frac{\partial u}{\partial y} = -\frac{1}{\rho}\frac{\partial p}{\partial x} + \nu\frac{\partial^2 u}{\partial x^2} + \nu\frac{\partial^2 u}{\partial y^2}$$

$$u\frac{\partial v}{\partial x} + v\frac{\partial v}{\partial y} = -\frac{1}{\rho}\frac{\partial p}{\partial y} + \nu\frac{\partial^2 u}{\partial x^2} + \nu\frac{\partial^2 v}{\partial y^2}$$

Using the order of magnitudes established above, the various terms in these two equations will be of the following order of magnitude:

$$\frac{U^2}{x} + \frac{U^2}{x} = -\frac{1}{\rho}\frac{\partial p}{\partial x} + \nu\frac{U}{x^2} + \nu\frac{U}{\delta^2}$$

$$\frac{\delta U^2}{x^2} + \frac{\delta U^2}{x^2} = -\frac{1}{\rho}\frac{\partial p}{\partial y} + \nu\frac{\delta U}{x^3} + \nu\frac{U}{x\delta}$$

No attempt has yet been made to estimate the order of magnitude of the pressure terms, so that they are carried along as they are.

In the first of these two equations, the two inertia terms are of the same order, but the second viscous term $(\nu\, \partial^2 u/\partial y^2)$ is seen to be much larger than the first viscous term $(\nu\, \partial^2 u/\partial x^2)$. Hence the latter viscous term may be neglected in boundary layers. Since fluid particles may be accelerated in boundary layers, and since strong viscous effects exist, the dominant viscous term is assumed to be of the same order of magnitude as the inertia terms. This gives

$$\frac{U^2}{x} \sim \nu\frac{U}{\delta^2}$$

or

$$\delta \sim \sqrt{\frac{\nu x}{U}}$$

That is, from purely order-of-magnitude considerations it may be deduced that the boundary-layer thickness will increase as $\sqrt{x}$. Furthermore, the condition

$\delta/x \ll 1$ becomes

$$\frac{x^2}{\delta^2} \sim \frac{Ux}{\nu} \gg 1$$

or

$$R_N = \frac{Ux}{\nu} \gg 1$$

That is, the boundary-layer assumption of $\delta/x \ll 1$ is equivalent to the condition $R_N \gg 1$, where the length scale used in the Reynolds number is $x$.

From the foregoing discussion, it is obvious that, provided the Reynolds number based on $x$ is large, the $x$ component of the Navier-Stokes equations may be approximated by the following equation:

$$u\frac{\partial u}{\partial x} + v\frac{\partial u}{\partial y} = -\frac{1}{\rho}\frac{\partial p}{\partial x} + \nu\frac{\partial^2 u}{\partial y^2}$$

Furthermore, from the order-of-magnitude balance for the $y$ component of the Navier-Stokes equations, it is evident that the inertia in the $y$ direction is of order $\delta/x$ smaller than that in the $x$ direction and so may be neglected by comparison. Also, the viscous terms in the $y$ direction are of order $\delta/x$ smaller than those which act in the $x$ direction, and so the former may be neglected. Thus the $y$ component of the Navier-Stokes equations becomes

$$0 = -\frac{1}{\rho}\frac{\partial p}{\partial y}$$

That is, $p$ is independent of the transverse coordinate $y$ in boundary layers, so that $p$ will be a function of $x$ only. Thus, in boundary layers, the continuity equation and the Navier-Stokes equations become

$$\frac{\partial u}{\partial x} + \frac{\partial v}{\partial y} = 0 \qquad (9.2a)$$

$$u\frac{\partial u}{\partial x} + v\frac{\partial u}{\partial y} = -\frac{1}{\rho}\frac{dp}{dx} + \nu\frac{\partial^2 u}{\partial y^2} \qquad (9.2b)$$

It will be noticed that the loss of the highest derivative in $x$ now makes the governing equations parabolic, whereas the Navier-Stokes equations are elliptic. This mathematical change has physical consequences which will be exposed in later sections.

Since the pressure $p$ is independent of the transverse coordinate $y$ in boundary layers, the pressure distribution along the boundary layer will be the same as that of the outer flow. But the outer flow is a potential flow, and so the Bernoulli equation is valid. Hence

$$\frac{p}{\rho} + \tfrac{1}{2}U^2 = \text{constant}$$

where the outer velocity $U$ will be a constant for flow over a plane surface, but in general it will be a function of $x$. Thus, from the Bernoulli equation

$$-\frac{1}{\rho}\frac{\partial p}{\partial x} = U\frac{dU}{dx}$$

Substituting this result into Eq. (9.2b) gives the following alternative form of the Prandtl boundary-layer equation:

$$u\frac{\partial u}{\partial x} + v\frac{\partial u}{\partial y} = U\frac{dU}{dx} + v\frac{\partial^2 u}{\partial y^2} \qquad (9.2c)$$

The boundary conditions which accompany any boundary-layer equations are the no-slip conditions on the surface and the condition that the outer-flow velocity is recovered far from the surface. That is, the following boundary conditions must be satisfied:

$$u(x,0) = 0 \qquad\qquad (9.3a)$$

$$v(x,0) = 0 \qquad\qquad (9.3b)$$

$$u(x,y) \to U(x) \qquad \text{as } y \to \infty \qquad (9.3c)$$

The last condition in effect matches the inner flow to the outer flow, so that the potential-flow solution must be known before the boundary-layer problem can be solved.

The alternative way of deriving the boundary-layer equations from the Navier-Stokes equations involves a limiting procedure similar to that which was used to extract the Stokes equations from the full Navier-Stokes equations. The Navier-Stokes equations are first written in terms of dimensionless variables, which results in a coefficient $1/R_N$ appearing in front of the viscous terms. Stretched coordinates $X = x$ and $Y = \sqrt{R_N}\,y$ are then introduced, which removes the coefficient $1/R_N$ from one of the viscous terms. If the limit $R_N \to \infty$ while $X$ and $Y$ are held fixed is now taken, the boundary-layer equations will be obtained. This approach is useful if higher approximations to the boundary-layer theory are required, that is, if an expansion type of solution is sought. However, the nature of the coordinate stretching is not obvious without appealing to the physical approach.

## 9.3 BLASIUS SOLUTION

An exact solution to the boundary-layer equations corresponding to a uniform flow over a flat surface was obtained by Blasius. The flow configuration for a flat boundary is shown in Fig. 9.3, where it is understood that $U$ is a constant and $\delta$ is a function of $x$. Since $U$ is constant, the pressure term in the boundary-layer equation is identically zero. Thus the continuity equation and the boundary-layer approximation to the Navier-Stokes equations are

$$\frac{\partial u}{\partial x} + \frac{\partial v}{\partial y} = 0$$

$$u\frac{\partial u}{\partial x} + v\frac{\partial u}{\partial y} = \nu\frac{\partial^2 u}{\partial y^2}$$

In order to reduce this pair of equations to a single equation, a stream function defined by $u = \partial\psi/\partial y$, $v = -\partial\psi/\partial x$ is introduced. This satisfies the continuity equation identically for all stream functions $\psi$ and yields the following form of the boundary-layer equation:

$$\frac{\partial\psi}{\partial y}\frac{\partial^2\psi}{\partial x\,\partial y} - \frac{\partial\psi}{\partial x}\frac{\partial^2\psi}{\partial y^2} = \nu\frac{\partial^3\psi}{\partial y^3}$$

Since this is a parabolic partial differential equation and since there is no geometric length scale in the problem, a similarity type of solution will be sought. Similarity solutions were discussed in Sec. 7.4 and, in the context of the problem in hand, were shown to be of the form

$$\psi(x, y) \sim f(\eta)$$

where

$$\eta \sim \frac{y}{x^n}$$

The value of $n$ for the case of a flat surface is $\frac{1}{2}$, so that the similarity variable $\eta$ is chosen to be

$$\eta = \frac{y}{\sqrt{\nu x/U}}$$

Here, the parameters $\nu$ and $U$ have been used to render the similarity variable dimensionless. For this choice of $\eta$, the $x$ component of velocity will have the following functional form:

$$u = \frac{\partial\psi}{\partial y} \sim \sqrt{\frac{U}{\nu x}} f'(\eta)$$

where the prime denotes differentiation with respect to $\eta$. But when $\eta = $ constant, $u$ should be constant, so that the proportionality factor in the equation $\psi(x, y) \sim f(\eta)$ should include a $\sqrt{x}$. Then, since the units of $\psi$ are a length squared divided by time, the dimensions will be correct if the proportionality factor also includes $\sqrt{\nu U}$. That is, a similarity solution of the following form is sought:

$$\psi(x, y) = \sqrt{\nu U x} f(\eta)$$

where

$$\eta = \frac{y}{\sqrt{\nu x/U}}$$

From these expressions, the various derivatives which appear in the boundary-layer equation may be evaluated as follows:

$$\frac{\partial \psi}{\partial x} = -\frac{U}{2}\frac{y}{x}f' + \tfrac{1}{2}\sqrt{\nu U}\,\frac{1}{x^{1/2}}f$$

$$= -\frac{1}{2}\sqrt{\frac{\nu U}{x}}\,\eta f' + \frac{1}{2}\sqrt{\frac{\nu U}{x}}\,f$$

$$\frac{\partial \psi}{\partial y} = U f'$$

$$\frac{\partial^2 \psi}{\partial x\,\partial y} = -\frac{U}{2}\sqrt{\frac{U}{\nu}}\,\frac{y}{x^{3/2}}f'' = -\frac{U}{2x}\eta f''$$

$$\frac{\partial^2 \psi}{\partial y^2} = U\sqrt{\frac{U}{\nu x}}\,f''$$

$$\frac{\partial^3 \psi}{\partial y^3} = \frac{U^2}{\nu x}f'''$$

Substituting these results into the equation for the stream function gives

$$-\frac{U^2}{2x}\eta f' f'' - \frac{U^2}{2x}(f - \eta f')f'' = \frac{U^2}{x}f'''$$

or

$$-\frac{U^2}{2x}ff'' = \frac{U^2}{x}f'''$$

Since $x$ may be canceled from this equation, the existence of a similarity type of solution is confirmed. That is, a solution of the assumed form exists provided the function $f$ satisfies the following conditions:

$$f''' + \tfrac{1}{2}ff'' = 0 \tag{9.4a}$$

$$f(0) = f'(0) = 0 \tag{9.4b}$$

$$f'(\eta) \to 1 \qquad \text{as } \eta \to \infty \tag{9.4c}$$

The boundary conditions (9.4b) and (9.4c) follow from the no-slip boundary conditions and the matching condition with the outer flow as described by Eqs. (9.3). From the solution to this problem the stream function may be obtained using the relationship

$$\psi(x, y) = \sqrt{\nu U x}\, f\!\left(\frac{y}{\sqrt{\nu x/U}}\right) \tag{9.4d}$$

The problem represented by Eqs. (9.4a), (9.4b), and (9.4c) is a well-posed problem. It is shown in the problems at the end of this chapter that the differential equation may be reduced in order. However, numerical integration is eventually required. In spite of this, the Blasius solution to the boundary-layer equations is considered to be exact, since the partial differential equation has been reduced to an ordinary differential equation which, together with the appropriate boundary conditions, may be solved numerically to a high degree of accuracy.

The results of interest which should be extracted from the solution are the shear-stress distribution along the surface, the drag acting on the surface, and the boundary-layer thickness. The shear stress on the surface is given by

$$\tau_0(x) = \mu \frac{\partial u}{\partial y}(x,0)$$

$$= \mu \frac{\partial^2 \psi}{\partial y^2}(x,0)$$

$$= \mu \sqrt{\frac{U^3}{\nu x}} f''(0)$$

Nondimensionalizing this surface shear stress by means of the kinetic energy in the free stream gives

$$\frac{\tau_0(x)}{\frac{1}{2}\rho U^2} = \frac{2f''(0)}{\sqrt{R_N}}$$

where the Reynolds number is based on the distance $x$ from the leading edge of the surface to the location under consideration. But the value of $f''(0)$ is found numerically to be 0.332, so that the shear-stress distribution along the surface will be given by the expression

$$\frac{\tau_0(x)}{\frac{1}{2}\rho U^2} = \frac{0.664}{\sqrt{R_N}} \tag{9.5a}$$

This result shows that the shear stress falls off as $\sqrt{x}$ along the surface.

The drag force acting on the surface may be evaluated by integrating the shear stress. That is, the drag force acting on the surface up to the station $x$ will be given by

$$F_D = \int_0^x \tau_0(\xi)\, d\xi$$

Thus the drag coefficient of the surface will be

$$C_D = \frac{F_D/x}{\frac{1}{2}\rho U^2} = \frac{1}{x}\int_0^x \frac{\tau_0(\xi)}{\frac{1}{2}\rho U^2}\, d\xi$$

Using the result obtained above for the surface shear-stress distribution gives

$$C_D = \frac{0.664}{x} \int_0^x \frac{d\xi}{\sqrt{R_N}}$$

$$C_D = \frac{1.328}{\sqrt{R_N}}$$

(9.5b)

Strictly speaking, the shear-stress distribution given by Eq. (9.5a) should not be used near the leading edge of the surface, since the boundary-layer assumptions are no longer valid there. However, any difference between the actual shear-stress and that given by Eq. (9.5a) is not likely to create any significant discrepancy because of the relatively short distance involved. The shear stress actually has a singularity at $x = 0$, but this singularity is integrable, so that the drag force is not singular. Indeed, Eq. (9.5b) shows that the drag force varies as $\sqrt{x}$, where $x$ is the point up to which the accumulated drag is being considered.

To obtain the boundary-layer thickness, it is observed from the numerical solution that $u = 0.99U$ when $\eta = 5.0$. Then, using the definition of $\eta$ and the fact that $y = \delta$ when $u = 0.99U$ gives

$$\frac{\delta}{\sqrt{\nu x / U}} = 5.0$$

hence

$$\frac{\delta}{x} = \frac{5.0}{\sqrt{R_N}}$$

(9.5c)

where, again, the length used in the Reynolds number is the distance $x$ to which the boundary-layer thickness applies. In the same way the following expressions are obtained from the numerical results for the displacement thickness and the momentum thickness:

$$\frac{\delta^*}{x} = \frac{1.72}{\sqrt{R_N}}$$

(9.5d)

$$\frac{\theta}{x} = \frac{0.664}{\sqrt{R_N}}$$

(9.5e)

These results show that the various boundary-layer thicknesses grow as $\sqrt{x}$ and that $\theta < \delta^* < \delta$.

## 9.4 FALKNER-SKAN SOLUTIONS

A whole family of similarity solutions to the boundary-layer equations were found by Falkner and Skan. These solutions are obtained by seeking general similarity-type solutions and interpreting the flow field for each solution so obtained.

Look for general similarity solutions of the form

$$u(x, y) = U(x)f'(\eta)$$

where

$$\eta = \frac{y}{\xi(x)}$$

Here, $U(x)$ is the outer flow and $\xi(x)$ is an unspecified function of $x$ which will be determined later. For this form of velocity the stream function must be

$$\psi(x, y) = U(x)\xi(x)f(\eta)$$

But, from Eq. (9.2c), the equation to be satisfied by $\psi$ is

$$\frac{\partial\psi}{\partial y}\frac{\partial^2\psi}{\partial x\,\partial y} - \frac{\partial\psi}{\partial x}\frac{\partial^2\psi}{\partial y^2} = U\frac{dU}{dx} + \nu\frac{\partial^3\psi}{\partial y^3}$$

The various terms which appear in this equation may be evaluated as follows:

$$\frac{\partial\psi}{\partial x} = \frac{dU}{dx}\xi f + U\frac{d\xi}{dx}f - U\xi\frac{y}{\xi^2}\frac{d\xi}{dx}f'$$

$$= \frac{dU}{dx}\xi f + U\frac{d\xi}{dx}f - U\frac{d\xi}{dx}\eta f'$$

$$\frac{\partial\psi}{\partial y} = Uf'$$

$$\frac{\partial^2\psi}{\partial x\,\partial y} = \frac{dU}{dx}f' - U\frac{y}{\xi^2}\frac{d\xi}{dx}f''$$

$$= \frac{dU}{dx}f' - \frac{U}{\xi}\frac{d\xi}{dx}\eta f''$$

$$\frac{\partial^2\psi}{\partial y^2} = \frac{U}{\xi}f''$$

$$\frac{\partial^3\psi}{\partial y^3} = \frac{U}{\xi^2}f'''$$

Substituting these results into the equation to be satisfied by $\psi$ gives

$$Uf'\left(\frac{dU}{dx}f' - \frac{U}{\xi}\frac{d\xi}{dx}\eta f''\right) - \left(\frac{dU}{dx}\xi f + U\frac{d\xi}{dx}f - U\frac{d\xi}{dx}\eta f'\right)\frac{U}{\xi}f'' = U\frac{dU}{dx} + \nu\frac{U}{\xi^2}f'''$$

$$U\frac{dU}{dx}(f')^2 - U\frac{dU}{dx}ff'' - U^2\frac{1}{\xi}\frac{d\xi}{dx}ff'' = U\frac{dU}{dx} + \nu\frac{U}{\xi^2}f'''$$

where the second term in the first bracket has been canceled with the third term in the second bracket. Combining the second and third terms of this equation gives

$$U\frac{dU}{dx}(f')^2 - \frac{U}{\xi}\frac{d}{dx}(U\xi)ff'' = U\frac{dU}{dx} + \nu\frac{U}{\xi^2}f'''$$

This equation may be put in standard form by multiplying by $\xi^2/(\nu U)$, giving

$$f''' + \left[\frac{\xi}{\nu}\frac{d}{dx}(U\xi)\right]ff'' + \left[\frac{\xi^2}{\nu}\frac{dU}{dx}\right]\{1 - (f')^2\} = 0$$

If a similarity solution exists, this should now be an ordinary differential equation for the function $f$ in terms of $\eta$. Thus the two coefficients inside the brackets should be constants at most, say $\alpha$ and $\beta$, respectively. That is, for a similarity solution we must have

$$\frac{\xi}{\nu}\frac{d}{dx}(U\xi) = \alpha$$

$$\frac{\xi^2}{\nu}\frac{dU}{dx} = \beta$$

where $\alpha$ and $\beta$ are constants. A convenient alternative to one of the above equations may be obtained as follows:

$$\frac{d}{dx}(U\xi^2) = 2U\xi\frac{d\xi}{dx} + \xi^2\frac{dU}{dx}$$

$$= 2\xi\left(U\frac{d\xi}{dx} + \xi\frac{dU}{dx}\right) - \xi^2\frac{dU}{dx}$$

$$= 2\xi\frac{d}{dx}(U\xi) - \xi^2\frac{dU}{dx}$$

i.e.,
$$\frac{d}{dx}(U\xi^2) = \nu(2\alpha - \beta)$$

This equation, together with either of the foregoing two equations, is sufficient to relate $U$ and $\xi$ to the constants $\alpha$ and $\beta$. In terms of $\alpha$ and $\beta$, the differential equation to be solved for the function $f$ is

$$f''' + \alpha ff'' + \beta\left[1 - (f')^2\right] = 0$$

The boundary conditions which accompany this differential equation are the same as for the flat surface. If the problem so obtained is a solvable one, then we have found an exact solution to the boundary-layer equations.

From the foregoing analysis and discussion, it is evident that exact solutions to the boundary-layer equations may be obtained by pursuing the following procedure:

1. Select values of the constants $\alpha$ and $\beta$.
2. Find the corresponding values of $U(x)$ and $\xi(x)$ from the relations

$$\frac{d}{dx}(U\xi^2) = \nu(2\alpha - \beta) \tag{9.6a}$$

$$\xi^2 \frac{dU}{dx} = \beta\nu \tag{9.6b}$$

3. Determine the function $f(\eta)$ which is the solution to the following problem:

$$f''' + \alpha f f'' + \beta\left[1 - (f')^2\right] = 0 \tag{9.6c}$$

$$f(0) = f'(0) = 0 \tag{9.6d}$$

$$f'(\eta) \to 1 \quad \text{as } \eta \to \infty \tag{9.6e}$$

4. The stream function for the flow field in the boundary layer is then given by

$$\psi(x, y) = U(x)\xi(x)f\left(\frac{y}{\xi(x)}\right) \tag{9.6f}$$

Having chosen the constants $\alpha$ and $\beta$, a particular flow configuration is being considered. This flow configuration will not be known a priori but will become evident when step 2 is completed. The function $U(x)$ is the outer-flow velocity, which is the potential-flow velocity for the geometry under considera-tion. Then, when $U(x)$ is established through step 2, comparison with the results of Chap. 4 will reveal the geometry of the problem. Since $\alpha$ and $\beta$ have been chosen, the problem to be satisfied by the function $f(\eta)$ is now explicit, so that a solution may be sought. This solution, together with the quantities $U(x)$ and $\xi(x)$, completely determines the stream function for the problem from which all properties of the flow field may be derived.

Several exact solutions to the boundary-layer equations may be obtained by the foregoing method. The solution corresponding to a flat surface, which has already been established, will be obtained from the Falkner-Skan solutions to illustrate the procedure and to verify the result. Some new solutions to the boundary-layer equations will then be established in the next few sections.

It should be noted that for $\alpha = 1$ numerical solutions to Eqs. (9.6) show that $f''(0) \to 0$ as $\beta$ is decreased. The value for which $f''(0) = 0$ is $\beta = -0.1988$, and for values of $\beta$ which are smaller than this value, $f'(\eta) > 1$ at some location. This corresponds to $u > U$, which is physically unreasonable. There-fore, for $\alpha = 1$, we must have $\beta > -0.1988$.

The solution corresponding to a flat surface is obtained by choosing $\alpha = \frac{1}{2}$ and $\beta = 0$ in the Falkner-Skan solutions. Then, from Eqs. (9.6a) and (9.6b),

$$\frac{d}{dx}(U\xi^2) = \nu$$

$$\xi^2 \frac{dU}{dx} = 0$$

Since $\xi(x)$ cannot be zero, the second of these equations shows that $U(x) = c$, where $c$ is a constant. Then the other equation shows that

$$\xi(x) = \sqrt{\frac{\nu x}{c}}$$

The fact that $U(x)$ is a constant in this case identifies the geometry as a flat surface rather than a curvilinear one which may be thought of as being stretched out into a plane. Using the values of $\alpha$ and $\beta$ chosen above, the problem to be solved by the function $f(\eta)$ is

$$f''' + \tfrac{1}{2}ff'' = 0$$

$$f(0) = f'(0) = 0$$

$$f'(\eta) \to 1 \quad \text{as } \eta \to \infty$$

The stream function is then given by

$$\psi(x, y) = \sqrt{c\nu x}\, f\!\left(\frac{y}{\sqrt{\nu x/c}}\right)$$

These results are seen to agree identically with the Blasius solution which is given by Eqs. (9.4).

## 9.5  FLOW OVER A WEDGE

The solution to the boundary-layer equations corresponding to flow over a wedge may be obtained from the Falkner-Skan equations by setting $\alpha = 1$ and keeping $\beta$ arbitrary. Then, from Eqs. (9.6a) and (9.6b) with $\alpha = 1$, $U(x)$ and $\xi(x)$ will be defined by the following equations:

$$\frac{d}{dx}(\xi^2 U) = \nu(2 - \beta)$$

$$\xi^2 \frac{dU}{dx} = \nu\beta$$

Integrating the first of these equations gives

$$\xi^2 U = \nu(2 - \beta)x$$

Dividing the second of the foregoing equations by this last result gives

$$\frac{1}{U}\frac{dU}{dx} = \frac{\beta}{2-\beta}\frac{1}{x}$$

This equation may be integrated directly to give

$$\log U = \frac{\beta}{2-\beta}\log x + \log c$$

where $c$ is an arbitrary constant. Hence the outer-flow velocity corresponding to our choice of $\alpha$ is

$$U(x) = cx^{\beta/(2-\beta)} \qquad (9.7a)$$

but

$$\xi^2\frac{dU}{dx} = \nu\beta$$

$$\therefore \quad \xi^2 c\frac{\beta}{2-\beta}x^{-2(1-\beta)/(2-\beta)} = \nu\beta$$

hence

$$\xi(x) = \sqrt{\frac{\nu(2-\beta)}{c}}\, x^{(1-\beta)/(2-\beta)} \qquad (9.7b)$$

Equation (9.7a) shows that the outer flow is that over a wedge of angle $\pi\beta$. This may be shown by using the potential flow for a sector whose angle, measured in the fluid, is $\pi/n$. The result, as given by Eq. (4.10), is

$$F(z) = Uz^n$$

$$\therefore \quad W(z) = nUz^{n-1}$$

that is,

$$u - iv = nU(x + iy)^{n-1}$$

Hence on the surface $y = 0$ the velocity components are

$$u = nUx^{n-1}$$

$$v = 0$$

That is, the velocity given by Eq. (9.7a) has the same form as that near the boundary of the flow in a sector of angle $\pi/n$. To find the angle of the wedge corresponding to Eq. (9.7a), the exponents of $x$ in these two expressions are equated. Hence

$$n - 1 = \frac{\beta}{2-\beta}$$

This gives the angle of the half wedge measured in the fluid. Then, from the symmetry of the flow field, the angle of the wedge will be $2(\pi - \pi/n)$. From the above equation, this angle is $\pi\beta$, which is shown in Fig. 9.4.

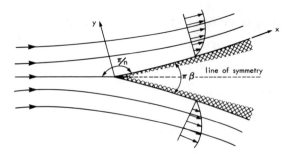

**FIGURE 9.4**
Boundary-layer flow over a wedge.

From Eqs. (9.6c), (9.6d), and (9.6e), the problem to be solved for the function $f$ is

$$f''' + ff'' + \beta\left[1 - (f')^2\right] = 0 \tag{9.7c}$$

$$f(0) = f'(0) = 0 \tag{9.7d}$$

$$f'(\eta) \to 1 \quad \text{as } \eta \to \infty \tag{9.7e}$$

This problem may be solved numerically. Having obtained the solution for $f(\eta)$, the stream function will be given by Eq. (9.6f), where $U(x)$ and $\xi(x)$ are given by Eqs. (9.7a) and (9.7b). This gives

$$\psi(x, y) = \sqrt{c(2 - \beta)\nu}\, x^{1(2 - \beta)} f\left(\frac{y}{\sqrt{(2 - \beta)\nu/c}}\, x^{-(1 - \beta)(2 - \beta)}\right) \tag{9.7f}$$

## 9.6 STAGNATION-POINT FLOW

Another exact solution to the boundary-layer equations which may be obtained from the Falkner-Skan similarity solution is that corresponding to a stagnation-point flow. The values of the constants $\alpha$ and $\beta$ which yield this solution are $\alpha = \beta = 1$. But this is equivalent to letting $\beta$ be unity in the solution for the flow over a wedge. Then the angle of the wedge becomes $\pi$, which means the flow impinges on a flat surface yielding a plane stagnation point.

The solution may be obtained by setting $\beta = 1$ in Eqs. (9.7). This gives

$$U(x) = cx \tag{9.8a}$$

$$\xi(x) = \sqrt{\frac{\nu}{c}} \tag{9.8b}$$

$$f''' + ff'' + 1 - (f')^2 = 0 \tag{9.8c}$$

$$f(0) = f'(0) = 0 \tag{9.8d}$$

$$f'(\eta) \to 1 \quad \text{as } \eta \to \infty \tag{9.8e}$$

$$\psi(x, y) = \sqrt{c\nu}\, xf\left(\frac{y}{\sqrt{\nu/c}}\right) \tag{9.8f}$$

It will be noticed that this is precisely the exact solution to the full Navier-Stokes equations which was obtained by Hiemenz for a stagnation point. This solution is given by Eqs. (7.7a), (7.7b), and (7.7c). Thus the exact solution to the boundary-layer equations is also an exact solution to the full Navier-Stokes equations in this instance.

## 9.7   FLOW IN A CONVERGENT CHANNEL

The boundary-layer solution for flow in a convergent channel may be obtained from the Falkner-Skan solution by choosing $\alpha = 0$ and $\beta = 1$. For these values of the constants, Eqs. (9.6a) and (9.6b) become

$$\frac{d}{dx}(U\xi^2) = -\nu$$

$$\xi^2\frac{dU}{dx} = \nu$$

Integrating the first of these equations gives

$$U\xi^2 = -\nu x$$

and dividing the second equation by this last result gives

$$\frac{1}{U}\frac{dU}{dx} = -\frac{1}{x}$$

Integrating this equation shows that the outer-flow velocity is of the form

$$U(x) = -\frac{c}{x} \tag{9.9a}$$

where $c$ is a constant. Then from one of the above results it follows that

$$\xi(x) = \sqrt{\frac{\nu}{c}}\,x \tag{9.9b}$$

Equation (9.9a) is the potential-flow velocity for flow in a convergent channel. That is, the solution obtained here corresponds to a boundary layer on the wall of a convergent channel in which the flow is directed inward to the apex of the channel walls. This flow configuration is shown in Fig. 9.5. It will be noted that

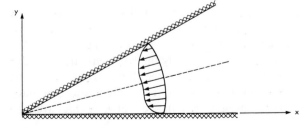

**FIGURE 9.5**
Boundary-layer flow on the wall of a convergent channel.

for $c < 0$, that is, for outward flow from the apex, Eq. (9.9b) shows that no solution exists. This may be interpreted as meaning that for flow in a divergent channel the adverse pressure gradient will cause the boundary layers to separate, and hence a reverse flow will result. This situation was encountered in Sec. 7.8 when the exact solution for viscous flow in convergent and divergent channels was studied.

For $\alpha = 0$ and $\beta = 1$, the problem to be satisfied by the function $f(\eta)$ is

$$f''' + 1 - (f')^2 = 0 \tag{9.9c}$$

$$f(0) = f'(0) = 0 \tag{9.9d}$$

$$f'(\eta) \to 1 \quad \text{as } \eta \to \infty \tag{9.9e}$$

Then, using Eq. (9.6f) and the results (9.9a) and (9.9b), the value of the stream function will be

$$\psi(x, y) = -\sqrt{c\nu}\, f\left(\frac{y}{\sqrt{\nu/c\, x}}\right) \tag{9.9f}$$

## 9.8   APPROXIMATE SOLUTION FOR A FLAT SURFACE

The foregoing solutions have all been exact in the sense that a similarity form of solution reduced the governing partial differential equations to a nonlinear ordinary differential equation which could be solved numerically to a high degree of accuracy. For situations where an exact solution does not exist, an approximate solution must be sought. One of the classical approximate methods which is widely used was introduced by von Kármán and refined by Pohlhausen. The basic procedure will be presented in this section in the context of boundary-layer flow on a flat surface, and the procedure will be generalized in the next section.

For flow over a flat surface the outer velocity $U$ is constant, so that the boundary-layer equations are

$$\frac{\partial u}{\partial x} + \frac{\partial v}{\partial y} = 0$$

$$u\frac{\partial u}{\partial x} + v\frac{\partial u}{\partial y} = \nu\frac{\partial^2 u}{\partial y^2}$$

Normally, a functional form of solution to these equations is sought which satisfies the equations identically at each point in space, that is, at each value of $x$ and $y$, and which tends to the appropriate values on the boundaries. If such a solution cannot be found, it may be possible to satisfy the basic equations on the average rather than at each and every point in the fluid. That is, if the boundary-layer equation is integrated with respect to $y$ across the boundary layer, the resulting equation will represent a balance between the average inertia forces and viscous forces for each $x$ location. Then a velocity distribution may be obtained which satisfies this averaged balance of forces but which does not satisfy the balance at each point across the boundary layer. The results which are extracted from such approximate solutions are found to be reasonably accurate in most instances.

Prior to integrating the boundary-layer equations, it is useful to recast them in a slightly different form. The term $u\,\partial u/\partial x$ may be rewritten as follows:

$$u\frac{\partial u}{\partial x} = \frac{\partial}{\partial x}(u^2) - u\frac{\partial u}{\partial x}$$

$$= \frac{\partial}{\partial x}(u^2) + u\frac{\partial v}{\partial y}$$

in which $\partial u/\partial x$ has been replaced by $-\partial v/\partial y$ from the continuity equation. Thus the boundary-layer form of the equation for the $x$ momentum may be written in the form

$$\frac{\partial}{\partial x}(u^2) + u\frac{\partial v}{\partial y} + v\frac{\partial u}{\partial y} = \nu\frac{\partial^2 u}{\partial y^2}$$

or

$$\frac{\partial}{\partial x}(u^2) + \frac{\partial}{\partial y}(uv) = \nu\frac{\partial^2 u}{\partial y^2}$$

This equation is still exact within the boundary-layer approximation. This local balance of forces will now be reduced to an average balance across the boundary layer by integrating with respect to $y$ from $y = 0$ to $y = \delta$.

$$\int_0^\delta \frac{\partial}{\partial x}(u^2)\,dy + [uv]_0^\delta = \nu\left[\frac{\partial u}{\partial y}\right]_0^\delta$$

But, $u(x,0) = v(x,0) = 0$ from the no-slip boundary condition and $u(x,\delta) = U$, which is the outer-flow velocity. Also, $\mu\,\partial u/\partial y = \tau_0$, the surface shear stress, when $y = 0$, and since the boundary-layer velocity profile should blend smoothly into the outer-flow velocity at $y = \delta$, $\partial u/\partial y = 0$ there. Hence the integrated boundary-layer equation becomes

$$\int_0^\delta \frac{\partial}{\partial x}(u^2)\,dy + Uv(x,\delta) = -\frac{\tau_0}{\rho}$$

The quantity $v(x,\delta)$ may be evaluated by integrating the continuity equation between the limits $y = 0$ and $y = \delta$. This gives

$$\int_0^\delta \frac{\partial u}{\partial x}\,dy + [v]_0^\delta = 0$$

$$\therefore \quad Uv(x,\delta) = -U\int_0^\delta \frac{\partial u}{\partial x}\,dy$$

Then the integrated boundary-layer equation may be written in the following form:

$$\int_0^\delta \frac{\partial}{\partial x}(u^2)\, dy - U\int_0^\delta \frac{\partial u}{\partial x}\, dy = -\frac{\tau_0}{\rho}$$

Finally, these two integrals involving derivatives may be expressed as derivatives of integrals through the rule of Leibnitz. For any function $f(x, y)$ this rule states

$$\int_{\alpha(x)}^{\beta(x)} \frac{\partial f}{\partial x}(x, y)\, dy = \frac{d}{dx}\int_{\alpha(x)}^{\beta(x)} f(x, y)\, dy - f(x, \beta)\frac{d\beta}{dx} + f(x, \alpha)\frac{d\alpha}{dx}$$

Using this rule, the integrals which appear above may be rewritten as follows:

$$\int_0^\delta \frac{\partial}{\partial x}(u^2)\, dy = \frac{d}{dx}\int_0^\delta u^2\, dy - U^2\frac{d\delta}{dx}$$

$$\int_0^\delta \frac{\partial u}{\partial x}\, dy = \frac{d}{dx}\int_0^\delta u\, dy - U\frac{d\delta}{dx}$$

Thus the integrated boundary-layer equation becomes

$$\frac{d}{dx}\int_0^\delta u^2\, dy - U\frac{d}{dx}\int_0^\delta u\, dy = -\frac{\tau_0}{\rho}$$

Since $U$ is a constant, it may be taken inside the derivative and the integral, and the two integrals may be combined. This gives

$$\frac{d}{dx}\int_0^\delta u(U - u)\, dy = \frac{\tau_0}{\rho}$$

This equation is known as the *momentum integral*, and it is valid for boundary-layer flow over a flat surface. Physically, this equation states that the rate of change of the momentum in the entire boundary layer at any value of $x$ is equal to the force produced by the shear stress at the surface at that location.

The manner in which the momentum integral is used is as follows: A form of velocity profile is first assumed, typically a polynomial in $y$. The arbitrary constants in this expression are used to match the required boundary conditions. These boundary conditions are

$$u(x, 0) = 0$$

$$u(x, \delta) = U$$

$$\frac{\partial u}{\partial y}(x, \delta) = 0$$

The first of these conditions is the no-slip boundary condition at the surface, the second condition matches the boundary-layer velocity to the outer-flow velocity, and the third condition ensures that the matching is smooth at $y = \delta$. It should be noted that all the higher derivatives should also be zero at $y = \delta$ for a smooth transition from the boundary layer to the outer flow. The number of conditions which can be satisfied, of course, depends upon the number of free parameters in the assumed velocity profile. It should further be noted that a series of boundary conditions should also be imposed at $y = 0$. The boundary-layer equation for this case is

$$u\frac{\partial u}{\partial x} + v\frac{\partial u}{\partial y} = \nu\frac{\partial^2 u}{\partial y^2}$$

Hence the no-slip boundary condition at $y = 0$ would automatically result in $\partial^2 u/\partial y^2 = 0$ at $y = 0$ if our velocity profile was the correct one. However, since we know that our assumed velocity profile is not the correct one, this boundary condition must be imposed separately. Likewise, by differentiating the boundary-layer equation, conditions for the third and higher derivatives will be obtained which should be imposed separately in our approximate solution. The number of boundary conditions out of this infinite array at $y = 0$ and $y = \delta$ which can be accommodated depends upon the number of free parameters which are available. Normally the three conditions mentioned above are included in the order of priority in which they are written down, then the condition $\partial^2 u/\partial y^2 = 0$ at $y = 0$ is imposed, then $\partial^2 u/\partial y^2 = 0$ at $y = \delta$, and so on.

Typically, the velocity profile is taken to be a polynomial in $y$, and the degree of this polynomial determines the number of boundary conditions which may be satisfied. For the case under consideration we propose the following form:

$$\frac{u}{U} = a_0 + a_1\frac{y}{\delta} + a_2\left(\frac{y}{\delta}\right)^2$$

Then three boundary conditions may be satisfied and since these boundary conditions all involve constants, the quantities $a_0$, $a_1$, and $a_2$ will be constants. Thus the velocity profile represented above will be similar at the various values of $x$ and so represents a similarity type of profile. The boundary conditions $u(x, 0) = 0$, $u(x, \delta) = U$, and $\partial u/\partial y(x, \delta) = 0$ give, respectively,

$$0 = a_0$$

$$1 = a_0 + a_1 + a_2$$

$$0 = a_1 + 2a_2$$

The solution to these equations is $a_0 = 0$, $a_1 = 2$, and $a_2 = -1$, so that the velocity profile becomes

$$\frac{u}{U} = 2\frac{y}{\delta} - \left(\frac{y}{\delta}\right)^2 \tag{9.10a}$$

Using the assumed velocity profile across the boundary layer will reduce the momentum integral to an ordinary differential equation for the boundary-layer thickness $\delta(x)$. The terms which appear in the momentum integral may be evaluated as follows:

$$\int_0^\delta u(U - u)\, dy = U^2 \int_0^\delta \frac{u}{U}\left(1 - \frac{u}{U}\right) dy$$

$$= U^2 \int_0^\delta \left[2\frac{y}{\delta} - \left(\frac{y}{\delta}\right)^2\right]\left[1 - 2\frac{y}{\delta} + \left(\frac{y}{\delta}\right)^2\right] dy$$

$$= \delta U^2 \int_0^1 (2\eta - \eta^2)(1 - 2\eta + \eta^2)\, d\eta$$

$$= \tfrac{2}{15} \delta U^2$$

$$\tau_0 = \mu \frac{\partial u}{\partial y}\bigg|_{y=0}$$

$$= \mu U \frac{\partial}{\partial y}\left[2\frac{y}{\delta} - \left(\frac{y}{\delta}\right)^2\right]_{y=0}$$

$$= \mu \frac{U}{\delta} \frac{\partial}{\partial \eta}(2\eta - \eta^2)\bigg|_{\eta=0}$$

$$= 2\mu \frac{U}{\delta}$$

Substituting these results into the momentum integral gives

$$\frac{d}{dx}\left(\tfrac{2}{15} \delta U^2\right) = \frac{2\nu U}{\delta}$$

$$\delta\, d\delta = 15\frac{\nu}{U}\, dx$$

Integrating this equation and setting $\delta = 0$ when $x = 0$ gives

$$\delta = \sqrt{30}\sqrt{\frac{\nu x}{U}}$$

In nondimensional form this result becomes

$$\frac{\delta}{x} = \frac{5.48}{\sqrt{R_N}} \tag{9.10$b$}$$

where the length scale which has been used in the Reynolds number is the distance $x$. Equation (9.10$a$) compares favorably with Eq. (9.5$c$) which is the exact solution for a flat surface. The relation $\tau_0 = 2\mu U/\delta$ shows that the shear stress on the surface is given by

$$\frac{\tau_0}{\tfrac{1}{2}\rho U^2} = \frac{0.73}{\sqrt{R_N}} \tag{9.10$c$}$$

This result also compares favorably with the exact solution which is given by Eq. (9.5*a*).

It is evident that the momentum-integral approach is capable of producing meaningful results, even when it is used in conjunction with a rather crude approximation to the form of the velocity profile. In the case under consideration here a second-degree polynomial was used. An even more crude representation of the velocity profile would be a straight line which matches only the boundary conditions $u(x, 0) = 0$ and $u(x, \delta) = U$. On the other hand, third-, or higher-degree polynomials could also be employed which would yield more accurate results. The second-order profile which was used here gives $\partial^2 u / \partial y^2 (x, 0) = -2U/\delta^2$ instead of zero. By employing a third-degree polynomial the correct velocity curvature could be imposed, which would yield more accurate results. This will be confirmed in the problems at the end of the chapter.

## 9.9 GENERAL MOMENTUM INTEGRAL

The momentum integral which was developed in the previous section for flat surfaces will be generalized here to include outer flows whose velocities are, in general, functions of $x$. The boundary layer may still be considered to be stretched out in a plane, provided the radius of curvature of the body is large compared with the boundary-layer thickness and centrifugal effects are negligible. In such cases the outer-flow velocity $U(x)$ will not be constant but will be defined by the potential-flow solution for the body under consideration.

Performing the same manipulation on the term $u \, \partial u / \partial x$ as was carried out in the previous section, the boundary-layer equations may be written in the form

$$\frac{\partial u}{\partial x} + \frac{\partial v}{\partial y} = 0$$

$$\frac{\partial}{\partial x}(u^2) + \frac{\partial}{\partial y}(uv) = U \frac{dU}{dx} + \frac{\mu}{\rho} \frac{\partial^2 u}{\partial y^2}$$

Integrating the second equation across the boundary layer and utilizing the boundary conditions $u(x, 0) = 0$, $u(x, \delta) = U$, $\mu \, \partial u / \partial y(x, 0) = \tau_0$, and $\partial u / \partial y(x, \delta) = 0$, gives

$$\int_0^\delta \frac{\partial}{\partial x}(u^2) \, dy + Uv(x, \delta) = \frac{dU}{dx} \int_0^\delta U \, dy - \frac{\tau_0}{\rho}$$

The outer-flow velocity $U(x)$ at the edge of the boundary layer depends upon $x$ only, and $dU/dx$ has been taken outside the integral while $U$ has been kept inside the integral in the pressure term. This is purely a matter of convenience which will permit two integrals to be combined in the subsequent analysis.

Integrating the continuity equation shows that

$$v(x, \delta) = -\int_0^\delta \frac{\partial u}{\partial x} \, dy$$

so that the momentum-integral equation becomes

$$\int_0^\delta \frac{\partial}{\partial x}(u^2) \, dy - U\int_0^\delta \frac{\partial u}{\partial x} \, dy = \frac{dU}{dx}\int_0^\delta U \, dy - \frac{\tau_0}{\rho}$$

Using Leibnitz's rule permits the order of the integration and the differentiation to be interchanged, yielding the following result:

$$\frac{d}{dx}\int_0^\delta u^2 \, dy - U^2\frac{d\delta}{dx} - U\frac{d}{dx}\int_0^\delta u \, dy + U^2\frac{d\delta}{dx} = \frac{dU}{dx}\int_0^\delta U \, dy - \frac{\tau_0}{\rho}$$

The second integral may be rewritten as follows:

$$U\frac{d}{dx}\int_0^\delta u \, dy = \frac{d}{dx}\int_0^\delta Uu \, dy - \frac{dU}{dx}\int_0^\delta u \, dy$$

Thus the momentum integral becomes

$$\frac{d}{dx}\int_0^\delta u^2 \, dy - \frac{d}{dx}\int_0^\delta Uu \, dy + \frac{dU}{dx}\int_0^\delta u \, dy = \frac{dU}{dx}\int_0^\delta U \, dy - \frac{\tau_0}{\rho}$$

The first and second integrals may now be combined and the third and fourth integrals may be combined to give

$$\frac{d}{dx}\int_0^\delta u(U - u) \, dy + \frac{dU}{dx}\int_0^\delta (U - u) \, dy = \frac{\tau_0}{\rho}$$

But the integrands of these two integrals are essentially zero for $y > \delta$, so that the upper limits of integration may be taken to be infinity. This gives

$$\frac{d}{dx}\left[ U^2\int_0^\infty \frac{u}{U}\left(1 - \frac{u}{U}\right) dy \right] + \frac{dU}{dx}U\int_0^\infty \left(1 - \frac{u}{U}\right) dy = \frac{\tau_0}{\rho}$$

Now the first integral is the momentum thickness $\theta$ and the second integral is the displacement thickness $\delta^*$, as may be seen from comparison with Eqs. (9.1c) and (9.1b), respectively. Then the momentum integral may be rewritten in the form

$$\frac{d}{dx}(U^2\theta) + \frac{dU}{dx}U\delta^* = \frac{\tau_0}{\rho}$$

Expanding the first derivative and dividing the entire equation by $U^2$ yields the following alternative form of the generalized momentum integral:

$$\frac{d\theta}{dx} + (2\theta + \delta^*)\frac{1}{U}\frac{dU}{dx} = \frac{\tau_0}{\rho U^2} \tag{9.11}$$

For any assumed form of velocity profile across the boundary layer, $\theta$, $\delta^*$, and $\tau_0$ may be evaluated from their definitions. Then Eq. (9.11) will provide an ordinary differential equation which may be solved for the boundary-layer thickness. The manner in which the solution is carried out, for the case of a fourth-order polynomial for $u$, will be covered in the next section.

## 9.10 KÁRMÁN-POHLHAUSEN APPROXIMATION

The general momentum integral, when used in conjunction with a fourth-order polynomial to represent the velocity profile, is known as the Kárman-Pohlhausen method. The velocity profile is taken to be of the form

$$\frac{u}{U} = a + b\eta + c\eta^2 + d\eta^3 + e\eta^4$$

where

$$\eta(x, y) = \frac{y}{\delta(x)}$$

The coefficients $a$, $b$, $c$, $d$, and $e$ will, in general, be functions of $x$, so that solutions which are not similar may be obtained. The foregoing velocity profile can satisfy five boundary conditions, and these are taken to be

$$u(x, 0) = 0$$

$$u(x, \delta) = U(x)$$

$$\frac{\partial u}{\partial y}(x, \delta) = 0$$

$$\frac{\partial^2 u}{\partial y^2}(x, 0) = -\frac{U(x)}{\nu}\frac{dU(x)}{dx}$$

$$\frac{\partial^2 u}{\partial y^2}(x, \delta) = 0$$

The fourth boundary condition comes from the boundary-layer form of the momentum equation and the no-slip boundary condition. In terms of the dimensionless velocity $u/U$ and the dimensionless coordinate $\eta = y/\delta$, these boundary conditions become

$$\frac{u}{U} = 0 \quad \text{and} \quad \frac{\partial^2(u/U)}{\partial\eta^2} = -\frac{\delta^2}{\nu}\frac{dU}{dx} = -\Lambda(x) \quad \text{on } \eta = 0$$

and

$$\frac{u}{U} = 1 \quad \text{and} \quad \frac{\partial(u/U)}{\partial\eta} = \frac{\partial^2(u/U)}{\partial\eta^2} = 0 \quad \text{on } \eta = 1$$

The quantity $\Lambda(x)$ which has been introduced here is a dimensionless variable which is a measure of the pressure gradient in the outer flow.

Imposing the foregoing boundary conditions on the assumed form of velocity profile gives the following set of algebraic equations for the unknown coefficients:

$$0 = a$$
$$-\Lambda = 2c$$
$$1 = a + b + c + d + e$$
$$0 = b + 2c + 3d + 4e$$
$$0 = 2c + 6d + 12e$$

The solution to this set of equations is

$$a = 0$$
$$b = 2 + \frac{\Lambda}{6}$$
$$c = -\frac{\Lambda}{2}$$
$$d = -2 + \frac{\Lambda}{2}$$
$$e = 1 - \frac{\Lambda}{6}$$

Thus the assumed form of velocity profile which satisfies the principal boundary conditions is

$$\frac{u}{U} = \left(2 + \frac{\Lambda}{6}\right)\eta - \frac{\Lambda}{2}\eta^2 - \left(2 - \frac{\Lambda}{2}\right)\eta^3 + \left(1 - \frac{\Lambda}{6}\right)\eta^4$$

It is advantageous to separate the right-hand side of this expression into terms which are independent of $\Lambda(x)$ and terms which depend upon $\Lambda(x)$. This gives

$$\frac{u}{U} = (2\eta - 2\eta^3 + \eta^4) + \frac{\Lambda}{6}(\eta - 3\eta^2 + 3\eta^3 - \eta^4)$$

$$\frac{u}{U} = 1 - (1 + \eta)(1 - \eta)^3 + \frac{\Lambda}{6}\eta(1 - \eta)^3$$

$$(9.12a)$$

Equation $(9.12a)$ is now in the form

$$\frac{u}{U} = F(\eta) + \Lambda G(\eta)$$

where the functions $F(\eta)$ and $G(\eta)$ are shown schematically in Fig. 9.6a. The function $F(\eta)$ is seen to be a monotonically increasing function of $\eta$ which ranges from zero at $\eta = 0$ to unity at $\eta = 1$. The function $G(\eta)$ increases from zero at $\eta = 0$ to a maximum of 0.0166 at $\eta = 0.25$, after which it drops off to zero at $\eta = 1$.

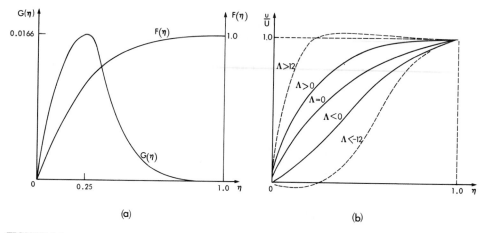

**FIGURE 9.6**
(a) Form of the functions $F(\eta)$ and $G(\eta)$ and (b) the velocity profiles for the various values of the pressure parameter $\Lambda(x)$.

Figure 9.6b shows the velocity profiles corresponding to various values of the pressure parameter $\Lambda$. For $\Lambda = 0$ the velocity profile corresponds to a flat surface in which the representation is a fourth-order polynomial. For values of $\Lambda$ which are greater than 12, the resulting velocity profiles show that $u/U > 1$. Since the boundary-layer velocity is not expected to exceed that of the outer flow locally, it is concluded that $\Lambda < 12$. Also, for values of $\Lambda$ which are less than $-12$, the velocity profiles show negative regions which correspond to reverse flow. Although reverse flows do occur physically, under these conditions the basic assumptions upon which the theory is based cannot be justified. Thus the parameter $\Lambda$ should be greater than $-12$. Combining these two results it is concluded that the parameter $\Lambda$ should lie in the range

$$-12 < \Lambda(x) < 12 \tag{9.12b}$$

Having established a suitable approximation to the velocity profile, the boundary-layer thicknesses and the surface shear stress may be evaluated. Substituting the velocity profile (9.12a) into Eq. (9.1b) gives the following expression for the displacement thickness in terms of the boundary-layer thickness:

$$\delta^* = \delta \int_0^1 \left(1 - \frac{u}{U}\right) d\eta$$

$$= \delta \int_0^1 \left[(1 + \eta)(1 - \eta)^3 - \frac{\Lambda}{6}\eta(1 - \eta)^3\right] d\eta \tag{9.12c}$$

$$\delta^* = \delta \left(\frac{3}{10} - \frac{\Lambda}{120}\right)$$

Similarly, Eq. (9.1c) yields the following expression for the momentum thickness:

$$\theta = \delta \int_0^1 \frac{u}{U}\left(1 - \frac{u}{U}\right) d\eta$$

$$= \delta \int_0^1 \left[1 - (1 + \eta)(1 - \eta)^3 + \frac{\Lambda}{6}\eta(1 - \eta)^3\right]$$

$$\times \left[(1 + \eta)(1 - \eta)^3 - \frac{\Lambda}{6}\eta(1 - \eta)^3\right] d\eta$$

$$\theta = \delta \left(\frac{37}{315} - \frac{\Lambda}{945} - \frac{\Lambda^2}{9,072}\right)$$

(9.12d)

The shear stress on the surface for this velocity distribution will be

$$\tau_0 = \mu \frac{U}{\delta} \left.\frac{\partial(u/U)}{\partial \eta}\right|_{\eta=0}$$

$$= \mu \frac{U}{\delta} \frac{\partial}{\partial \eta}\left[1 - (1 + \eta)(1 - \eta)^3 + \frac{\Lambda}{6}\eta(1 - \eta)^3\right]\Bigg|_{\eta=0}$$

(9.12e)

$$\tau_0 = \mu \frac{U}{\delta}\left(2 + \frac{\Lambda}{6}\right)$$

The foregoing expressions relate the displacement and momentum thicknesses and the surface shear stress to the boundary-layer thickness, which as yet, is unknown. These relations follow purely from the velocity profile under consideration. The additional relation which is required to determine the absolute value of these various quantities is supplied by the momentum integral.

Multiplying the momentum integral [Eq. (9.11)] by $U\theta/\nu$ gives the following additional relation connecting $\theta$, $\delta^*$, and $\tau_0$:

$$\frac{U\theta}{\nu}\frac{d\theta}{dx} + (2\theta + \delta^*)\frac{\theta}{\nu}\frac{dU}{dx} = \frac{\tau_0\theta}{\mu U}$$

or

$$\frac{1}{2}U\frac{d}{dx}\left(\frac{\theta^2}{\nu}\right) + \left(2 + \frac{\delta^*}{\theta}\right)\frac{\theta^2}{\nu}\frac{dU}{dx} = \frac{\tau_0\theta}{\mu U}$$

Expressions for the various quantities which appear in this equation will now be established as functions of the pressure parameter $\Lambda(x)$. Recall

$$\Lambda = \frac{\delta^2}{\nu}\frac{dU}{dx}$$

hence

$$\frac{\theta^2}{\nu}\frac{dU}{dx} = \frac{\theta^2}{\delta^2}\Lambda$$

But $\theta/\delta$ may be evaluated in terms of $\Lambda$ through Eq. (9.12$d$). This gives

$$\frac{\theta^2}{\nu}\frac{dU}{dx} = \left(\frac{37}{315} - \frac{\Lambda}{945} - \frac{\Lambda^2}{9,072}\right)^2 \Lambda$$

or

$$\frac{\theta^2}{\nu}\frac{dU}{dx} = K(x)$$

where

$$K(x) = \left(\frac{37}{315} - \frac{\Lambda}{945} - \frac{\Lambda^2}{9,072}\right)\Lambda$$

The term $\delta^*/\theta$ which appears in the momentum integral may be similarly evaluated using Eqs. (9.12$c$) and (9.12$d$). Thus

$$\frac{\delta^*}{\theta} = \frac{\left(\dfrac{3}{10} - \dfrac{\Lambda}{120}\right)}{\left(\dfrac{37}{315} - \dfrac{\Lambda}{945} - \dfrac{\Lambda^2}{9,072}\right)}$$

or

$$\frac{\delta^*}{\theta} = f(K)$$

where

$$f(K) = \frac{\left(\dfrac{3}{10} - \dfrac{\Lambda}{120}\right)}{\left(\dfrac{37}{315} - \dfrac{\Lambda}{945} - \dfrac{\Lambda^2}{9,072}\right)}$$

The function $f$ depends on $\Lambda$ and hence upon $x$. However, $K$ is also a function of $x$, so that $f$ may be considered to be a function of $K$. The other parameter which appears in the momentum integral is $\tau_0\theta/(\mu U)$. Multiplying Eqs. (9.12$e$) and (9.12$d$) gives the following expression for this parameter:

$$\frac{\tau_0\theta}{\mu U} = g(K)$$

where

$$g(K) = \left(2 + \frac{\Lambda}{6}\right)\left(\frac{37}{315} - \frac{\Lambda}{945} - \frac{\Lambda^2}{9,072}\right)$$

These results will now be substituted into the momentum integral. For the time being the leading term in the momentum integral will be retained in its existing form. Substituting from the above results for $\delta^*/\theta$, $\theta^2(dU/dx)/\nu$, and $\tau_0\theta(\mu U)$ then gives

$$\tfrac{1}{2}U\frac{d}{dx}\left(\frac{\theta^2}{\nu}\right) + [2 + f(K)]K = g(K)$$

where
$$K = \frac{\theta^2}{\nu} \frac{dU}{dx}$$

It is now proposed to take $Z = \theta^2/\nu$ as a new dependent variable so that
$$K = Z \frac{dU}{dx}$$

and the momentum integral becomes
$$U \frac{dZ}{dx} = 2\{g(K) - [2 + f(K)]K\}$$

or
$$U \frac{dZ}{dx} = H(K)$$

where
$$H(K) = 2\{g(K) - [2 + f(K)]K\}$$

Then, substituting for $g(K)$ and $K$ from their definitions shows that $H(K)$ is related to $\Lambda(x)$ through the following identity:

$$
\begin{aligned}
H(K) = 2 &\left\{ \left(2 + \frac{\Lambda}{6}\right)\left(\frac{37}{315} - \frac{\Lambda}{945} - \frac{\Lambda^2}{9{,}072}\right) \right.\\
&\left. - \left[2 + \frac{\left(\dfrac{3}{10} - \dfrac{\Lambda}{120}\right)}{\left(\dfrac{37}{315} - \dfrac{\Lambda}{945} - \dfrac{\Lambda^2}{9{,}072}\right)}\right]\left(\frac{37}{315} - \frac{\Lambda}{945} - \frac{\Lambda^2}{9{,}072}\right)^2 \Lambda \right\}\\
= 2&\left(\frac{37}{315} - \frac{\Lambda}{945} - \frac{\Lambda^2}{9{,}072}\right)\left[2 - \frac{116}{315}\Lambda + \left(\frac{2}{945} + \frac{1}{120}\right)\Lambda^2 + \frac{2}{9{,}072}\Lambda^3\right]
\end{aligned}
$$

where the quantity $K$ is related to $\Lambda$ by the expression

$$K = \left(\frac{37}{315} - \frac{\Lambda}{945} - \frac{\Lambda^2}{9{,}072}\right)^2 \Lambda$$

From these two expressions, both $K$ and $H(K)$ may be evaluated for any value of the pressure parameter $\Lambda(x)$. Thus a curve of $H(K)$ as a function of $K$ may be constructed. The form of this curve is shown in Fig. 9.7. The momentum integral has been reduced to the ordinary differential equation $U\,dZ/dx = H(K)$ where the functional form of the quantity $H$ is sufficiently complex that this integral cannot be evaluated explicitly. However, it may be seen from Fig. 9.7 that the function $H$ is approximately linear in $K$ over the range of interest.

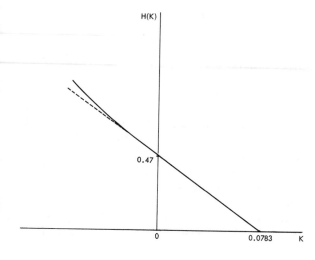

**FIGURE 9.7**
Exact form of the function $H(K)$
(solid line) and straight-line approximation (dotted line).

Thus the function $H$ may be approximated by the equation

$$H(K) = 0.47 - 6K$$

Then the momentum integral becomes

$$U\frac{dZ}{dx} = 0.47 - 6K$$

$$= 0.47 - 6Z\frac{dU}{dx}$$

or

$$\frac{1}{U^5}\frac{d}{dx}(ZU^6) = 0.47$$

In this form the momentum integral may be expressed in terms of the following
quadrature:

$$Z(x) = \frac{0.47}{U(x)^6}\int_0^x U(\xi)^5 \, d\xi$$

Then, since $Z = \theta^2/\nu$, the value of $\theta$ will be

$$\theta^2(x) = \frac{0.47\nu}{U(x)^6}\int_0^x U(\xi)^5 \, d\xi \qquad (9.12f)$$

For any given boundary shape the approximate solution to the boundary-
layer equations may be obtained as follows: For the specified boundary shape
the potential-flow problem should be solved to yield the outer velocity $U(x)$.
Then Eq. (9.12f) may be used to evaluate the momentum thickness $\theta(x)$. The

pressure parameter $\Lambda(x)$ may then be evaluated from the relation

$$\left( \frac{37}{315} - \frac{\Lambda}{945} - \frac{\Lambda^2}{9{,}072} \right)^2 \Lambda = \frac{\theta^2}{\nu} \frac{dU}{dx} \tag{9.12g}$$

Having established $\Lambda$, the boundary-layer thickness $\delta$ may be evaluated from Eq. (9.12d), and the displacement thickness $\delta^*$ may then be obtained from Eq. (9.12c). The velocity distribution across the boundary layer will be given by Eq. (9.12a), and the shear stress at the surface will be given by Eq. (9.12e). In practice it is difficult to evaluate the quantity $\Lambda(x)$ from Eq. (9.12g) unless $\Lambda = $ constant. It is therefore much simpler to choose specific functions $\Lambda(x)$ and use the foregoing equations to determine the outer-flow velocity and hence the nature of the boundary shape.

As an example, the Kármán-Pohlhausen approximation will be applied to the case of flow over a flat surface. For a flat surface the outer-flow velocity $U$ will be constant, so that Eq. (9.12f) gives

$$\theta^2 = 0.47 \frac{\nu x}{U}$$

or
$$\frac{\theta}{x} = \frac{0.686}{\sqrt{R_N}} \tag{9.13a}$$

Since $U = $ constant, $dU/dx = 0$, so that Eq. (9.12g) will have the solution $\Lambda = 0$. Then, from Eq. (9.12d),

$$\theta = \tfrac{37}{315}\delta$$

so that, from Eq. (9.13a), the boundary-layer thickness will be

$$\frac{\delta}{x} = \frac{5.84}{\sqrt{R_N}} \tag{9.13b}$$

Equations (9.12c) and (9.12e) may now be employed to evaluate the displacement thickness and the surface shear stress. This gives

$$\frac{\delta^*}{x} = \frac{1.75}{\sqrt{R_N}} \tag{9.13c}$$

$$\frac{\tau_0}{\tfrac{1}{2}\rho U^2} = \frac{0.686}{\sqrt{R_N}} \tag{9.13d}$$

These results compare favorably with the results obtained from the Blasius solution which are given by Eqs. (9.5). The principal result which is of physical interest is the surface shear stress. The exact solution has a coefficient of 0.664, whereas the Kármán-Pohlhausen approximate solution has a coefficient of 0.686. Thus the shear-stress distribution obtained here is within 3.5 percent of the exact solution. It should also be noted that there is considerable improvement with the fourth-order polynomial velocity distribution used here over the second-order polynomial analysis which yielded a coefficient of 0.73 as was established in Eq. (9.10c).

## 9.11 BOUNDARY-LAYER SEPARATION

It is known from experimental observations that boundary layers have a tendency to separate from the surface over which they flow to form a wake behind the body as shown in Fig. 9.1. The existence of such wakes leads to large streamwise pressure differentials across the body, which results in a substantial pressure drag or form drag. Indeed, for bluff bodies such as circular cylinders, the form drag constitutes almost all the total drag at Reynolds numbers or $10^4$ or higher. That is, the shear stress along the surface of a cylinder produces a drag force which is negligible compared with the form drag for large Reynolds numbers. For lifting bodies such as airfoils, separation of the boundary layer can destroy the bound vortex on the body, thus destroying the lift which the airfoil generates. This is the so-called "stall condition."

A simple qualitative explanation for the existence of boundary-layer separation on a bluff body may be given as follows: The pressure gradient along a boundary layer is determined by that of the outer flow, as was established earlier. Then, if a region of adverse pressure gradient exists in the outer flow, this pressure gradient will exert itself along the surface of the body near which the fluid velocity is small. The momentum contained in the fluid layers which are adjacent to the surface will be insufficient to overcome the force exerted by teh pressure gradient, so that a region of reverse flow will exist. That is, at some point the adverse pressure gradient will cause the fluid layers adjacent to the surface to flow in a direction opposite to that of the outer flow. Such a flow configuration means that the boundary layer has separated from the surface and is deflected over the reverse-flow region.

Figure 9.8 shows the qualitative form of the velocity profile in a boundary layer in the vicinity of the separation point. Prior to separation the velocity gradient at the surface is positive, so that the shear stress there opposes the outer-flow field. After separation the velocity gradient at the surface is negative,

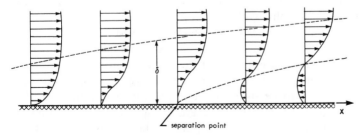

**FIGURE 9.8**
Velocity profiles in a boundary layer in the vicinity of separation.

so that the shear stress has changed its sign and direction. This observation leads to the classical definition of a separation point as a point at which the shear stress vanishes. That is, separation is said to occur at the point where the velocity gradient vanishes.

$$\frac{\partial u}{\partial y}(x,0) = 0 \qquad \text{for separation} \qquad (9.14)$$

Using this definition of separation, it may be shown that separation can occur only in a region of adverse pressure gradient. Along the surface $y = 0$ the boundary-layer equations reduce to

$$0 = -\frac{dp}{dx} + \mu \frac{\partial^2 u}{\partial y^2}$$

owing to the no-slip boundary condition. Thus the curvature of the velocity profile is proportional to the pressure gradient along the surface. Then if $dp/dx$ is negative, the curvature of the velocity profile is negative and will remain negative at the surface just as it is at the edge of the boundary layer. That is, separation will not occur in a region of favorable pressure gradient. On the other hand, if $dp/dx$ is positive, the curvature of the velocity profile will be positive at the surface. Since $\partial^2 u/\partial y^2$ must still be negative at the edge of the boundary layer, the velocity profile must go through an inflection point some-where between $y = 0$ and $y = \delta$. Such a velocity profile may lead to separation if the curvature at $y = 0$ is sufficiently positive to yield a reverse-flow configura-tion as shown in Fig. 9.8. Thus it may be concluded that separation can occur in a region of positive pressure gradient.

Calculating the location of the separation point is not an easy matter. The obvious way of proceeding is first to solve the potential-flow problem for the body in question. The pressure so obtained could then be substituted into the boundary-layer equations, which could then be solved by either an exact solution or an approximate solution. From the solution to the boundary-layer equations the location of the point of zero shear stress could then be located. The obvious difficulty with such a procedure is that as soon as the boundary layer separates, the pressure distribution will differ from that predicted by the potential-flow solution, since the latter applies to a different streamline config-uration from that which exists physically.

There are two principal approaches which are used to overcome the difficulty outlined above. The approach which was used by Hiemenz involved determining the pressure distribution around the body in question experimen-tally. The resulting pressure curve may then be represented analytically by a polynomial which permits it to be used in the boundary-layer equations. The results obtained by this method show good agreement with experimental obser-vations. However, the disadvantage of this approach lies in the fact that the pressure distribution must be established experimentally for each body shape and for each Reynolds number of interest. Measuring the pressure distribution around circular cylinders is not difficult, since a single pressure tap may be rotated to sense the pressure at different angles from the front stagnation point. On the other hand, measuring the pressure distribution around noncircular cylinders is not such a simple matter.

The second approach which is used to determine analytically the location of the separation point is to modify the potential-flow model from which the pressure distribution is obtained. Several flow models exist, each of which takes into account the separated configuration of the outer flow. The difficulty with this approach is that empirical constants exist in the potential-flow model and experimental results must be consulted to establish these constants.

From the foregoing discussion it is evident that the subject of boundary-layer separation is one which is not well understood analytically. Indeed, it is still not clear whether or not the boundary-layer equations are regular at separation. One school of thought claims that the boundary-layer equations are regular at separation by virtue of the appropriate pressure distribution. Some recent results even question the validity of the condition (9.14) at separation. Evidence suggests that the location of the point where $dp/dx$ vanishes, that where $\partial u/\partial y$ vanishes, and that where separation occurs are all distinct. However, no length scales could be established so that all these points could possible concur within the appropriate macroscopic length scales. What is known is that boundary layers will separate in adverse pressure gradients, so that the magnitude and extent of such pressure gradients should be minimized. This means that bodies should be streamlined rather than bluff and should be oriented at small angles of attack. Also, it is known that sharp corners which bend away from the fluid become separation points, so that such corners should be avoided if separation is to be delayed as far as possible.

## 9.12 STABILITY OF BOUNDARY LAYERS

Like any fluid-flow situation, boundary layers may become unstable. Usually, instabilities in boundary layers manifest themselves in turbulence. That is, a laminar boundary layer which becomes unstable usually becomes a turbulent boundary layer. The properties of laminar and turbulent boundary layers are quite different. For example, the angle to the location of the separation point on a circular cylinder, measured from the front stagnation point, is about 82° for a laminar boundary layer and about 108° for a turbulent boundary layer. This significant change in location of the separation points results in an appreciable drop in the drag coefficient as shown in Fig. 8.2. It is therefore of some interest to investigate the stability of boundary layers.

The basis of our stability calculation will be to introduce a small disturbance into the boundary-layer variables and determine whether this disturbance grows or decays with time. If the disturbance grows with time, the boundary layer will be classified as unstable, and if the disturbance grows with time, the boundary layer will be classified as stable. Intermediate to these two situations

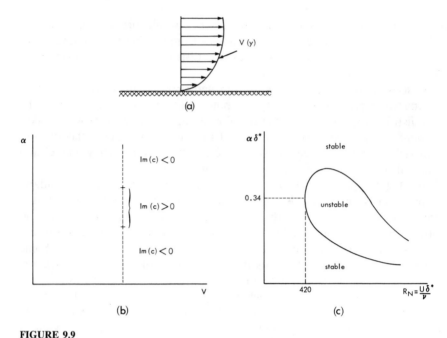

**FIGURE 9.9**
(*a*) Undisturbed boundary-layer velocity profile, (*b*) stability-calculation results for fixed *V*, and (*c*) stability diagram.

is the case of marginal stability in which the disturbance neither grows nor decays.

Figure 9.9*a* shows the velocity profile in a narrow strip of a boundary layer. For such a narrow strip the velocity in the horizontal direction may be considered to be a funciton of *y* only, say *V*(*y*), and the vertical velocity may be considered to be zero. The undisturbed boundary-layer velocity *V*(*y*) acts in the horizontal direction, although the symbol *V* has been used. This symbol has been employed to avoid confusion with the outer-flow velocity *U*(*x*) at the edge of the boundary layer.

A small, but arbitrary, disturbance is introduced to this boundary-layer velocity profile so that the velocity components and the pressure become

$$u(x, y, t) = V(y) + u'(x, y, t)$$

$$v(x, y, t) = 0 + v'(x, y, t)$$

$$p(x, y, t) = p_0(x) + p'(x, y, t)$$

where $|u'/V|$, $|v'/V|$, and $|p'/p_0|$ are all small compared with unity. Substituting these instantaneous local values of the velocity components and the pressure into the continuity and the Navier-Stokes equations gives

$$\frac{\partial u'}{\partial x} + \frac{\partial v'}{\partial y} = 0$$

$$\frac{\partial u'}{\partial t} + (V + u')\frac{\partial u'}{\partial x} + v'\left(\frac{dV}{dy} + \frac{\partial u'}{\partial y}\right)$$

$$= -\frac{1}{\rho}\left(\frac{dp_0}{dx} + \frac{\partial p'}{\partial x}\right) + \nu\left(\frac{\partial^2 u'}{\partial x^2} + \frac{d^2 V}{dy^2} + \frac{\partial^2 u'}{\partial y^2}\right)$$

$$\frac{\partial v'}{\partial t} + (V + u')\frac{\partial v'}{\partial x} + v'\frac{\partial v'}{\partial y} = -\frac{1}{\rho}\frac{\partial p'}{\partial y} + \nu\left(\frac{\partial^2 v'}{\partial x^2} + \frac{\partial^2 v'}{\partial y^2}\right)$$

As a special case, when the perturbation is zero, the foregoing equations reduce to

$$0 = -\frac{1}{\rho}\frac{dp_0}{dx} + \nu\frac{d^2 V}{dy^2}$$

Hence these terms may be removed from the equation of $x$ momentum. Furthermore, since the perturbation is assumed to be small, products of all primed quantities may be neglected as being small. Thus the linearized equations governing the motion of the disturbance are

$$\frac{\partial u'}{\partial x} + \frac{\partial v'}{\partial y} = 0$$

$$\frac{\partial u'}{\partial t} + V\frac{\partial u'}{\partial x} + v'\frac{dV}{dy} = -\frac{1}{\rho}\frac{\partial p'}{\partial x} + \nu\left(\frac{\partial^2 u'}{\partial x^2} + \frac{\partial^2 u'}{\partial y^2}\right)$$

$$\frac{\partial v'}{\partial t} + V\frac{\partial v'}{\partial x} = -\frac{1}{\rho}\frac{\partial p'}{\partial y} + \nu\left(\frac{\partial^2 v'}{\partial x^2} + \frac{\partial^2 v'}{\partial y^2}\right)$$

These three equations may be reduced to two by introducing a perturbation stream function defined by

$$u' = \frac{\partial \psi}{\partial y} \qquad v' = -\frac{\partial \psi}{\partial x}$$

In terms of this stream function the governing equations become

$$\frac{\partial^2 \psi}{\partial y \, \partial t} + V\frac{\partial^2 \psi}{\partial x \, \partial y} - \frac{\partial \psi}{\partial x}\frac{dV}{dy} = -\frac{1}{\rho}\frac{\partial p'}{\partial x} + \nu\left(\frac{\partial^3 \psi}{\partial x^2 \, \partial y} + \frac{\partial^3 \psi}{\partial y^3}\right)$$

$$-\frac{\partial^2 \psi}{\partial x \, \partial t} - V\frac{\partial^2 \psi}{\partial x^2} = -\frac{1}{\rho}\frac{\partial p'}{\partial y} - \nu\left(\frac{\partial^3 \psi}{\partial x^3} + \frac{\partial^3 \psi}{\partial x \, \partial y^2}\right)$$

Finally, by forming the mixed derivative $\partial^2 p'/\partial x\,\partial y$, these two equations may be reduced to one which may be written in the form

$$\left(\frac{\partial}{\partial t} + V\frac{\partial}{\partial x}\right)\left(\frac{\partial^2 \psi}{\partial y^2} + \frac{\partial^2 \psi}{\partial x^2}\right) - \frac{d^2 V}{dy^2}\frac{\partial \psi}{\partial x} = \nu\left(\frac{\partial^4 \psi}{\partial y^4} + 2\frac{\partial^4 \psi}{\partial x^2\,\partial y^2} + \frac{\partial^4 \psi}{\partial x^4}\right)$$

The stream function for the disturbance must satisfy this linear, fourth-order, partial differential equation.

Since the disturbance under consideration is arbitrary in form, it may be Fourier-analyzed in the $x$ direction. That is, the perturbation stream function may be represented by the following Fourier integral:

$$\psi(x, y, t) = \int_0^\infty \Psi(y)e^{i\alpha(x-ct)}\,d\alpha$$

where $\alpha$ is real and positive. The time variation has been taken to be $e^{-i\alpha ct}$, so that if the imaginary part of $c$ is positive, the disturbance will grow, and if it is negative, the disturbance will decay with time. For $c = 0$, the disturbance will be neutrally stable. Substituting the above expression for the stream function into the governing equation yields the following integro-differential equation:

$$\int_0^\infty \left[(-i\alpha c + i\alpha V)(\Psi'' - \alpha^2\Psi) - i\alpha\Psi V''\right]e^{i\alpha(x-ct)}\,d\alpha$$

$$= \int_0^\infty \left[\nu(\Psi'''' - 2\alpha^2\Psi'' + \alpha^4\Psi)\right]e^{i\alpha(x-ct)}\,d\alpha$$

where the primes denote derivatives with respect to $y$. Since this equation should be valid for any arbitrary disturbance whatsoever, it should be valid for each individual value of the inverse wavelength $\alpha$. Thus the integrand in the above equation should vanish. This gives

$$(V - c)(\Psi'' - \alpha^2\Psi) - V''\Psi = \frac{\nu}{i\alpha}(\Psi'''' - 2\alpha^2\Psi'' + \alpha^4\Psi) \qquad (9.15a)$$

Equation (9.15a) is known as the *Orr-Sommerfeld* equation. The boundary conditions which accompany this equation may be derived from the condition that the disturbance should vanish at the surface $y = 0$ and at the edge of the boundary layer. Thus

$$u'(x, 0, t) = v'(x, 0, t) = 0$$
$$u'(x, y, t) = v'(x, y, t) \to 0 \qquad \text{as } y \to \infty$$

In terms of the stream function $\Psi(y)$ these boundary conditions become

$$\Psi(0) = \Psi'(0) = 0 \qquad\qquad (9.15b)$$
$$\Psi(y) = \Psi'(y) \to 0 \qquad \text{as } y \to \infty \qquad\qquad (9.15c)$$

Solutions to the Orr-Sommerfeld equation are obtained as follows. For a given undisturbed velocity profile and disturbance wavelength, both $V(y)$ and $\alpha$ will be known. Then Eqs. (9.15a), (9.15b), and (9.15c) represent an eigenvalue problem for the time coefficient $c$. Then if each possible wavelength in turn is treated, results of the form indicated in Fig. 9.9$b$ may be established. That is, regions which are stable (corresponding to the imaginary part of $c$ being negative) and regions which are unstable (corresponding to the imaginary part of $c$ being positive) may be identified. Then, by considering all possible values of the undisturbed boundary-layer velocity which are less than the outer-flow velocity, a stability diagram may be constructed. That is, by considering all possible values of $V(y)$ in the range $0 \le V(y) \le U(x)$, the stability boundaries for that particular $x$ location may be established. Figure 9.9$c$ shows the results of carrying out such a procedure for flow over a flat surface. Here the Reynolds number has been based on the displacement thickness of the boundary layer and the inverse wavelength $\alpha$ has been nondimensionalized by the same quantity. It may be seen that the lower Reynolds number for which an instability may occur is 420. Thus

$$\frac{U\delta^*_{\text{crit}}}{\nu} = 420 \tag{9.16}$$

Hence an arbitrary disturbance which has a Fourier component whose wavelength is such that $\alpha\delta^* = 0.34$ will lie on the stability boundary. Thus it may be expected that for Reynolds numbers in excess of 420, arbitrary disturbances will

**PLATE 2**
Flow around a snowmobile. (*Photograph courtesy of National Research Council of Canada.*)

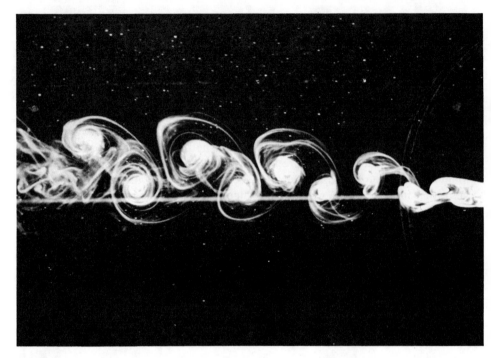

**PLATE 3**
Vortex street generated by a circular cylinder. (*Photograph courtesy of National Research Council of Canada.*)

be unstable. Such instabilities will manifest themselves in the form of turbulence at Reynolds numbers slightly larger than this critical value.

## PROBLEMS

**9.1.** It was observed in Sec. 9.2 that the boundary-layer equations are parabolic. Show that they may be put in the form of the one-dimensional diffusion or heat conduction equation by taking $h = p + \rho \mathbf{u} \cdot \mathbf{u}/2$ as a new dependent variable with $\xi = x$ and $\eta = \psi$ as new independent variables.

**9.2.** The Blasius solution for flow over a flat surface involves solving a third-order, nonlinear, ordinary differential equation. It will be noticed that this equation is invariant to the following transformations:

(*a*) $f \rightarrow f$, $\eta \rightarrow \eta + $ constant

(*b*) $f \rightarrow f/a$, $\eta \rightarrow a\eta$

Show that (*a*) enables the Blasius equation to be reduced to a second-order equation by taking $F = df/d\eta$ as a new dependent variable and $f$ as a new independent variable. Then show that (*b*) enables the resulting second-order equation to be reduced to a first-order equation by taking $G = (dF/df)/f$ as a new dependent variable and $\xi = F/f^2$ as a new independent variable. Find the resulting ordinary differential equation for $G(\xi)$.

**9.3.** In Sec. 9.7 a similarity solution to the boundary-layer equations reduced them to a third-order, nonlinear, ordinary differential equation. Show that this equation may be integrated to give

$$f'(\eta) = 3 \tanh^2 \left( \frac{\eta}{\sqrt{2}} + 1.146 \right) - 2$$

where the prime denotes differentiation with respect to $\eta$.

**9.4.** Fig. 9.10 illustrates a two-dimensional jet entering a reservoir which contains a stationary fluid. A solution is sought to the laminar boundary layer equations for this situation. Assuming that there is no pressure gradient along the jet, look for a similarity solution for the stream function of the following form:

$$\psi(x, y) = 6\alpha\nu x^{1/3} f(\eta)$$

where
$$\eta = \alpha \frac{y}{x^{2/3}}$$

In the above expressions, $\alpha$ is a dimensional constant and $\nu$ is the kinematic viscosity of the fluid. Obtain an expression for the function $f(\eta)$ in this solution and the boundary conditions which it has to satisfy. From the solution for $f(\eta)$, obtain the solution for the stream function $\psi(x, y)$.

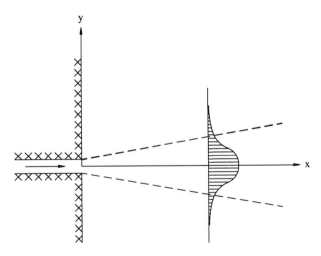

**FIGURE 9.10**
Jet entering a reservoir.

**9.5.** Use the momentum integral and the velocity profile

$$\frac{u}{U} = a + b\frac{y}{\delta}$$

to evaluate the boundary-layer thicknesses $\delta$, $\delta^*$, and $\theta$ and the surface shear stress $\tau_0$ for flow over a flat surface.

**9.6.** Repeat Prob. 9.4 using the velocity distribution

$$\frac{u}{U} = a + b\frac{y}{\delta} + c\left(\frac{y}{\delta}\right)^2 + d\left(\frac{y}{\delta}\right)^3$$

**9.7.** Repeat Prob. 9.4 using the velocity distribution

$$\frac{u}{U} = \sin\left(\frac{\pi y}{2\delta}\right)$$

**9.8.** Use the Kármán-Pohlhausen approximation to obtain a solution for the boundary layer which develops on a surface for which the outer flow velocity is defined by the expression

$$U(x) = Ax^{1/6}$$

where $A$ is a constant. From the solution evaluate the boundary-layer thicknesses $\delta$, $\delta^*$, and $\theta$ and the surface shear stress $\tau_0$.

**9.9.** Figure 9.11 shows a viscous, incompressible liquid flowing down a vertical surface. A boundary layer develops near the vertical surface and grows to approach the free surface. Taking into account the force due to gravity, write down the boundary-layer equations for this flow configuration. From these equations obtain the corresponding momentum integral. Hence, by employing a second-order polynomial for the velocity distribution, obtain an expression for the boundary-layer thickness $\delta(x)$.

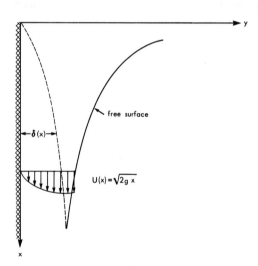

**FIGURE 9.11**
Liquid flow down a vertical surface.

# CHAPTER
# 10

# BUOYANCY-DRIVEN FLOWS

There exists a large class of fluid flows in which the motion is caused by buoyancy in the fluid. Buoyancy is the force which is experienced in a fluid due to a variation of density in the presence of a gravitational field. According to the definition of an incompressible fluid, as was presented in Sec. 1.6, variations in the density normally mean that the fluid is compressible, rather than incompressible. That being the case, one might expect that the material content of this chapter would be presented in Part IV of the text, rather than Part III. The rationale for this apparent contradiction is discussed below.

For many of the fluid flows of the type mentioned above, the density variation is important only in the body-force term of the Navier-Stokes equations. In all other places in which the density appears in the governing equations, the variation of density leads to an insignificant effect. That is, compressibility of the fluid is not a prime consideration. However, viscous effects are of first-order importance. Buoyancy results in a force acting on the fluid, and the fluid would accelerate continuously if it were not for the existence of the viscous forces. The viscous forces oppose the buoyancy forces and cause the fluid to move with a velocity distribution which creates a balance between the opposing buoyancy and viscous forces. Therefore, if buoyancy-driven flows are to be classified as being *viscous flows of incompressible fluids* or *compressible flows of inviscid fluids*, the former is the more appropriate classification.

The situation depicted above occurs in *natural convection*. The other type of convection is *forced convection* in which the fluid moves under the influence of forces other than the buoyancy force. Since density variations exist in buoyancy-driven flows, the density is no longer a known quantity. This means that the continuity and Navier-Stokes equations no longer constitute a complete set of equations from which the solution to a flow problem may be obtained. The energy equation is required in order to yield a complete set of equations, and this adds to the complication of solutions to this class of problems.

The equations which are the most commonly used to solve buoyancy-driven flow problems employ the *Boussinesq approximation*. This is the first topic which is addressed in this chapter. The balance of the chapter is devoted to a presentation of some of the solutions of the governing equations, as defined by the Boussinesq approximation.

## 10.1 THE BOUSSINESQ APPROXIMATION

The equations governing the flow of an incompressible fluid in which gravity provides the only significant body force are written below.

$$\nabla \cdot \mathbf{u} = 0$$

$$\rho \frac{\partial \mathbf{u}}{\partial t} + \rho(\mathbf{u} \cdot \nabla)\mathbf{u} = -\nabla p + \mu \nabla^2 \mathbf{u} - \rho g \mathbf{e}_z$$

Here, $\mathbf{e}_z$ is the unit vector acting in the positive $z$ direction, and it is assumed that gravity acts in the negative $z$ direction. In the absence of any motion, these equations reduce to the following form:

$$0 = -\nabla p_0 - \rho_0 g \mathbf{e}_z \qquad (10.1)$$

where $p_0$ and $\rho_0$ are, respectively, the pressure and the density distributions which exist under static equilibrium. Then we may adopt the following notation for the pressure, density, and velocity distributions in the fluid during convective motion,

$$p = p_0 + p^*$$
$$\rho = \rho_0 + \rho^*$$
$$\mathbf{u} = 0 + \mathbf{u}^*$$

In the above, $p^*$ is the pressure in the fluid relative to the static value, $\rho^*$ is the density measured relative to the static value, and $\mathbf{u}^*$ is the velocity of the fluid during the convective motion. Substituting these values into the equations quoted above yields the following result:

$$\nabla \cdot \mathbf{u}^* = 0$$

$$(\rho_0 + \rho^*)\frac{\partial \mathbf{u}^*}{\partial t} + (\rho_0 + \rho^*)(\mathbf{u}^* \cdot \nabla)\mathbf{u}^* = -\nabla p^* + \mu \nabla^2 \mathbf{u}^* - \rho^* g \mathbf{e}_z$$

In the above, the equation of static equilibrium, Eq. (10.1), has been subtracted.

The equations presented above are exact for an incompressible fluid which has a density variation, or stratification, throughout it. The Boussinesq approximation consists of neglecting any variation of density except in the gravitational term. The latter term is of prime importance since it represents the force which causes the motion which is being represented. However, the variation of density is assumed to have only a minor effect on the inertia force. This may be considered to be reasonable where relatively small density differences exist over moderate distances. Then, considering $\rho$ to be constant, the Boussinesq approximation to the governing equations is

$$\nabla \cdot \mathbf{u} = 0 \tag{10.2a}$$

$$\rho \frac{\partial \mathbf{u}}{\partial t} + \rho (\mathbf{u} \cdot \nabla) \mathbf{u} = -\nabla p + \mu \nabla^2 \mathbf{u} - \Delta \rho \, g \mathbf{e}_z \tag{10.2b}$$

In Eqs. (10.2a) and (10.2b) it is understood that the pressure $p$ is measured relative to the static pressure distribution. The quantity $\Delta \rho$ is the density difference relative to the static distribution and it is positive when the density is greater than the static value.

Strictly speaking, the equations presented above are valid only for a fluid in which the density varies, but which is incompressible. However, the idea behind the Boussinesq approximation may be extended to include compressible fluids too. Provided that the variation in density is small, it may be assumed that in buoyancy-driven flows the variation in density is negligible in all of the terms in the governing equations except the gravitational term. This means that the variation in density may be neglected in the continuity equation as well as in the equations of dynamics.

## 10.2 THERMAL CONVECTION

In thermal convection the density variation is caused by temperature variations in the fluid. This is to be contrasted with the case of density variations which are caused by salinity variations in water, etc. In thermal convection the density is usually expressed in terms of the temperature by the following relationship,

$$\rho = \rho_0 [1 - \beta (T - T_0)] \tag{10.3}$$

In the above, $\beta$ is the coefficient of thermal expansion of the fluid and $T_0$ is the temperature of the fluid which exists at static equilibrium.

The representation of the density given by Eq. (10.3) is valid for moderate departures of the temperature $T$ from the static value $T_0$ for an incompressible fluid. In general, the thermal equation of state may be written in the form $\rho = \rho(p, T)$. Hence it follows, without invoking the condition of incompressibility, that

$$\rho = \rho_0 + (p - p_0) \frac{\partial \rho}{\partial p} (p_0, T_0) + (T - T_0) \frac{\partial \rho}{\partial T} (p_0, T_0)$$

In the above, only the linear terms in the pressure difference and the temperature difference have been retained in this Taylor series expansion. Now if it is assumed that compressible effects are negligible, the second term on the right-hand side of this equation will be negligible. This is equivalent to saying that the density is a function of the temperature only, rather than being a function of both the pressure and the density. The third term on the right-hand side of the above equation may be evaluated for the case of an ideal gas, for which $\rho = p/RT$, giving the result

$$\rho = \rho_0 + (T - T_0)\left(\frac{-p}{RT_0^2}\right)$$

$$= \rho_0 - \rho_0\frac{(T - T_0)}{T_0}$$

This is the same form as Eq. (10.3) and it shows that for an ideal gas the thermal expansion coefficient assumes the value $\beta = 1/T_0$. In general, the value of $\beta$ is determined experimentally and it is a property of the fluid in the same sense as the viscosity is a property.

From Eq. (10.3) it follows that $\Delta\rho = -\rho_0\beta(T - T_0) = -\rho\beta(T - T_0)$, where the density $\rho$ is assumed to be constant and equal to the value which exists when there is no motion. Then, substituting this value into Eq. (10.2$b$) yields the following form of the equations governing the motion which results when thermal convection occurs:

$$\mathbf{\nabla} \cdot \mathbf{u} = 0 \tag{10.4}$$

$$\rho\frac{\partial\mathbf{u}}{\partial t} + \rho(\mathbf{u} \cdot \mathbf{\nabla})\mathbf{u} = -\nabla p + \mu\nabla^2\mathbf{u} + \rho g\beta(T - T_0)\mathbf{e}_z \tag{10.5}$$

Equations (10.4) and (10.5) constitute four scalar equations for five unknown quantities. The unknown quantities are the velocity vector $\mathbf{u}$, the pressure $p$, and the temperature $T$. Then, in order to achieve a closed mathematical system, the thermal energy equation must be employed. This means that the dynamics of the system, and its thermodynamics, are no longer independent of each other. That is, permitting the density of the fluid to vary with its temperature in the buoyancy term has coupled the system's dynamics and thermodynamics.

The appropriate form of the equation of conservation of energy was derived in Prob. (3.1) and it is given by Eq. (3.6) which is rewritten below.

$$\rho\frac{\partial h}{\partial t} + \rho(\mathbf{u} \cdot \mathbf{\nabla})h = \frac{\partial p}{\partial t} + (\mathbf{u} \cdot \mathbf{\nabla})p + \mathbf{\nabla} \cdot (k\nabla T) + \Phi \tag{10.6}$$

where $$h = h(\rho, T) \tag{10.7}$$

In the above equations, $h$ is the enthalpy of the fluid and $\Phi$ is the dissipation function. In accordance with the Boussinesq approximation, the density is assumed to be constant.

In general, the enthalpy $h$ is a function of the pressure $p$ and the temperature $T$. However, if we restrict our discussion to ideal gases, it follows that $h$ may be considered to be a function of $T$ only. Then for cases where $h$ may be considered to be a function of $T$ only, including all fluids which are ideal gases, Eqs. (10.4)–(10.7) may be rewritten in the following form:

$$\nabla \cdot \mathbf{u} = 0 \tag{10.8}$$

$$\rho \frac{\partial \mathbf{u}}{\partial t} + \rho(\mathbf{u} \cdot \nabla)\mathbf{u} = -\nabla p + \mu \nabla^2 \mathbf{u} + \rho g \beta (T - T_0)\mathbf{e}_z \tag{10.9}$$

$$\rho C_p \frac{\partial T}{\partial t} + \rho C_p(\mathbf{u} \cdot \nabla)T = \frac{\partial p}{\partial t} + (\mathbf{u} \cdot \nabla)p + \nabla \cdot (k\,\nabla T) + \Phi \tag{10.10}$$

In the above equations, the density $\rho$ is assumed to be constant, the pressure $p$ is measured relative to the static value. The quantity $C_p$ is the specific heat at constant pressure and $\Phi$ is the dissipation function.

## 10.3 BOUNDARY LAYER APPROXIMATIONS

Buoyancy-driven flows which comply with the general Boussinesq approximation are governed by Eqs. (10.4)–(10.7). For the case of thermal convection in which the density may be considered to be a function of the temperature only, the simplified form of the governing equations is given by Eqs. (10.8)–(10.10). In this section we further simplify the governing equations by applying the boundary layer approximation to the latter set of equations and by assuming that the fluid properties remain constant.

In the interests of consistency with Chap. 9, we consider two-dimensional, steady flow in the x-y plane, in which the main flow is in the x direction. Since the flow is buoyancy driven, this requires that we adopt the configuration illustrated in Fig. 10.1. This is the same situation as depicted in Fig. 9.3 except that there is no externally-driven flow and the coordinate system has been rotated through an angle of 90°. In Fig. 10.1, the quantity $\delta_T$ is the *thermal boundary layer thickness* which is assumed to be of the same order of magnitude as the boundary layer thickness, $\delta$.

For boundary-layer-like flows, the dynamic equations are approximated in the same way that they were in the previous chapter. That is, the equations of the dynamics are the same as those derived in Chap. 9, and given by Eqs. (9.2a) and (9.2b), except that the buoyancy term which appears in Eq. (10.9) acts in the x direction. Then it remains to arrive at a consistent version of the energy equation [Eq. (10.10)]. Writing this equation explicitly for steady, two-dimensional flows for which the viscosity coefficient and the thermal conductivity are constant, yields the following result:

$$\rho C_p \left( u \frac{\partial T}{\partial x} + v \frac{\partial T}{\partial y} \right) = \left( u \frac{\partial p}{\partial x} + v \frac{\partial p}{\partial y} \right) + k \left( \frac{\partial^2 T}{\partial x^2} + \frac{\partial^2 T}{\partial y^2} \right) + \Phi$$

where

$$\Phi = 2\mu \left[ \left( \frac{\partial u}{\partial x} \right)^2 + \left( \frac{\partial v}{\partial y} \right)^2 \right] + \mu \left( \frac{\partial u}{\partial y} + \frac{\partial v}{\partial x} \right)^2$$

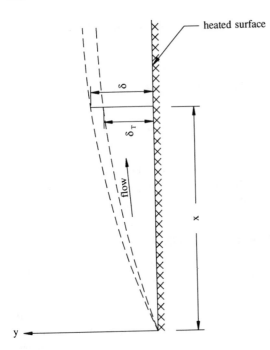

**FIGURE 10.1**
Development of thermal and momentum boundary layers on a vertical heated surface.

The following observations may be made regarding the various terms which appear in this equation.

1. The convective terms on the left-hand side of this equation are both of the same order of magnitude—as was the case for the convection of momentum in the boundary layer equations.

2. In the first bracketed term on the right-hand side, the pressure gradient across the boundary layer is negligibly small. This fact is verified by the $y$ component of the momentum equation in the boundary layer equations.

3. In the heat conduction term, the component involving the second derivative with respect to $y$ is considerably larger than that with respect to $x$. This is the same approximation as was made with the viscous terms in the boundary layer equations.

4. For moderate velocities induced by thermal convection, the dissipation of energy by the action of viscosity is negligibly small. That is, $\Phi$ may be neglected.

Applying these observations and assumptions to the energy equation, as written above, results in the following reduced form:

$$\rho C_p \left( u \frac{\partial T}{\partial x} + v \frac{\partial T}{\partial y} \right) = u \frac{\partial p}{\partial x} + k \frac{\partial^2 T}{\partial y^2}$$

Combining this result with the continuity and momentum equations results in the following set of equations for buoyancy-driven thermal convection according to the boundary layer approximation:

$$\frac{\partial u}{\partial x} + \frac{\partial v}{\partial y} = 0 \qquad (10.11a)$$

$$u\frac{\partial u}{\partial x} + v\frac{\partial u}{\partial y} = -\frac{1}{\rho}\frac{dp}{dx} + v\frac{\partial^2 u}{\partial y^2} + g\beta(T - T_0) \qquad (10.11b)$$

$$u\frac{\partial T}{\partial x} + v\frac{\partial T}{\partial y} = \frac{1}{\rho C_p}\left(u\frac{dp}{dx}\right) + \kappa\frac{\partial^2 T}{\partial y^2} \qquad (10.11c)$$

In Eq. (10.11$c$) the quantity $\kappa = k/\rho C_p$ is the thermal diffusivity of the fluid.

Eqs. (10.11$a$) and (10.11$b$) are the same Eqs. (9.2$a$) and (9.2$b$), except that the buoyancy term exists in Eq. (10.11$b$). This additional term involves the local value of the temperature of the fluid which, in general, is unknown. This requires the inclusion of the energy equation in order to yield a closed set of equations. Equation (10.11$c$) is the form of the energy equation which is consistent with the boundary layer approximation and which is valid for moderate temperature differentials from the ambient value. Equations (10.11$a$), (10.11$b$), and (10.11$c$) are to be solved subject to the no-slip boundary condition on the surface $y = 0$, and subject to the condition that the velocity should be zero far from the heated surface. In addition, either the temperature of the heated surface or the heat flux on its surface must be specified.

## 10.4   VERTICAL ISOTHERMAL SURFACE

In this section we apply the equations derived above to the flow which is induced by a vertical surface which is isothermal at a temperature which is elevated relative to the ambient. The situation is as depicted in Fig. 10.1 in which the temperature of the vertical surface is everywhere $T_s$ while that far from the surface is $T_0$, both of which are constants. For such a configuration there is negligible pressure gradient in the $x$ direction. Then, from Eqs. (10.11), the mathematical problem to be solved becomes

$$\frac{\partial u}{\partial x} + \frac{\partial v}{\partial y} = 0 \qquad (10.12a)$$

$$u\frac{\partial u}{\partial x} + v\frac{\partial u}{\partial y} = v\frac{\partial^2 u}{\partial y^2} + g\beta(T - T_0) \qquad (10.12b)$$

$$u\frac{\partial T}{\partial x} + v\frac{\partial T}{\partial y} = \kappa\frac{\partial^2 T}{\partial y^2} \qquad (10.12c)$$

The boundary conditions which accompany these differential equations are

$$u(x,0) = 0 \qquad\qquad (10.13a)$$

$$v(x,0) = 0 \qquad\qquad (10.13b)$$

$$u(x,y) \to 0 \qquad \text{as } y \to \infty \qquad\qquad (10.13c)$$

$$T(x,0) = T_s \qquad\qquad (10.13d)$$

$$T(x,y) \to T_0 \qquad \text{as } y \to \infty \qquad\qquad (10.13e)$$

The first two of these conditions are the usual no-slip boundary condition while the third condition ensures that the effect of the heated surface does not extend far from the surface. The last two conditions specify that the temperature in the fluid is $T_s$ at the vertical surface and $T_0$ far from the surface.

In order to facilitate the solution to Eqs. (10.12), two changes will be made. First, the stream function $\psi(x, y)$ will be introduced as was done in Chapter 9. This will permit Eqs. (10.12$a$) and (10.12$b$) to be replaced by a single equation involving the stream function. Second, the temperature $T(x, y)$ will be replaced by a dimensionless temperature difference $\theta(x, y)$ which is defined as follows:

$$\theta(x, y) = \frac{T(x, y) - T_0}{T_s - T_0}$$

This dimensionless temperature varies in value between zero, far from the surface, to unity at the surface. This makes it preferable to the alternate dimensionless temperature defined by the quantity $\beta(T - T_0)$. In terms of these new variables, Eqs. (10.12) become

$$\frac{\partial \psi}{\partial y}\frac{\partial^2 \psi}{\partial x \partial y} - \frac{\partial \psi}{\partial x}\frac{\partial^2 \psi}{\partial y^2} = \nu \frac{\partial^3 \psi}{\partial y^3} + g\beta(T_s - T_0)\theta \qquad (10.14a)$$

$$\frac{\partial \psi}{\partial y}\frac{\partial \theta}{\partial x} - \frac{\partial \psi}{\partial x}\frac{\partial \theta}{\partial y} = \kappa \frac{\partial^2 \theta}{\partial y^2} \qquad (10.14b)$$

Following the methods employed in Chapter 9, we now look for a similarity solution to this problem of the following form:

$$\psi(x, y) = C_1 x^m f(\eta)$$

and
$$\theta(x, y) = F(\eta)$$

where
$$\eta(x, y) = C_2 \frac{y}{x^n}$$

In the above, $m$ and $n$ are undetermined exponents, although not necessarily integers, and $C_1$ and $C_2$ are constants whose values will render the functions $f$, $F$, and $\eta$ dimensionless. From the definitions of these quantities, the various derivatives which appear in the differential equations are evaluated as follows:

$$\frac{\partial \psi}{\partial x} = C_1 x^{m-1}(mf - n\eta f')$$

$$\frac{\partial \psi}{\partial y} = C_1 c_2 x^{m-n} f'$$

$$\frac{\partial^2 \psi}{\partial x \, \partial y} = C_1 C_2 x^{m-n-1}\{(m-n)f' - n\eta f''\}$$

$$\frac{\partial^2 \psi}{\partial y^2} = C_1 C_2^2 x^{m-2n} f''$$

$$\frac{\partial^3 \psi}{\partial y^3} = C_1 C_2^3 x^{m-3n} f'''$$

$$\frac{\partial \theta}{\partial x} = -nx^{-1}\eta F'$$

$$\frac{\partial \theta}{\partial y} = C_2 x^{-n} F'$$

$$\frac{\partial^2 \theta}{\partial y^2} = C_2^2 x^{-2n} F''$$

The primes in the above expressions represent derivatives with respect to the similarity variable $\eta$. Substituting these expressions into Eqs. (10.14a) and (10.14b) produces the following equations:

$$C_1^2 C_2^2 x^{2m-2n-1}\{(m-n)(f')^2 - mff''\} = \nu C_1 C_2^3 x^{m-3n} f''' + g\beta(T_s - T_0)F$$

$$- mC_1 x^{m-n-1} fF' = \kappa C_2 x^{-2n} F''$$

If these equations are to be reduced to a pair of ordinary differential equations, the powers of $x$ in the first equation must be zero, and the powers of $x$ on each side of the second equation must be equal. That is, the following relations must exist:

$$2m - 2n - 1 = 0$$

$$m - 3n = 0$$

$$m - n - 1 = -2n$$

Although it is generally not possible to satisfy three conditions with only two quantities, the three equations above are satisfied by the values

$$m = \tfrac{3}{4} \qquad n = \tfrac{1}{4}$$

Using these values, the ordinary differential equations derived above simplify to the following form:

$$f''' + \frac{C_1}{4\nu C_2}\left[3ff'' - 2(f')^2\right] + \frac{g\beta(T_s - T_0)}{\nu C_1 C_2^3}F = 0$$

$$F'' + \frac{3}{4}\frac{C_1}{\kappa C_2}fF' = 0$$

Having selected the values for the exponents $m$ and $n$ in order to produce a similarity solution, it is now possible to define explicitly the constants $C_1$ and $C_2$ in such a way that the functions $f$, $F$, and $\eta$ are dimensionless. The quantities which are available for this purpose are $\nu$, $g$, and $\kappa$. Then it is sufficient, from dimensionality considerations, to define $C_1$ and $C_2$ as follows:

$$C_1 \sim \nu\left(\frac{g}{\nu^2}\right)^{1/4}$$

$$C_2 \sim \left(\frac{g}{\nu^2}\right)^{1/4}$$

However, by including *dimensionless* constants of proportionality in the definitions of these two quantities, permits two of the coefficients which appear in the differential equations to be normalized to unity. Noting that the quantity $\beta(T_s - T_0)$ is dimensionless, we define the constants $C_1$ and $C_2$ as follows:

$$C_1 = \frac{\nu}{4}\left(\frac{4g\beta(T_s - T_0)}{\nu^2}\right)^{1/4}$$

$$C_2 = \left(\frac{4g\beta(T_s - T_0)}{\nu^2}\right)^{1/4}$$

With this choice of values for the constants $C_1$ and $C_2$ the differential equations for the functions $f$ and $F$ become

$$f''' + 3ff'' - 2(f')^2 + F = 0 \qquad\qquad (10.15a)$$

$$F'' + 3P_r fF' = 0 \qquad\qquad (10.15b)$$

In the above, the parameter $P_r = \nu/\kappa$ is the *Prandtl number*. Numerically, the Prandtl number is about 0.7 for air and about 7.0 for water. In terms of the functions $f$ and $F$ the boundary conditions defined by Eqs. (10.13) become

$$f(0) = f'(0) = 0 \qquad\qquad (10.15c)$$

$$f'(\eta) \to 0 \qquad \text{as } \eta \to \infty \qquad\qquad (10.15d)$$

$$g(0) = 1 \qquad\qquad (10.15e)$$

$$g(\eta) \to 0 \qquad \text{as } \eta \to \infty \qquad\qquad (10.15f)$$

Once the solution to the ordinary differential system defined by Eqs. (10.15) has been obtained, the corresponding solution for the stream function and the dimensionless temperature are given by the following relations:

$$\psi(x, y) = \frac{\nu}{4}\left(\frac{4g\beta(T_s - T_0)}{\nu^2}\right)^{1/4} x^{3/4} f(\eta) \qquad (10.16a)$$

$$\theta(x, y) = F(\eta) \qquad (10.16b)$$

where $\qquad \eta(x, y) = \left(\frac{4g\beta(T_s - T_0)}{\nu^2}\right)^{1/4} \frac{y}{x^{1/4}} \qquad (10.16c)$

The problem defined by Eqs. (10.15) was solved by Pohlhausen for $P_r = 0.733$. The physical result which is of greatest interest is the rate at which convective heat transfer takes place between the vertical surface and the ambient fluid. The result so obtained is usually quoted in the following nondimensional form:

$$N_u = 0.359(G_r)^{1/4} \qquad (10.17)$$

where $\qquad N_u = \dfrac{hl}{k}$

and $\qquad G_r = \dfrac{gl^3(T_s - T_0)}{\nu^2 T_0}$

The parameters in Eq. (10.17) are the *Nusselt number* $N_u$, which is the non-dimensional heat transfer rate, and the *Grashof number* $G_r$, which is the nondimensional temperature differential which drives the convection. In these quantities, $h$ is the rate of heat transfer per unit area per unit time, $k$ is the thermal conductivity of the ambient fluid, and $l$ is the length of surface over which the heat transfer takes place.

## 10.5  LINE SOURCE OF HEAT

Figure 10.2 shows the physical situation which exists when a line source of heat is immersed in an otherwise stationary fluid. The situation is similar to that of the vertical surface, except that there is no physical surface involved, and no characteristic temperature differential.

The equations governing the motion which is induced in this situation will be the same as those of Sec. 10.4, and they are given by Eqs. (10.12). However, the boundary conditions are different in this case. Since there is no physical surface in the present case, the conditions (10.13a) and (10.13d) are no longer relevant. The first of these conditions must be replaced by a statement that the $x$ axis is a line of symmetry and the second condition by a statement which ensures that the total heat rising from the source is the same at all streamwise

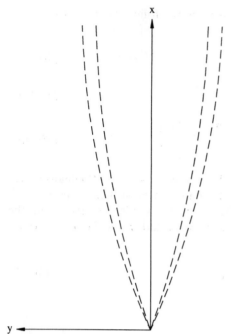

**FIGURE 10.2**
Thermal convection from a line source of heat.

locations. These new conditions are expressed by the following equations:

$$\frac{\partial u}{\partial y}(x,0) = 0$$

$$\int_{-\infty}^{\infty} \rho u C_p (T - T_0) \, dy = Q$$

In the above, $Q$ is the value of the total amount of heat which leaves the source per unit time per unit length of source.

With the changes noted above, the problem to be solved consists of the differential system defined by Eqs. (10.12), subject to the following boundary conditions:

$$\frac{\partial u}{\partial y}(x,0) = 0 \tag{10.18a}$$

$$v(x,0) = 0 \tag{10.18b}$$

$$\int_{-\infty}^{\infty} \rho u C_p (T - T_0) \, dy = Q \tag{10.18c}$$

$$\frac{\partial T}{\partial y}(x,0) = 0 \tag{10.18d}$$

$$T(x,y) \to 0 \quad \text{as } y \to \pm\infty \tag{10.18e}$$

As in the case of the isothermal surface, we recast the differential equations in terms of the stream function and a dimensionless temperature. The former is defined to satisfy the continuity equation as before, but the dimensionless temperature has to be redefined. The surface temperature no longer needs normalizing to unity, so that the appropriate definition of the dimensionless temperature in this case is

$$\theta(x, y) = \beta\{T(x, y) - T_0\}$$

In terms of the stream function and the new dimensionless temperature, Eqs. (10.12) reduce to the following form:

$$\frac{\partial \psi}{\partial y}\frac{\partial^2 \psi}{\partial x \partial y} - \frac{\partial \psi}{\partial x}\frac{\partial^2 \psi}{\partial y^2} = \nu\frac{\partial^3 \psi}{\partial y^3} + g\theta \qquad (10.19a)$$

$$\frac{\partial \psi}{\partial y}\frac{\partial \theta}{\partial x} - \frac{\partial \psi}{\partial x}\frac{\partial \theta}{\partial y} = \kappa\frac{\partial^2 \theta}{\partial y^2} \qquad (10.19b)$$

In order to obtain solutions to these equations we seek a similarity solution which is suggested by that which was found in the previous section, but in which the dimensionless temperature also has a coefficient which is a function of $x$. The form of the solution which is sought is

$$\psi(x, y) = C_1 x^m f(\eta)$$

where

$$\eta(x, y) = C_2\frac{y}{x^n}$$

and

$$\theta(x, y) = C_3 x^r F(\eta)$$

In these expressions, $m$, $n$, and $r$ are undetermined exponents and the quantities $C_1$, $C_2$, and $C_3$ are constants which render the functions $f$, $F$, and $\eta$ dimensionless. It is not to be assumed that any of these quantities have the same values as they did in Sec. 10.4. The derivatives of the stream function are the same in Sec. 10.4, and the derivatives of the dimensionless temperature are

$$\frac{\partial \theta}{\partial x} = C_3 x^{r-1}(rF - n\eta F')$$

$$\frac{\partial \theta}{\partial y} = C_2 C_3 x^{r-n}F'$$

$$\frac{\partial^2 \theta}{\partial y^2} = C_2^2 C_3 x^{r-2n}F''$$

Using these results for the differentials, Eqs. (10.19) reduce to the following form:

$$C_1^2 C_2^2 x^{2m-2n-1}\{(m-n)(f')^2 - mff''\} = \nu C_1 C_2^3 x^{m-3n} f''' + gC_3 x^r F$$

$$C_1 C_2 C_3 x^{m-n+r-1}(rf'F - mfF') = \kappa C_2^2 C_3 x^{r-2n} F''$$

For a similarity solution to exist, the $x$ dependence in these equations must cancel. This leads to the following equations relating the exponents $m$, $n$, and $r$,

$$2m - 2n - 1 = m - 3m$$
$$2m - 2n - 1 = r$$
$$m - n + r - 1 = r - 2n$$

The first two of these relations come from the momentum equation while the last relation comes from the energy equation. It will be seen that the first and the last equations are the same, so that the requirement of reducing the partial differential equations to ordinary differential equations is met by satisfying the first two of the equations presented above. Rewriting these equations shows that the similarity condition is met provided

$$m = \tfrac{1}{4}(3 + r) \qquad n = \tfrac{1}{4}(1 - r)$$

It will be noted that for the special case $r = 0$, the solution obtained in Sec. 10.4 is recovered. In order to determine the value of $r$ for the case under consideration, the condition given by Eq. (10.18c) must be invoked. This condition specifies the following:

$$Q = \int_{-\infty}^{\infty} \rho u C_p (T - T_0)\, dy$$

$$= \int_{-\infty}^{\infty} \rho \frac{\partial \psi}{\partial y} C_p \frac{\theta}{\beta} \frac{x^n}{C_2}\, d\eta$$

In the above, it has been noted that

$$dy = \frac{x^n}{C_2}\, d\eta + nx^{n-1} \frac{1}{C_2} \eta\, dx$$

However, the integration indicated above is carried out in a plane $x = $ constant, so that $dy$ will be proportional to $x^n\, d\eta$. Substituting the values established for the quantities in the integrand produces the result,

$$Q = C_1 C_3 \frac{\rho C_p}{\beta} x^{m+r} \int_{-\infty}^{\infty} f'F\, d\eta \qquad (10.20)$$

Since the quantity $Q$ should be independent of $x$, it follows that $(m + r)$ should be zero. This additional requirement, coupled with the results obtained above, leads to the following values for the exponents $m$, $n$, and $r$.

$$m = \tfrac{3}{5} \qquad n = \tfrac{2}{5} \qquad r = -\tfrac{3}{5}$$

For these values of the exponents, the differential equations for the functions $f$

and $F$ become

$$f''' + \frac{C_1}{5\nu C_2}\left[3ff'' - (f')^2\right] + \frac{gC_3}{\nu C_1 C_2^3}F = 0 \qquad (10.21a)$$

$$F'' + \frac{3C_1}{5\kappa C_2}\frac{d}{d\eta}(fF) = 0 \qquad (10.21b)$$

In order to render the functions $f$, $F$, and $\eta$ dimensionless, we choose

$$C_1 \sim \nu\left(\frac{g}{\nu^2}\right)^{1/5}$$

$$C_2 \sim \left(\frac{g}{\nu^2}\right)^{1/5}$$

$$C_3 \sim \left(\frac{g}{\nu^2}\right)^{-1/5}$$

As was the case in the previous section, we may include dimensionless constants of proportionality in the definitions of the above quantities. The purpose of doing this is to simplify the parameters which appear in the resulting differential equations and boundary conditions. The differential equations are given above, and the condition which may be considerably simplified through normalization is given by Eq. (10.20). In the latter context, it is noted that the following quantity is dimensionless,

$$\frac{\rho\nu C_p}{\beta Q}$$

With this observation, the following definitions of the constants $C_1$, $C_2$, and $C_3$ are adopted in order to simplify the coefficients in the resulting problem,

$$C_1 = \nu\left(\frac{\beta Q}{\rho\nu C_p}\frac{g}{\nu^2}\right)^{1/5}$$

$$C_2 = \frac{1}{5}\left(\frac{\beta Q}{\rho\nu C_p}\frac{g}{\nu^2}\right)^{1/5}$$

$$C_3 = \left(\frac{\rho^4\nu^4 C_p^4}{\beta^4 Q^4}\frac{g}{\nu^2}\right)^{-1/5}$$

Using these definitions, the differential equations, as given by Eqs. (10.21), reduce to the following form:

$$f''' + 3ff'' - (f')^2 + F = 0 \qquad (10.22a)$$

$$F'' + 3P_r\frac{d}{d\eta}(fF) = 0 \qquad (10.22b)$$

The boundary conditions which accompany this differential system are given by Eqs. (10.18). In terms of the new variables and parameters these equations become

$$f(0) = f''(0) = 0 \tag{10.23a}$$

$$\int_{-\infty}^{\infty} f'F \, d\eta = 1 \tag{10.23b}$$

$$F'(0) = 0 \tag{10.23c}$$

$$F(\eta) \to 0 \quad \text{as } \eta \to \pm\infty \tag{10.23d}$$

The solutions to Eqs. (10.22) are of the form

$$f(\eta) = A \tanh \alpha\eta$$

$$F(\eta) = B \operatorname{sech}^2 \alpha\eta$$

This form of solution satisfies Eqs. (10.23a), (10.23c), and (10.23d) for all finite values of the constants $A$, $B$, and $\alpha$. Then, substitution of the assumed form of solutions into Eqs. (10.22a), (10.22b), and (10.23b) produces restrictions on the values of $A$, $B$, and $\alpha$. These restrictions are, respectively,

$$\alpha = \tfrac{5}{6}A \quad \text{and} \quad B = \tfrac{50}{27}A^4$$

$$\alpha = 3P_r A$$

$$B = \tfrac{3}{4}A$$

These four conditions cannot be satisfied by the constants $A$, $B$, and $\alpha$ alone, and the solution only exists for a particular value of the Prandtl number $P_r$. The solution to the above equations is

$$A = \left(\tfrac{81}{200}\right)^{1/5}$$

$$B = \tfrac{3}{4}\left(\tfrac{200}{81}\right)^{1/5}$$

$$\alpha = \tfrac{5}{6}\left(\tfrac{81}{200}\right)^{1/5}$$

$$P_r = \tfrac{5}{18}$$

In summary, a similarity solution has been found for a particular value of the Prandtl number only, and the solution is as follows:

$$\psi(x, y) = \frac{6\alpha}{5}\nu\left(\frac{\beta Q}{\rho\nu C_p}\frac{g}{\nu^2}\right)^{1/5} x^{3/5} \tanh \alpha\eta \tag{10.24a}$$

$$\theta(x, y) = \frac{5}{8\alpha}\left(\frac{\beta^4 Q^4}{\rho^4\nu^4 C_p^4}\frac{\nu^2}{g}\right)^{1/5} x^{-3/5} \operatorname{sech}^2 \alpha\eta \tag{10.24b}$$

where $$\eta(x, y) = \frac{1}{5}\left(\frac{\beta Q}{\rho \nu C_p}\frac{g}{\nu^2}\right)^{1/5}\frac{y}{x^{2/5}} \tag{10.24c}$$

and $$\alpha = \frac{5}{6}\left(\frac{81}{200}\right)^{1/5} \tag{10.24d}$$

The above solution is valid for a Prandtl number of $P_r = 5/18$. It shows that the centerline temperature $(T(x, 0) - T_0)$ varies as $x^{-3/5}$.


## 10.6  POINT SOURCE OF HEAT

A solution analogous to that obtained in Sec. 10.5 may be obtained for the case of a point source of heat. The physical situation which will exist is illustrated in Fig. 10.2, it being understood that in the present case there will be angular symmetry about the $x$ axis. In recognition of this fact, the preferred coordinate system involves circular cylindrical coordinates in which the coordinates $y$ and $z$ are replaced by $R$ and $\theta$. Under these circumstances, the coordinate system will be $(R, \theta, x)$, which is different from the usual situation in which the axis of symmetry is the $z$ axis.

Noting that there will be no $\theta$ dependence due to the symmetry already noted, the governing equations, which are described by Eqs. (10.12), may be rewritten in terms of the preferred coordinate system as follows:

$$\frac{\partial}{\partial x}(Ru) + \frac{\partial}{\partial R}(Ru_R) = 0$$

$$u\frac{\partial u}{\partial x} + u_R\frac{\partial u}{\partial R} = \frac{\nu}{R}\frac{\partial}{\partial R}\left(R\frac{\partial u}{\partial R}\right) + g\beta(T - T_0)$$

$$u\frac{\partial T}{\partial x} + u_R\frac{\partial T}{\partial R} = \kappa\frac{1}{R}\frac{\partial}{\partial R}\left(R\frac{\partial T}{\partial R}\right)$$

In the above, $u$ is the velocity in the $x$ direction and $u_R$ is the velocity in the radial direction, perpendicular to the $x$ axis. In order to facilitate obtaining a solution to this set of differential equations, a Stokes stream function and a dimensionless temperature are introduced as follows:

$$Ru = \frac{\partial \psi}{\partial R} \quad \text{and} \quad Ru_R = -\frac{\partial \psi}{\partial x}$$

$$\theta = \beta(T - T_0)$$

In terms of these new dependent variables, the differential equations quoted above assume the following form:

$$\frac{1}{R}\frac{\partial \psi}{\partial R}\frac{\partial}{\partial x}\left(\frac{1}{R}\frac{\partial \psi}{\partial R}\right) - \frac{1}{R}\frac{\partial \psi}{\partial x}\frac{\partial}{\partial R}\left(\frac{1}{R}\frac{\partial \psi}{\partial R}\right)$$

$$= \frac{\nu}{R}\frac{\partial}{\partial R}\left[R\frac{\partial}{\partial R}\left(\frac{1}{R}\frac{\partial \psi}{\partial R}\right)\right] + g\beta(T - T_0)$$

$$\frac{\partial \psi}{\partial R}\frac{\partial T}{\partial x} - \frac{\partial \psi}{\partial x}\frac{\partial T}{\partial R} = \kappa\frac{\partial}{\partial R}\left(R\frac{\partial T}{\partial R}\right)$$

The boundary conditions which accompany these differential equations are the following:

$$\frac{\partial u}{\partial R}(x,0) = \frac{\partial}{\partial R}\left(\frac{1}{R}\frac{\partial \psi}{\partial R}\right)_{R=0} = 0 \qquad (10.25a)$$

$$u_R(x,0) = -\frac{1}{R}\left(\frac{\partial \psi}{\partial x}\right)_{R=0} = 0 \qquad (10.25b)$$

$$\int_0^\infty \rho u C_p(T - T_0)2\pi R\,dR = \int_0^\infty \rho\frac{\partial \psi}{\partial R}C_p\frac{\theta}{\beta}2\pi\,dR = Q \qquad (10.25c)$$

$$\frac{\partial T}{\partial R}(x,0) = \frac{1}{\beta}\left(\frac{\partial \theta}{\partial R}\right)_{R=0} = 0 \qquad (10.25d)$$

$$T(x,0) \to 0 \quad \text{and} \quad \theta(x,0) \to 0 \quad \text{as } R \to \pm\infty \qquad (10.25e)$$

Solutions to the differential system are sought of the following form:

$$\psi(x,R) = C_1 x^m f(\eta)$$

where

$$\eta(x,R) = C_2\frac{R}{x^n}$$

and

$$\theta(x,R) = C_3 x^r F(\eta)$$

Substitution of these assumed forms of solution into the differential system shows that a similarity solution exists for the following values of the exponents $m$, $n$, and $r$,

$$m = 1 \quad \text{and} \quad 4n + r = 1$$

The third equation which is required to define the solution is obtained from the condition defined by Eq. (10.25c). In terms of the new variables, this condition becomes

$$2\pi\rho C_1 C_3\frac{C_p}{\beta}x^{m+r}\int_0^\infty f'F\,d\eta = Q \qquad (10.26)$$

Since the quantity $Q$, the heat leaving the source per unit time, must be independent of $x$, the additional requirement is that $(m + r) = 0$. This results in the following values of the exponents for a similarity solution to exist:

$$m = 1 \qquad n = \tfrac{1}{2} \qquad r = -1$$

For these values of the exponents $m$, $n$, and $r$, the differential equations for the stream function and the dimensionless temperature become

$$f''' - \left(1 - \frac{C_1}{\nu}f\right)\frac{d}{d\eta}\left(\frac{f'}{\eta}\right) + \frac{gC_3}{\nu C_1 C_2^4}\eta F = 0$$

$$F' + \frac{C_1}{\kappa\eta}fF = 0$$

Also, the values of the constants $C_1$, $C_2$, and $C_3$ which preserve the correct dimensions of the stream function and the dimensionless temperature are

$$C_1 \sim \nu$$

$$C_2 \sim \left(\frac{g}{\nu^2}\right)^{1/6}$$

$$C_3 \sim \left(\frac{g}{\nu^2}\right)^{-1/3}$$

For a point source, as we have here, the quantity $Q$ has the dimensions of quantity of heat per unit time. Then a dimensionless parameter for this case is

$$\frac{\rho\nu C_p}{\beta Q}\left(\frac{\nu^2}{g}\right)^{1/3}$$

Then, in order to simplify the coefficients in the differential equations and in Eq. (10.26) we choose the following values for the constants:

$$C_1 = \nu$$

$$C_2 = \left(\frac{\beta Q}{\rho\nu C_p}\right)^{1/4}\left(\frac{g}{\nu^2}\right)^{1/4}$$

$$C_3 = \frac{\beta Q}{\rho\nu C_p}$$

Using these values for the constants of proportionality in the expressions for the stream function and the dimensionless temperature, the system reduces to the following ordinary differential system:

$$f''' - (1 - f)\frac{d}{d\eta}\left(\frac{1}{\eta}f'\right) + \eta F = 0 \qquad (10.27a)$$

$$F' + P_r\frac{1}{\eta}fF = 0 \qquad (10.27b)$$

In arriving at Eq. (10.27b), the energy equation has been integrated once and the boundary condition (10.25$d$) has been employed. The boundary conditions which accompany these two differential equations are

$$f(0) = f'(0) = F'(0) = 0 \qquad (10.28a)$$

$$\int_0^\infty f'F\,d\eta = \frac{1}{2\pi} \qquad (10.28b)$$

Closed form solutions to the problem posed above exist, and they will be explored in the problems at the end of the chapter.

## 10.7 STABILITY OF HORIZONTAL LAYERS

When a horizontal layer of fluid is heated from below, or cooled from above, a buoyancy force exists which can result in convective motion. However, if the buoyancy force is not sufficiently large, no motion occurs. This situation may be qualitatively explained as follows.

Consider a horizontal layer of fluid as shown in Fig. 10.3. The fluid is at rest, and heat is passing through the fluid by conduction from the lower surface to the upper surface. For simplicity, the two horizontal surfaces are considered to be isothermal, although they have different temperatures. Under these circumstances, the buoyancy force will tend to cause the fluid to rise from the lower surface, resulting in natural convection.

Suppose that while the fluid is still at rest a small-amplitude disturbance is introduced. It may be that the viscous forces which act on the disturbing motion exceed the buoyancy force which causes any convection which may arise. Under these circumstances the disturbance will decay and the motion will cease. On the other hand, if the buoyancy force exceeds the viscous forces, the disturbance

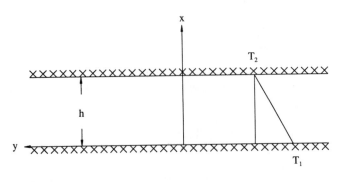

**FIGURE 10.3**
Horizontal layer of fluid heated from below.

will grow and convective motion will result. These observations suggest that a stability analysis of the situation depicted could identify the existence of a minimum value of the buoyancy force below which no motion will exist. The situation described above may be analyzed quantitatively in a manner similar to that which was used in Sec. 9.12 which dealt with the stability of boundary layers. However, the governing equations will be different in the current case, because of the existence of heat addition. The relevant equations are also different from those which were used in the previous few sections since the disturbance mentioned above will not, in general, satisfy the assumptions of the boundary layer approximation.

The situation depicted in Fig. 10.3 involves heat conduction in a stationary fluid. A small-amplitude disturbance is assumed to be introduced into this situation. The equations which govern the motion involved in this disturbance will be unsteady and three-dimensional. Following the Boussinesq approximation, we consider variations in density to be important only in the gravitational term. We further consider the fluid properties to be constant and in the energy equation we neglect the viscous dissipation of energy and the effects of pressure variations in the transfer of energy. Then, using the density variation defined by Eq. (10.3), the equations governing the motion associated with the disturbance will be

$$\nabla \cdot \mathbf{u} = 0$$

$$\frac{\partial \mathbf{u}}{\partial t} + (\mathbf{u} \cdot \nabla)\mathbf{u} = -\frac{1}{\rho}\nabla p + \nu \nabla^2 \mathbf{u} - g[1 - \beta(T - T_0)]\mathbf{e}_x$$

$$\frac{\partial T}{\partial t} + (\mathbf{u} \cdot \nabla)T = \kappa \nabla^2 T$$

In the above equations, the pressure is measured relative to its absolute value; that is, it is no longer measured relative to the static value. But, from Fig. 10.3, the static temperature distribution, $T_s(x)$, is represented by

$$T_s(x) = T_1 - (T_1 - T_2)\frac{x}{h}$$

Before the disturbance is introduced, the velocity vector $\mathbf{u}$ in the above equations will be zero. Then, using the temperature distribution specified above, the equations reduce to the following form:

$$0 = -\frac{1}{\rho}\frac{dp_s}{dx} - g\left\{1 - \beta\left[(T_1 - T_0) - (T_1 - T_2)\frac{x}{h}\right]\right\}$$

The pressure distribution which exists in the stationary state has been labelled $p_s$ and, as before, the density is understood to be evaluated at the reference temperature $T_0$.

When the disturbance is introduced, the field variables are assumed to be perturbed in the following manner:

$$\mathbf{u}(x, y, z, t) = 0 + \mathbf{u}'(x, y, z, t)$$

$$p(x, y, z, t) = p_0(x) + p'(x, y, z, t)$$

$$T(x, y, z, t) = T_s(x) + T'(x, y, z, t)$$

Here, the primed quantities are, by assumption, small perturbations caused by the disturbance. Then products of primed quantities may be neglected. Thus the linearized form of these equations is

$$\nabla \cdot \mathbf{u}' = 0$$

$$\frac{\partial \mathbf{u}'}{\partial t} = -\frac{1}{\rho} \nabla p' + \nu \nabla^2 \mathbf{u}' + g\beta T' \mathbf{e}_x$$

$$\frac{\partial T'}{\partial t} - u' \frac{(T_1 - T_2)}{h} = \kappa \nabla^2 T'$$

The pressure may be eliminated from this system of equations by taking the curl of the momentum equation. Then, it is proposed to take the curl of the resulting equation and to use the identity

$$\nabla \times (\nabla \times \mathbf{u}') = \nabla(\nabla \cdot \mathbf{u}') - \nabla^2 \mathbf{u}' = -\nabla^2 \mathbf{u}'$$

in which the continuity equation has been utilized. In this way the momentum equation becomes

$$\left( \nabla^2 - \frac{1}{\nu} \frac{\partial}{\partial t} \right) \nabla^2 \mathbf{u}' = -\frac{g\beta}{\nu} \left( \mathbf{e}_x \nabla^2 - \nabla \frac{\partial}{\partial x} \right) T'$$

The $y$ and $z$ components of velocity may now be eliminated by taking the dot product of this equation with the unit vector $\mathbf{e}_x$. Thus the problem reduces to the following two equations:

$$\left( \nabla^2 - \frac{1}{\nu} \frac{\partial}{\partial t} \right) \nabla^2 u' = -\frac{g\beta}{\nu} \left( \nabla^2 - \frac{\partial}{\partial x^2} \right) T'$$

$$\left( \nabla^2 - \frac{1}{\kappa} \frac{\partial}{\partial t} \right) T' = -\frac{(T_1 - T_2)}{\kappa h} u'$$

The disturbance which is represented by the perturbation quantities $u'$ and $T'$ is arbitrary in its form. Therefore, it may be represented by Fourier integrals in the $y$ and $z$ directions. Thus we represent the velocity and temperature perturbations by the following expressions:

$$u'(x, y, z, t) = \int_{-\infty}^{\infty} \int_{-\infty}^{\infty} U'(x, t) e^{i(k_y y + k_z z)} \, dk_y \, dk_z$$

$$T'(x, y, z, t) = \int_{-\infty}^{\infty} \int_{-\infty}^{\infty} \theta'(x, t) e^{i(k_y y + k_z z)} \, dk_y \, dk_z$$

Substituting these expressions into the equations which govern the disturbance and using the fact that the result must be valid for all wavelengths of disturbance, results in the following two differential equations:

$$\left( \frac{\partial^2}{\partial x^2} - k^2 - \frac{1}{\nu} \frac{\partial}{\partial t} \right) \left( \frac{\partial^2}{\partial x^2} - k^2 \right) U' = \frac{g\beta}{\nu} k^2 \theta'$$

$$\left( \frac{\partial^2}{\partial x^2} - k^2 - \frac{1}{\kappa} \frac{\partial}{\partial t} \right) \theta' = -\frac{(T_1 - T_2)}{\kappa h} U'$$

where

$$k^2 = k_y^2 + k_z^2$$

We next use the fact that the coefficients in the above equations are constants. Then we can seek solutions to the differential equations of the following form:

$$U'(x, t) = U(x) e^{\sigma(\kappa / h^2) t}$$

$$\theta'(x, t) = \theta(x) e^{\sigma(\kappa / h^2) t}$$

In the above, the parameter $\sigma$ has been made dimensionless by dividing the time by the quantity $h^2 / \kappa$, which is the time required for heat to diffuse across the fluid layer. Substituting this representation of the disturbance into the two governing equations gives

$$\left( D^2 - \alpha^2 - \frac{\sigma}{P_r} \right) (D^2 - \alpha^2) U = \frac{g\beta}{\nu h^2} \alpha^2 \theta \qquad (10.29a)$$

$$(D^2 - \alpha^2 - \sigma) \theta = -\frac{(T_1 - T_2)}{\kappa h} U \qquad (10.29b)$$

where

$$\alpha = hk$$

and

$$D = h \frac{d}{dx}$$

In these equations, $\alpha$ is a dimensionless wave number and $D$ is the dimensionless derivative with respect to $x$. Eliminating the temperature $\theta$ between these two equations yields the following stability equation,

$$\left[ (D^2 - \alpha^2)(D^2 - \alpha^2 - \sigma) \left( D^2 - \alpha^2 - \frac{1}{P_r} \sigma \right) + \alpha^2 R_a \right] U = 0 \quad (10.30)$$

where

$$R_a = \frac{g h^3 \beta (T_1 - T_2)}{\kappa \nu}$$

The parameter $R_a = P_r G_r$ is the *Rayleigh number*, where $P_r$ is the Prandtl number and $G_r$ is the Grashof number. It is a measure of the strength of the buoyancy force which tries to initiate convective motion.

For the configuration depicted in Fig. 10.3, the boundary conditions require that the velocity and the temperature perturbations vanish at the boundaries, $x = 0$ and $x = h$. The first of these conditions requires that $U = 0$, while second condition requires that

$$D^2 \left( D^2 - 2\alpha^2 - \frac{\sigma}{P_r} \right) U = 0$$

This latter result follows from Eq. (10.29a) and the fact that $U$ itself vanishes at the boundaries. In addition, the no-slip condition at the boundaries requires that not only $u'$ but also $v'$ and $w'$ must vanish on the boundaries. With reference to the continuity equation, this condition will be satisfied if $DU$ vanishes on the boundaries. Putting these boundary conditions together produces the following set of conditions which are to be satisfied:

$$U = DU = D^2 \left( D^2 - 2\alpha^2 - \frac{\sigma}{P_r} \right) U = 0 \quad \text{on} \quad \frac{x}{h} = 0, 1 \quad (10.31)$$

The problem posed by Eqs. (10.30) and (10.31) represents an eigenvalue problem. For given values of the Rayleigh number $R_a$, the wave number of the disturbance $\alpha$, and the Prandtl number $P_r$, the eigenvalue will be the time coefficient $\sigma$. That is, for given values of $R_a$, $\alpha$, and $P_r$, there will be a value of the quantity $\sigma$ which satisfies the conditions specified above. As the value of the wave number is varied, different values of $\sigma$ will be obtained. The largest real value of $\sigma$ will define the Fourier component of the disturbance which is the fastest growing.

It was stated earlier that there was qualitative reason to expect that there was a minimum value of the buoyancy force for convection to start. If this is so, there will be a minimum value of the Rayleigh number below which no convection will take place. In order to identify this minimum value, we note that the situation which will exist in such a case will correspond to the wavelength of the fastest-growing component having $\sigma = 0$. All other components will be decaying. Then, at the onset of instability the time coefficient in the above equations will be zero. For this situation Eqs. (10.30) and (10.31) become

$$\left[ (D^2 - \alpha^2)^3 + \alpha^2 R_a \right] U = 0 \quad (10.32a)$$

$$U = DU = (D^2 - \alpha^2)^2 U = 0 \quad \text{on} \quad \frac{x}{h} = 0, 1 \quad (10.32b)$$

The eigenvalue is now the Rayleigh number $R_a$, which can be determined from the above equations for any given value of the wave number $\alpha$. Then, the minimum value of $R_a$, with respect to $\alpha$, will be the *critical Rayleigh number*. This corresponds to the magnitude of smallest temperature gradient for which all disturbances, that is all possible wave numbers, will decay in time rather than grow in time and produce convective motion.

The problem posed by Eqs. (10.32) has a solution which yields a value of 1707.8 for the critical Rayleigh number. When one of the boundaries is free, the appropriate boundary condition is that the surface be free of stress. In this case the value of 1100.7 is obtained for the critical Rayleigh number. For two free boundaries, the value of the critical Rayleigh number is 657.5.

## PROBLEMS

**10.1.** It was stated in Sec. 10.6 that a similarity solution existed to the problem posed by a point source of heat in which

$$\psi(x, R) = C_1 x^m f(\eta)$$

$$\eta(x, R) = C_2 \frac{R}{x^n}$$

$$\theta(x, R) = C_3 x^r F(\eta)$$

Carry out the analysis of this solution by substituting this assumed form of solution into the differential equations for the stream function and the temperature to verify that a similarity solution to these equations exists provided

$$m = 1 \qquad 4n + r = 1$$

Then, by considering Eq. (10.26), show that a solution to the convection problem exists in which

$$m = 1 \qquad n = \tfrac{1}{2} \qquad r = -1$$

**10.2.** The problem posed by convection from a point source of heat was reduced in Sec. 10.6 to the differential equations (10.27) with the boundary conditions (10.28). Assume that a solution to this problem exists for the case $P_r = 1$ in which

$$f(\eta) = A \frac{\eta}{a + \eta^2}$$

$$F(\eta) = B \frac{1}{\left(a + \eta^2\right)^3}$$

In the above, the quantities $A$, $B$, and $a$ are unspecified constants.
(a) Show that the above satisfies Eq. (10.27b) provided $A = 6$.
(b) Show the above satisfies Eqs. (10.27a) and (10.28b) provided

$$a = 6\sqrt{2\pi} \qquad \text{and} \qquad B = \frac{\left(6\sqrt{2\pi}\right)^3}{3\pi}$$

**10.3.** Show that for a point source of heat in a fluid for which $P_r = 2$, a solution exists of the following form:

$$f(\eta) = A\frac{\eta^2}{a + \eta^2}$$

$$F(\eta) = B\frac{1}{\left(a + \eta^2\right)^4}$$

Find the values of the constants $A$, $B$, and $a$ which satisfy Eqs. (10.27a), (10.27b), and (10.28b).

**10.4.** The problem of marginal stability of a layer of fluid which is heated from below is represented by Eqs. (10.32a) and (10.32b). The general solution to Eq. (10.32a) will be of the form

$$U(x) = C_1e^{-\gamma_1 x} + C_2e^{-\gamma_2 x} + C_3e^{-\gamma_3 x} + C_4e^{\gamma_1 x} + C_5e^{\gamma_2 x} + C_6e^{\gamma_3 x}$$

(a) Find the values of the constants $\gamma_i$ which satisfy the differential equation, Eq. (10.32a).

(b) The existence of a nontrivial solution which satisfies the boundary conditions (10.32b) leads to a certain determinant being zero. Find this determinant but do not attempt to solve the problem of setting the determinant to zero.

**10.5.** Replace the boundary conditions defined by Eq. (10.32b) for the case of two free surfaces at $x = 0$ and $x = h$.

## FURTHER READING—PART III

The topic of laminar viscous flows is fairly well covered in texts, especially the boundary layer section of the material. Listed below are some books which adequately cover and extend the material which was treated in Part III of this book.

Batchelor, G. K.: "An Introduction to Fluid Dynamics," Cambridge University Press, London, 1967.

Gebhart, Benjamin, Yogesh Jaluria, Roop L. Mahajan, and Bahgat Sammakia: "Buoyancy-Induced Flows and Transport," Hemisphere Publishing Corporation, New York, 1988.

Rosenhead, L. (ed.): "Laminar Boundary Layers," Oxford University Press, London, 1963.

Schlichting, Hermann: "Boundary-Layer Theory," 6th ed., McGraw-Hill Book Company, New York, 1968.

Van Dyke, Milton: "Perturbation Methods in Fluid Dynamics," Academic Press, Inc., New York, 1964.

Yih, Chia-Shun: "Fluid Mechanics," McGraw-Hill Book Company, New York, 1969.

# COMPRESSIBLE
# FLOW
# OF INVISCID
# FLUIDS

In this part of the book some phenomena which are associated with the compressibility of fluids will be uncovered and some methods of solving the governing equations will be established. In order to do this, the viscosity of the fluid will again be neglected, but owing to the high speeds associated with most compressible effects, the inertia of the fluid will be retained. That is, the fluids under consideration and the flow fields associated with them will be considered to be such that viscous effects are negligible but such that compressible effects are important.

Part IV of the book encompasses Chaps. 11, 12, and 13. Chapter 11 deals with the propagation of disturbances in compressible fluids and shows how shock waves are formed. This is followed by a treatment of both normal and oblique shock waves. Chapter 12 deals with one-dimensional flow situations and shows how pressure signals react upon reaching interfaces between different fluids and also solid boundaries. Nonadiabatic flows, including heat addition and friction, are also included. The final chapter, Chap. 13, deals with multidimensional flow fields, both subsonic and supersonic. These include the Prandtl-Glauert rule for subsonic flow and Ackeret's theory for supersonic flow.

## Governing Equations and Boundary Conditions

When the density of the fluid is not constant, the equations of continuity and momentum conservation are no longer sufficient to permit a solution for the velocity and pressure fields to be obtained. This is because the density, which is now a dependent variable, appears in these equations. To close the system of equations, the conservation of thermal energy must be utilized. Thus, from Eqs. (1.3$a$), (1.9$a$), and (1.11), the equations governing the motion of an inviscid fluid in which there are no body forces are

$$\frac{\partial \rho}{\partial t} + \nabla \cdot (\rho \mathbf{u}) = 0 \tag{IV.1}$$

$$\rho \frac{\partial \mathbf{u}}{\partial t} + \rho (\mathbf{u} \cdot \nabla) \mathbf{u} = -\nabla p \tag{IV.2}$$

$$\rho \frac{\partial e}{\partial t} + \rho (\mathbf{u} \cdot \nabla) e = -p \nabla \cdot \mathbf{u} + \nabla \cdot (k \nabla T) \tag{IV.3$a$}$$

In addition, equations of state must be included. These equations will be of the general form

$$p = p(\rho, T)$$
$$e = e(\rho, T)$$

The foregoing set of equations represents seven scalar equations for the seven unknowns $\mathbf{u}$, $p$, $\rho$, $e$, and $T$.

Two useful alternative forms of the thermal-energy equation exist. One of these was derived in the problems at the end of Chap. 3 and is given by Eq. (3.6). This equation, which introduces the enthalpy $h$ of the fluid in preference to the internal energy $e$, is

$$\rho \frac{\partial h}{\partial t} + \rho (\mathbf{u} \cdot \nabla) h = \frac{\partial p}{\partial t} + (\mathbf{u} \cdot \nabla) p + \nabla \cdot (k \nabla T) \tag{IV.3$b$}$$

It should be noted that Eq. (IV.3$b$) follows directly from Eq. (IV.3$a$) without further approximation. If the form (IV.3$b$) is employed, the caloric equation of state for $e$ should be replaced by the following caloric equation of state for $h$:

$$h = h(p, T)$$

The second alternative form of the thermal-energy equation is obtained from Eq. (IV.3$a$) as a special case. For situations in which heat conduction is negligible, Eq. (IV.3a) may be written in the form

$$\rho \frac{De}{Dt} = -p \nabla \cdot \mathbf{u}$$

If, in addition, the fluid is a perfect gas, it follows from the results of thermodynamics that

$$e = e(T)$$

$$\frac{de}{dT} = C_v$$

and

$$p = \rho RT$$

Thus, the thermal-energy equation may be written in the following form:

$$\rho C_v \frac{DT}{Dt} = -p \nabla \cdot \mathbf{u}$$

It should be noted that $De/Dt = (de/dT)\, DT/Dt = C_v\, DT/Dt$, so that the above result is valid even if $C_v$ is not constant.

Using the continuity equation, $\nabla \cdot \mathbf{u}$ may be replaced by $-(D\rho/Dt)/\rho$ in the above equation. Also, $T$ may be replaced by $p/(\rho R)$ from the thermal equation of state. Thus, the energy equation may be rewritten as follows:

$$\rho C_v \frac{D}{Dt}\left(\frac{p}{\rho R}\right) = \frac{p}{\rho}\frac{D\rho}{Dt}$$

$$\frac{\rho C_v}{R}\left(\frac{1}{\rho}\frac{Dp}{Dt} - \frac{p}{\rho^2}\frac{D\rho}{Dt}\right) = \frac{p}{\rho}\frac{D\rho}{Dt}$$

$$\frac{1}{p}\frac{Dp}{Dt} = \left(\frac{R + C_v}{C_v}\right)\frac{1}{\rho}\frac{D\rho}{Dt}$$

$$= \frac{\gamma}{\rho}\frac{D\rho}{Dt}$$

The last result follows from the thermodynamic relations

$$C_p - C_v = R$$

and $\gamma = C_p/C_v$. The thermal-energy equation is now in the form of logarithmic derivatives which may be combined as follows:

$$\frac{D}{Dt}(\log p) = \frac{D}{Dt}(\log \rho^\gamma)$$

$$\frac{D}{Dt}\left(\log \frac{p}{\rho^\gamma}\right) = 0 \qquad\qquad\qquad (IV.3c)$$

$$\therefore \quad \frac{p}{\rho^\gamma} = \text{constant along each streamline}$$

The above result will be recognized as the isentropic law for thermodynamic processes. This is compatible with the assumptions of an inviscid fluid in which heat conduction is negligible. The latter assumption means that the flow is adiabatic, and the absence of viscosity eliminates any irreversible losses. Equation (IV.3c) states that the quantity $p/\rho^\gamma$ is constant along each streamline, which means that the entropy is constant along each streamline. But if the flow originates in a region where the entropy is constant everywhere, then the constant in Eq. (IV.3c) will be the same from streamline to streamline. That is, $p/\rho^\gamma$ will be constant everywhere for adiabatic flow of a perfect gas which originates in an isentropic-flow field or reservoir.

The boundary conditions which accompany the foregoing equations may specify the velocity and the temperature or the heat flux. Since inviscid fluids are again being considered, the no-slip boundary condition cannot be imposed at rigid boundaries as was the case in Part III. Rather, the condition $\mathbf{u} \cdot \mathbf{n} = U$ which was used in Part II must again be employed, for the same reason as before.

This chapter establishes the relationships which exist for shock waves which occur in supersonic flow. First, the propagation of infinitesimal internal waves is studied, which establishes the speed of sound in a gas. It is then shown how this acoustical result is modified in the case of finite-amplitude disturbances. That is, it is shown how nonlinear effects grow to cause a shock wave to form. The remainder of the chapter is devoted to the study of steady flows involving standing shock waves.

The famous Rankine-Hugoniot relations for a normal shock wave are first derived. These relations show, among other things, that the flow through a shock wave is nonisentropic. From the second law of thermodynamics it is then shown that shock waves can occur only in supersonic flow and that, in the case of a normal shock wave, the downstream Mach number will be less than unity. This is followed by derivation of the working equations for both normal shock waves and oblique shock waves. That is, relationships are established which permit the conditions downstream of a shock wave to be calculated if the upstream conditions are known and, in the case of oblique shock waves, the angle of the boundary which is inducing the shock wave, relative to the flow direction.

## 11.1 PROPAGATION OF INFINITESIMAL DISTURBANCES

By studying the equations of motion for a small-amplitude internal disturbance in a gas, the speed at which such disturbances propagate may be established.

This speed is, of course, the speed of sound, since sound is a small-amplitude disturbance. Then consider a perfect gas which is originally at rest and through which a one-dimensional or plane disturbance is traveling. It will be assumed that this disturbance travels at a sufficiently fast speed that heat conduction may be neglected. That is, it is assumed that the flow is adiabatic. Then, from Eqs. (IV.1), (IV.2), and (IV.3c), the fluid variables must satisfy the following conditions:

$$\frac{\partial \rho}{\partial t} + \frac{\partial}{\partial x}(\rho u) = 0$$

$$\frac{\partial u}{\partial t} + u\frac{\partial u}{\partial x} = -\frac{1}{\rho}\frac{\partial p}{\partial x}$$

$$\frac{p}{\rho^\gamma} = \text{constant}$$

The flow field under consideration is isentropic, so that the pressure may be considered to be a function of one thermodynamic variable only, say the density. That is, $p$ may be considered to be $p(\rho)$ only where the particular function which applies is defined by the energy equation which is written above. Then the pressure term in the Euler equation may be rewritten as follows:

$$\frac{\partial p}{\partial x} = \frac{dp}{d\rho}\frac{\partial \rho}{\partial x}$$

Using this relation, the continuity and momentum equations may be rewritten as follows:

$$\frac{\partial \rho}{\partial t} + u\frac{\partial \rho}{\partial x} + \rho\frac{\partial u}{\partial x} = 0$$

$$\frac{\partial u}{\partial t} + u\frac{\partial u}{\partial x} + \frac{1}{\rho}\frac{dp}{d\rho}\frac{\partial \rho}{\partial x} = 0$$

So far, the above equations are exact within the assumptions of one-dimensional motion of an inviscid fluid in which the flow is adiabatic. In order to utilize the assumption of a small-amplitude disturbance, the field variables will now be written in terms of their undisturbed values plus a perturbation which is caused by the passage of the disturbance. The undisturbed velocity is zero, and the undisturbed pressure and density will be denoted by the constants $p_0$ and $\rho_0$, respectively. Then the instantaneous field variables may be written as follows:

$$p = p_0 + p'$$

$$\rho = \rho_0 + \rho'$$

$$u = 0 + u'$$

Substituting these expressions into the two equations derived above gives

$$\frac{\partial \rho'}{\partial t} + u' \frac{\partial \rho'}{\partial x} + (\rho_0 + \rho') \frac{\partial u'}{\partial x} = 0$$

$$\frac{\partial u'}{\partial t} + u' \frac{\partial u'}{\partial x} + \frac{1}{(\rho_0 + \rho')} \frac{dp}{d\rho} \frac{\partial \rho'}{\partial x} = 0$$

The quantities $\rho'/\rho_0$, $p'/p_0$, and $u'$ will be small for a small-amplitude disturbance, and so products of all primed quantities may be neglected as being quadratically small. The meaning of the statement $u'$ is small will be clarified later. Thus the linearized form of the foregoing equations is

$$\frac{\partial \rho'}{\partial t} + \rho_0 \frac{\partial u'}{\partial x} = 0$$

$$\frac{\partial u'}{\partial t} + \frac{1}{\rho_0} \left( \frac{dp}{d\rho} \right)_0 \frac{\partial \rho'}{\partial x} = 0$$

It has been considered that the quantity $dp/d\rho$ has been expanded in a Taylor series and the quantity $(dp/d\rho)_0$ is the leading term in such an expansion. The meaning of the subscript zero is that the quantity $dp/d\rho$ should be evaluated in the undisturbed gas.

From these equations it follows that:

$$\rho_0 \frac{\partial^2 u'}{\partial x\, \partial t} = - \frac{\partial^2 \rho'}{\partial t^2} = - \left( \frac{dp}{d\rho} \right)_0 \frac{\partial^2 \rho'}{\partial x^2}$$

so that the equation to be satisfied by the density perturbation is

$$\frac{\partial^2 \rho'}{\partial t^2} - \left( \frac{dp}{d\rho} \right)_0 \frac{\partial^2 \rho'}{\partial x^2} = 0$$

Likewise, by eliminating $\rho'$, the equation governing the velocity perturbation is found to be

$$\frac{\partial^2 u'}{\partial t^2} - \left( \frac{dp}{d\rho} \right)_0 \frac{\partial^2 u'}{\partial x^2} = 0$$

Thus both the density perturbation and the velocity perturbation will have the same functional form, so that $u'$ may be considered to be a function of $\rho'$ only. That is, whatever the dependence of $\rho'$ is on $x$ and $t$, $u'$ will have the same form of dependence, so that a simple relationship must exist between $u'$ and $\rho'$. The foregoing partial differential equations will be recognized as being one-dimensional wave equations. Thus, the solution for $\rho'$ will be of the form

$$\rho'(x,t) = f\left( x - \sqrt{\left( \frac{dp}{d\rho} \right)_0}\, t \right) + g\left( x + \sqrt{\left( \frac{dp}{d\rho} \right)_0}\, t \right)$$

where $f$ and $g$ are any differentiable functions of their arguments. The first term in this expression represents a wave traveling in the positive $x$ direction with velocity $\sqrt{(dp/d\rho)_0}$ and the second term represents a wave traveling in the negative $x$ direction with the same velocity. Thus the speed with which the density perturbation travels, and also that with which the velocity perturbation travels, is $\sqrt{(dp/d\rho)_0}$. Then, since the disturbance was assumed to be small and since sound is a small disturbance, this will be the speed with which sound travels. That is, if $a_0$ denotes the speed of sound in a quiescent gas, it follows that

$$a_0 = \sqrt{\left(\frac{dp}{d\rho}\right)_0}$$

The foregoing result may be put in a different form by evaluating the indicated derivative through use of the isentropic relationship and the ideal-gas law. From Eq. (IV.3c),

$$\frac{p}{\rho^\gamma} = \frac{p_0}{\rho_0^\gamma}$$

$$\frac{dp}{d\rho} = \gamma\rho^{\gamma-1}\frac{p_0}{\rho_0^\gamma}$$

$$= \gamma\frac{p}{\rho}$$

Hence, employing the gas law $p = \rho RT$ gives

$$\frac{dp}{d\rho} = \gamma RT$$

Thus the speed of sound in a quiescent gas may be written

$$a_0 = \sqrt{\gamma RT_0} = \sqrt{\gamma\frac{p_0}{\rho_0}} \tag{11.1a}$$

where $T_0$ is the temperature in the undisturbed gas. This familiar result shows that the speed of sound in a gas may be considered to be a function of the temperature of the gas only and that the speed increases as the square root of the temperature of the gas.

It is now possible to be more precise concerning the assumption which was made earlier that the perturbation velocity $u'$ is small. A quantitative interpretation of this assumption may be obtained from our original linearized form of the momentum equation together with the solution just obtained. The linearized form of the momentum equation which was used above is

$$\frac{\partial u'}{\partial t} + \frac{a_0^2}{\rho_0}\frac{\partial \rho'}{\partial x} = 0$$

But it was shown that, for a wave traveling in the positive $x$ direction, the solution for $u'$ was $f(x - a_0 t)$, so that

$$\frac{\partial u'}{\partial t} = -a_0 f'(x - a_0 t) = -a_0 \frac{\partial u'}{\partial x}$$

where $f'$ is the derivative of $f$ with respect to its argument. Thus the linearized form of the momentum equation may be written in the form

$$-a_0 \frac{\partial u'}{\partial x} + \frac{a_0^2}{\rho_0} \frac{\partial \rho'}{\partial x} = 0$$

or

$$\frac{\partial u'}{\partial x} = \frac{a_0}{\rho_0} \frac{\partial \rho'}{\partial x}$$

Integrating this equation with respect to $x$ and noting that $u' = 0$ when $\rho' = 0$ gives the following algebraic relation between the velocity and the density perturbations:

$$\frac{u'}{a_0} = \frac{\rho'}{\rho_0} \tag{11.1b}$$

Equation (11.1b) shows that the meaning of the assumption $u'$ is small is that $u'/a_0 \ll 1$, since it was already assumed that $\rho'/\rho_0 \ll 1$. This result also exposes the simple relationship which exists between $u'$ and $\rho'$ which was deduced to exist.

Another result which may be deduced from Eq. (11.1b) concerns a fundamental difference between compression waves and expansion waves. For compression waves, the density perturbation $\rho'$ will be positive. Then Eq. (11.1b) shows that the velocity perturbation $u'$ will also be positive. That is, the fluid velocity behind a compression wave will be such that the fluid particles tend to follow the wave as shown in Fig. 11.1a. On the other hand, $\rho'$ will be negative for an expansion wave, so that Eq. (11.1b) shows that $u'$ will also be negative. That is, the fluid behind an expansion wave will tend to move away

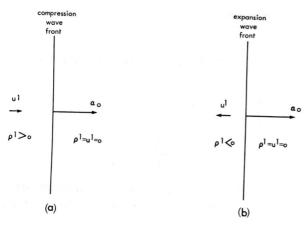

FIGURE 11.1
Fluid velocity induced by (a) a compression wave front and (b) an expansion wave front.

from the wave front as shown in Fig. 11.1. This fundamental difference between compression waves and expansion waves will be discussed further in later sections.

## 11.2 PROPAGATION OF FINITE DISTURBANCES

Consider, again, the passage of a plane wave through an otherwise quiescent fluid, but this time no assumption will be made about the infinitesimal nature of the wave amplitude. By retaining the effects of finite amplitude, the phenomena which are associated with finite-amplitude disturbances may be exposed. It will be shown that finite-amplitude waves do not propagate undisturbed but that they form shock waves.

The continuity and momentum equations are the same as those which were the starting point of the previous section.

$$\frac{\partial \rho}{\partial t} + u \frac{\partial \rho}{\partial x} + \rho \frac{\partial u}{\partial x} = 0$$

$$\frac{\partial u}{\partial t} + u \frac{\partial u}{\partial x} = -\frac{1}{\rho} \frac{\partial p}{\partial x}$$

In the previous section it was found that, for infinitesimal waves, $u$ was a function of $\rho$ only and $p$ was a function of $\rho$ only. Although the flow will contain finite-amplitude effects here, it will be assumed that $u$ and $p$ will again be functions of $\rho$ only. Then, from $u = u(\rho)$, it follows that

$$\frac{\partial \rho}{\partial t} = \frac{d\rho}{du} \frac{\partial u}{\partial t}$$

$$\frac{\partial \rho}{\partial x} = \frac{d\rho}{du} \frac{\partial u}{\partial x}$$

Also, from $p = p(\rho)$ only,

$$\frac{\partial p}{\partial x} = \frac{dp}{d\rho} \frac{d\rho}{du} \frac{\partial u}{\partial x}$$

Thus the continuity and momentum equations may be rewritten in the following form:

$$\frac{d\rho}{du} \left[ \frac{\partial u}{\partial t} + u \frac{\partial u}{\partial x} \right] + \rho \frac{\partial u}{\partial x} = 0$$

$$\frac{\partial u}{\partial t} + u \frac{\partial u}{\partial x} = -\frac{1}{\rho} \frac{dp}{d\rho} \frac{d\rho}{du} \frac{\partial u}{\partial x}$$

The bracketed term in the first equation also appears in the second equation and so may be readily eliminated between these two equations. The resulting relation is

$$\rho \frac{du}{d\rho} \frac{\partial u}{\partial x} = \frac{1}{\rho} \frac{dp}{d\rho} \frac{d\rho}{du} \frac{\partial u}{\partial x}$$

Canceling $\partial u / \partial x$ from this equation and solving for $du$ gives

$$du = \pm \sqrt{\frac{dp}{d\rho} \frac{d\rho}{\rho}}$$

For convenience, the quantity $dp/d\rho$ will be denoted by $a^2$, although no physical significance will be attached to the quantity $a$ at this time. However, it is known that $a \rightarrow a_0$ as the amplitude becomes infinitesimal. In terms of this quantity $a$, the above equation becomes

$$\frac{du}{a} = \pm \frac{d\rho}{\rho}$$

The analogous equation which was obtained in the previous section was $du/a_0 = d\rho/\rho_0$ for a forward-running wave. Thus, in order that the result obtained here may reduce to the linear result for weak waves, the plus sign must be associated with a forward-running wave and the minus sign should be associated with a backward-running wave. This gives a fluid-particle velocity which follows a compression wave and moves away from an expansion wave as before. The foregoing relation shows that for a forward-running wave

$$\frac{du}{a} = \frac{d\rho}{\rho} \qquad (11.2a)$$

This result will be used in the momentum equation as follows:

$$\frac{\partial u}{\partial t} + u \frac{\partial u}{\partial x} = -\frac{1}{\rho} \frac{\partial p}{\partial x}$$

$$= -\frac{1}{\rho} \frac{dp}{d\rho} \frac{d\rho}{du} \frac{\partial u}{\partial x}$$

$$= -\frac{1}{\rho} a^2 \frac{\rho}{a} \frac{\partial u}{\partial x}$$

That is, the momentum equation for a forward-running wave may be written in the following form:

$$\frac{\partial u}{\partial t} + (u + a) \frac{\partial u}{\partial x} = 0$$

Solutions to this equation are of the form

$$u(x, t) = f[x - (u + a)t] \qquad (11.2b)$$

where $f$ is any differentiable function. It should be noted that in this instance both $u$ and $a$ are functions of the two independent variables $x$ and $t$.

The foregoing solution represents a wave traveling in the positive $x$ direction with velocity

$$U = u + a$$

The wave speed $U$ may be related to the speed of an infinitesimal wave, that is, to the speed of sound $a_0$, by use of the isentropic law $p/\rho^\gamma = p_0/\rho_0^\gamma$. From the definition of the quantity $a$, it follows that

$$a = \sqrt{\frac{dp}{d\rho}}$$

$$= \sqrt{\gamma \rho^{\gamma-1} \frac{p_0}{\rho_0^\gamma}}$$

$$= \sqrt{\gamma \frac{p_0}{\rho_0} \left(\frac{\rho}{\rho_0}\right)^{(\gamma-1)/2}}$$

$$= a_0 \left(\frac{\rho}{\rho_0}\right)^{(\gamma-1)/2}$$

where the definition of the speed of sound has been employed from Eq. (11.1a). Using this result and Eq. (11.2a), the local value of the fluid velocity may be related to the local speed of sound. From Eq. (11.2a).

$$du = a \frac{d\rho}{\rho}$$

$$= \frac{a_0}{\rho_0^{(\gamma-1)/2}} \rho^{(\gamma-3)/2} \, d\rho$$

where the relation between $a$ and $a_0$ established above has been employed. This equation may be integrated to give

$$u = \frac{a_0}{\rho_0^{(\gamma-1)/2}} \frac{\rho^{(\gamma-1)/2}}{(\gamma-1)/2} + \text{constant}$$

Using the fact that when $u = 0$, $\rho = \rho_0$ shows the value of the constant of integration is $2a_0/(\gamma - 1)$, so that the expression for $u$ becomes

$$u = \frac{2}{\gamma - 1} \left[ a_0 \left(\frac{\rho}{\rho_0}\right)^{(\gamma-1)/2} - a_0 \right]$$

$$= \frac{2}{\gamma - 1} (a - a_0)$$

Here the relation which was established between $a$ and $a_0$ has again been used. That is, the quantity $a$ is related to the local fluid velocity $u$ in the following way:

$$a = a_0 + \frac{\gamma - 1}{2}u$$

This result shows that $a > a_0$ for $u > 0$ and that the difference between $a$ and $a_0$ is proportional to the local fluid velocity $u$. Using this result, the speed of propagation of a finite-amplitude disturbance may be evaluated. It was shown from Eq. (10.2b) that such a disturbance travels with velocity

$$U(x, t) = a + u$$
$$\therefore \quad U(x, t) = a_0 + \frac{\gamma + 1}{2}u \qquad (11.2c)$$

where the relation between $a$ and $u$ established above has been used. Equation (11.2c) shows that the speed of propagation of a finite-amplitude disturbance is greater than the speed of sound for $u > 0$ and that it is no longer constant but depends upon the value of the local fluid velocity.

Since the propagation speed defined by Eq. (11.2c) depends upon both $x$ and $t$, it is not an equilibrium speed. That is, the speed at which a finite-amplitude signal travels will change continuously according to Eq. (11.2c). It is instructive to deduce the manner in which a given wave front will change its characteristics as a result of this fact. In time $\tau$, Eq. (11.2c) shows that a disturbance will travel a distance $L$ which is given by the expression

$$L = \left(a_0 + \frac{\gamma + 1}{2}u\right)\tau$$

Then, relative to an observer who is moving at the speed of sound $a_0$, the distance traveled by the wave will be

$$S = \frac{\gamma + 1}{2}u\tau$$

That is, relative to the observer the wave will travel a distance which is dependent upon the magnitude and the sign of the local fluid velocity in the disturbance. Thus regions of high local velocity will travel faster than regions of low local velocity. Then a smooth disturbance of arbitrary form will develop as shown in Fig. 11.2.

At time $\tau_1$ a smooth velocity profile is considered to be traveling in the positive $x$ direction. Then, at some later time $\tau_2 > \tau_1$, the regions of higher velocity will have advanced further, relative to an observer moving at constant velocity $a_0$, than the regions of lower velocity. At time $\tau_3 > \tau_2$ the wave front is shown to be vertical as higher-velocity regions continue to advance faster than the slower regions. Finally, at time $\tau_4 > \tau_3$, the higher-velocity regions are shown as having overtaken the portion of the signal which is moving at the sonic velocity $a_0$. It is seen that this is an impossible configuration, since three values of $u$ exist at a given location. Thus it is concluded that the wave front will steepen as indicated until the situation depicted at time $\tau_3$ is reached. At this

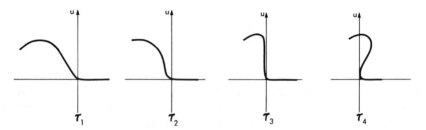

**FIGURE 11.2**
Progression of a finite-amplitude disturbance.

stage a sharp discontinuity in the field variables exists which is called a "shock wave." For times greater than $\tau_3$ this sharp wave front or shock wave will propagate in an equilibrium configuration.

To summarize, if a smooth, finite-amplitude compression wave is generated, it will travel in a nonequilibrium configuration. Different parts of the wave will travel at different speeds in such a way that the wave front will steepen as it progresses. Eventually, the steepening of the wave front will reach the point where the changes in velocity, pressure, etc., take place across a very narrow region. That is, a shock wave has been formed, and this shock wave will continue to travel at an equilibrium speed.

It should be noted that, in the foregoing argument, the fluid velocity $u$ was taken to be positive, which corresponds to a compression wave. For an expansion wave $u$ will be negative for a forward-running wave so that, according to Eq. (11.2$c$), the wave front will move more slowly than the speed of sound. Also, the more intense parts of the wave move the most slowly, so that the wave front will spread out rather than steepen. That is, compression waves steepen as they propagate but expansion waves will spread out.

## 11.3 RANKINE-HUGONIOT EQUATIONS

In the previous section it was shown how shock waves develop from finite-amplitude compression waves. In this section the variation of some of the fluid properties across a shock wave will be established. In particular, the Rankine-Hugoniot equations relate the density ratio across a shock wave to the pressure ratio and the fluid-velocity ratio.

Shock waves are very thin compared with most macroscopic length scales, so that they are conveniently approximated as line discontinuities in the fluid properties. For purposes of analysis it is convenient to adopt a frame of reference in which the shock wave is stationary and in which fluid approaches the shock wave in one state and leaves in another state. Figure 11.3 shows such a situation in which the incoming velocity, pressure, and density of the fluid are, respectively, $u_1$, $p_1$, and $\rho_1$. The corresponding outgoing values are $u_2$, $p_2$, and

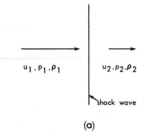

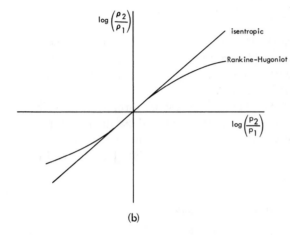

(b)

**FIGURE 11.3**
(*a*) Shock-wave configuration and (*b*) results from the Rankine-Hugoniot and the isentropic relations.

$p_2$. Since the shock wave is oriented normal to the velocity vector, it is called a "normal shock wave."

The quantities $u_1$, $p_1$, $\rho_1$, $u_2$, $p_2$, and $\rho_2$ will be related to each other through the equations of mass, momentum, and energy conservation. Since the shock wave represents a discontinuity in the fluid properties, differential equations cannot be used across it. Thus either the differential equations must be integrated to yield algebraic equations or the governing equations must be rederived in algebraic form.

Adopting the latter procedure, the equation of mass conservation may be readily written down by inspection from Fig. 11.3*a*.

$$\rho_1 u_1 = \rho_2 u_2 \tag{11.3a}$$

Multiplying these mass flow rates by the corresponding velocity magnitudes gives the change in momentum across the shock wave as $\rho_2 u_2^2 - \rho_1 u_1^2$. This change in momentum must be caused by the pressure force $p_1 - p_2$ per unit area, so that the equation of momentum conservation becomes

$$\rho_1 u_1^2 + p_1 = \rho_2 u_2^2 + p_2 \tag{11.3b}$$

Finally, the energy balance may be established as follows: The internal energy per unit mass will be

$$C_v T = C_p \frac{p}{\rho R}$$

$$= \frac{\gamma}{\gamma - 1} \frac{p}{\rho}$$

Here the ideal-gas law has been used together with the identity $C_p - C_v = R$. The total energy per unit mass will be the sum of the kinetic and internal components, so that the equation of energy conservation is

$$\frac{1}{2}u_1^2 + \frac{\gamma}{\gamma - 1} \frac{p_1}{\rho_1} = \frac{1}{2}u_2^2 + \frac{\gamma}{\gamma - 1} \frac{p_2}{\rho_2} \qquad (11.3c)$$

It should be noted that in deriving Eq. (11.3c) it has been implicitly assumed that the flow is adiabatic, although it has not been assumed that it is isentropic. Since shock waves involve high speeds and since heat conduction is a slow process, the adiabatic condition is well justified.

Equations (11.3a), (11.3b), and (11.3c) represents three equations in the six quantities $u_1$, $p_1$, $\rho_1$, $u_2$, $p_2$, and $\rho_2$. Hence, two of these quantities may be eliminated, leaving an equation connecting the remaining four quantities. The quantities $u_1$ and $u_2$ will be eliminated as follows: Dividing Eq. (11.3b) by Eq. (11.3a) gives

$$u_1 + \frac{p_1}{\rho_1 u_1} = u_2 + \frac{p_2}{\rho_2 u_2}$$

$$\therefore \quad u_2 - u_1 = \frac{p_1 - p_2}{\rho_1 u_1}$$

where $p_2/\rho_2 u_2$ has been rewritten $p_2/\rho_1 u_1$, which follows from the continuity equation. Multiplying the above equation by $u_2 + u_1$ gives

$$u_2^2 - u_1^2 = \frac{p_1 - p_2}{\rho_1}\left(1 + \frac{u_2}{u_1}\right)$$

But, from the continuity equation, $u_2/u_1 = \rho_1/\rho_2$, so that

$$u_2^2 - u_1^2 = (p_1 - p_2)\left(\frac{1}{\rho_1} + \frac{1}{\rho_2}\right)$$

The left-hand side of this equation may be replaced by a function of the pressures and densities only through use of the energy equation (11.3c). The resulting equation is

$$\frac{2\gamma}{\gamma - 1}\left(\frac{p_1}{\rho_1} - \frac{p_2}{\rho_2}\right) = (p_1 - p_2)\left(\frac{1}{\rho_1} + \frac{1}{\rho_2}\right)$$

This is the required equation which relates the pressures and densities only across the shock wave. Solving this equation for the density ratio results in the following alternative form of the above equation:

$$\frac{\rho_2}{\rho_1} = \frac{p_1 + (\gamma + 1)/(\gamma - 1)p_2}{(\gamma + 1)/(\gamma - 1)p_1 + p_2}$$

From the continuity equation $\rho_2/\rho_1 = u_1/u_2$, so that, combining this result with the above equation, the following conditions will apply across a normal shock wave:

$$\frac{\rho_2}{\rho_1} = \frac{1 + (\gamma + 1)/(\gamma - 1)(p_2/p_1)}{(\gamma + 1)/(\gamma - 1) + p_2/p_1} = \frac{u_1}{u_2} \qquad (11.4)$$

Equations (11.4) are called the *Rankine-Hugoniot equations*, and they relate the density ratio across a shock wave to the pressure ratio and the velocity ratio.

In the derivation of the Rankine-Hugoniot equations it was not assumed that the flow was isentropic, and indeed, it will now be shown that it is not isentropic. If the flow had been isentropic, the density ratio across the shock wave would have been

$$\frac{\rho_2}{\rho_1} = \left(\frac{p_2}{p_1}\right)^{1/\gamma}$$

Thus in a plot of $\log(\rho_2/\rho_1)$ versus $\log(p_2/p_1)$ the isentropic law will be a straight line of slope $1/\gamma$. The corresponding curve obtained from Eqs. (11.4) is a curved line as shown in Fig. 11.3b.

From the foregoing results it may be concluded that shock waves depart from the isentropic law unless $p_2/p_1$ and $\rho_2/\rho_1$ are close to unity. That is, unless the shock wave is very weak, it will not be isentropic.

## 11.4 CONDITIONS FOR NORMAL SHOCK WAVES

It will be shown in this section that, as a consequence of the second law of thermodynamics, only that portion of Fig. 11.3b which lies in the first quadrant has physical significance. This restriction will be shown to result in the restriction that the upstream Mach number $M_1$ must be greater than unity for shock waves to occur and the resulting downstream Mach number $M_2$ will be less than unity.

Using the results from thermodynamics which are quoted in Appendix E, the entropy difference across a shock wave $s_2 - s_1$ will be given by

$$s_2 - s_1 = C_p \log\left(\frac{T_2}{T_1}\right) - R \log\left(\frac{p_2}{p_1}\right)$$

Using the ideal-gas law, the temperature ratio in the above equation may be eliminated in favor of the pressure and density ratios. Thus the entropy change may be rewritten as follows:

$$s_2 - s_1 = C_p \log \left( \frac{p_2}{p_1} \frac{\rho_1}{\rho_2} \right) - R \log \left( \frac{p_2}{p_1} \right)$$

$$= (C_p - R) \log \left( \frac{p_2}{p_1} \right) - C_p \log \left( \frac{\rho_2}{\rho_1} \right)$$

But $C_p - R = C_v$, so that the entropy change across the shock wave, which will be denoted as $\Delta s$, may be evaluated from the following equation:

$$\frac{\Delta s}{C_v} = \log \left( \frac{p_2}{p_1} \right) - \gamma \log \left( \frac{\rho_2}{\rho_1} \right)$$

Using the above result, the entropy change and the density ratio will be compared, for a given pressure ratio, for two processes. The first process will be a shock wave which must obey the Rankine-Hugoniot equations and the second process will be a hypothetical isentropic one for the same pressure ratio as the shock wave. Then, from the above equation, the entropy changes in each of these processes will be

$$\left( \frac{\Delta s}{C_v} \right)_{R-H} = \log \left( \frac{p_2}{p_1} \right) - \gamma \log \left( \frac{\rho_2}{\rho_1} \right)_{R-H}$$

$$0 = \log \left( \frac{p_2}{p_1} \right) - \gamma \log \left( \frac{\rho_2}{\rho_1} \right)_I$$

where the subscript $R - H$ indicates the entropy change and the density ratio for a shock wave in which the pressure ratio is $p_2/p_1$ and the subscript $I$ denotes the density ratio for an isentropic process which spans the same pressure ratio. Subtracting these two equations to eliminate the common pressure ratio gives

$$\left( \frac{\Delta s}{C_v} \right)_{R-H} = \gamma \left[ \log \left( \frac{\rho_2}{\rho_1} \right)_I - \log \left( \frac{\rho_2}{\rho_1} \right)_{R-H} \right]$$

But the second law of thermodynamics requires that $\Delta s \geq 0$, so that

$$\log \left( \frac{\rho_2}{\rho_1} \right)_I \geq \log \left( \frac{\rho_2}{\rho_1} \right)_{R-H}$$

Figure 11.3b shows that this inequality can be satisfied only in the first quadrant of the diagram, which corresponds to $\log (\rho_2/\rho_1) > 0$ and $\log (p_2/p_1) > 0$. That is, in order to satisfy the second law of thermodynamics

$$\frac{\rho_2}{\rho_1} \geq 1 \qquad\qquad (11.5a)$$

which means that the gas must be compressed as it goes through a shock wave.

The continuity equation then shows that

$$\frac{u_1}{u_2} \geq 1 \tag{11.5b}$$

so that the fluid is slowed down as it passes through a shock wave.

The conditions (11.5a) and (11.5b) may be put into the more meaningful condition $M_1 \geq 1$. In order to achieve this alternative formulation, it is first necessary to derive a relationship which is known as the Prandtl or Meyer relation. In deriving this region, the subscript $*$ will be used to denote the value of a variable when $M = u/a = 1$, where $u$ is the fluid velocity and $a$ is the local value of the speed of sound. Then it follows that $u_* = a_*$.

The starting point in the derivation is the equation which is obtained by dividing the momentum equation (11.3b) by the continuity equation (11.3a).

$$u_1 + \frac{p_1}{\rho_1 u_1} = u_2 + \frac{p_2}{\rho_2 u_2}$$

Using the definition of the speed of sound to introduce $a_1^2 = \gamma p_1/\rho_1$ and $a_2^2 = \gamma p_2/\rho_2$ gives

$$u_1 + \frac{a_1^2}{\gamma u_1} = u_2 + \frac{a_2^2}{\gamma u_2}$$

$$\therefore \quad u_1 - u_2 = \frac{a_2^2}{\gamma u_2} - \frac{a_1^2}{\gamma u_1}$$

The right-hand side of this equation may be replaced by an equivalent expression which is obtained from the energy equation in the form

$$\frac{u_1^2}{2} + \frac{a_1^2}{\gamma - 1} = \frac{u_2^2}{2} + \frac{a_2^2}{\gamma - 1} = \frac{\gamma + 1}{2(\gamma - 1)} a_*^2$$

where the fact that $u_{1*} = u_{2*} = a_{1*} = a_{2*} = a_*$ has been used. Thus the velocity difference $u_1 - u_2$ may be rewritten in the following form:

$$u_1 - u_2 = \frac{1}{\gamma u_2}\left(\frac{\gamma + 1}{2}a_*^2 - \frac{\gamma - 1}{2}u_2^2\right) - \frac{1}{\gamma u_1}\left(\frac{\gamma + 1}{2}a_*^2 - \frac{\gamma - 1}{2}u_1^2\right)$$

This equation simplifies considerably to the form

$$u_1 u_2 = a_*^2 \tag{11.6}$$

which is the Prandtl or Meyer relation.

The above result will be used in the conditions which were established for a normal shock wave to obtain an alternative form of these conditions. Multiplying both the numerator and the denominator of the inequality (11.5b) by $u_1$

gives

$$\frac{u_1^2}{u_1 u_2} \geq 1$$

Then, using Eq. (11.6),

$$\frac{u_1^2}{a_*^2} \geq 1$$

The left-hand side of this inequality may be evaluated from the energy equation as follows:

$$\frac{u_1^2}{2} + \frac{a_1^2}{\gamma - 1} = \frac{\gamma + 1}{2(\gamma - 1)} a_*^2$$

Dividing this equation by $u_1^2$ gives

$$\frac{1}{2} + \frac{1}{\gamma - 1} \frac{1}{M_1^2} = \frac{\gamma + 1}{2(\gamma - 1)} \frac{a_*^2}{u_1^2}$$

$$\therefore \quad \frac{u_1^2}{a_*^2} = \frac{(\gamma + 1) M_1^2}{2 + (\gamma - 1) M_1^2}$$

Substituting this expression into the condition for a shock wave gives

$$\frac{(\gamma + 1) M_1^2}{2 + (\gamma - 1) M_1^2} \geq 1$$

which reduces to

$$M_1 \geq 1 \qquad (11.7a)$$

That is, a shock wave can occur only if the incoming flow is supersonic. Furthermore, in view of the Prandtl or Meyer relation [Eq. (11.6)], the inequality (11.7a) implies that

$$M_2 \leq 1 \qquad (11.7b)$$

To summarize, in order that the second law of thermodynamics may not be violated, a normal shock wave can occur only in supersonic flow, and the resulting downstream flow field will be subsonic. That is, the fluid will be compressed as it passes through the shock wave.

## 11.5  NORMAL-SHOCK-WAVE EQUATIONS

The results of the last two sections were intended to establish the fundamental phenomena of shock waves and the principal consequences of the existence of shock waves. However, the relationships established in these sections are not suitable for evaluating the conditions downstream of a shock wave in terms of the upstream conditions. It will be recalled that the three conservation equations connect six quantities, three upstream values and three downstream values. Then, it should be possible to eliminate any two of the downstream conditions and so obtain an equation which relates the remaining downstream condition to the three upstream conditions. In this way equations may be established for each of the downstream quantities in terms of the upstream conditions, which are presumably known. Rather than considering the velocity to be one of the quantities, the Mach number $M$ will be considered. Thus for supersonic flow in which a shock wave exists, the known quantities may be considered to be $p_1$, $\rho_1$, and $M_1$, while the unknown downstream quantities will be $p_2$, $\rho_2$, and $M_2$.

To evaluate downstream Mach number $M_2$ the energy equation involving the upstream conditions and the sonic conditions is employed as follows:

$$\frac{u_1^2}{2} + \frac{a_1^2}{\gamma - 1} = \frac{\gamma + 1}{2(\gamma - 1)} a_*^2$$

hence

$$\frac{a_*^2}{u_1^2} = 2\frac{\gamma - 1}{\gamma + 1}\left[\frac{1}{2} + \frac{1}{(\gamma - 1)M_1^2}\right]$$

Similarly, from the energy equation involving the downstream conditions and the sonic conditions, it follows that

$$\frac{a_*^2}{u_2^2} = 2\frac{\gamma - 1}{\gamma + 1}\left[\frac{1}{2} + \frac{1}{(\gamma - 1)M_2^2}\right]$$

These expressions will be used in the Prandtl or Meyer equation [Eq. (11.6)] as follows:

$$u_1 u_2 = a_*^2$$

$$\therefore \quad \frac{a_*^2}{u_1^2}\frac{a_*^2}{u_2^2} = 1$$

$$\left(\frac{\gamma - 1}{\gamma + 1}\right)^2\left[1 + \frac{2}{(\gamma - 1)M_1^2}\right]\left[1 + \frac{2}{(\gamma - 1)M_2^2}\right] = 1$$

Solving this equation for $M_2$ gives

$$M_2^2 = \frac{1 + [(\gamma - 1)/2]M_1^2}{\gamma M_1^2 - (\gamma - 1)/2} \qquad (11.8a)$$

That is, the downstream Mach number is a function only of the upstream Mach number and the specific-heat ratio of the gas. The variation of $M_2$ with $M_1$, as defined by Eq. (11.8a), is shown schematically in Fig. 11.4a. It will be seen that as the upstream Mach number increases, the downstream Mach number decreases. As $M_1 \to \infty$, Eq. (11.8a) shows that $M_2^2 \to (\gamma - 1)/2\gamma$, which defines the asymptotic limit.

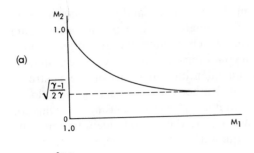

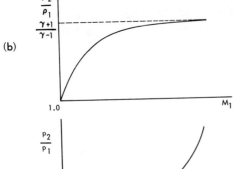

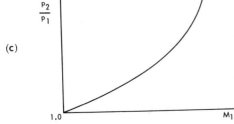

**FIGURE 11.4**
Conditions downstream of a normal shock wave; (*a*) the Mach number, (*b*) the density, and (*c*) the pressure.

The density ratio across the shock wave will be obtained by first evaluating the velocity ratio, then using the continuity equation. From Eq. (11.6),

$$\frac{u_2}{u_1} = \frac{a_*^2}{u_1^2}$$

$$= 2\frac{\gamma - 1}{\gamma + 1}\left[\frac{1}{2} + \frac{1}{(\gamma - 1)M_1^2}\right]$$

where the last equality was established earlier in this section. Simplifying the right-hand side of this equation gives

$$\frac{u_2}{u_1} = \frac{(\gamma - 1)M_1^2 + 2}{(\gamma + 1)M_1^2}$$

But, from the continuity equation, $u_2/u_1 = \rho_1/\rho_2$, so that the expression for the density ratio across the shock wave is

$$\frac{\rho_2}{\rho_1} = \frac{(\gamma + 1)M_1^2}{(\gamma - 1)M_1^2 + 2} \tag{11.8b}$$

The form of the density ratio as a function of the upstream Mach number is shown in Fig. 11.4b. The density ratio is a monotonically increasing function of $M_1$ and reaches an asymptote which, as shown by Eq. (11.8b), is $(\gamma + 1)/(\gamma - 1)$.

The pressure ratio across a normal shock wave may be readily evaluated from the Rankine-Hugoniot equations [Eqs. (11.4)] and the density ratio as given by Eq. (11.8b). This gives

$$\frac{1 + [(\gamma + 1)/(\gamma - 1)](p_2/p_1)}{(\gamma + 1)/(\gamma - 1) + p_2/p_1} = \frac{(\gamma + 1)M_1^2}{(\gamma - 1)M_1^2 + 2}$$

Solving this equation for the pressure ratio gives

$$\frac{p_2}{p_1} = 1 + \frac{2\gamma}{\gamma + 1}(M_1^2 - 1) \tag{11.8c}$$

The form of this result is shown in Fig. 11.4c. It will be seen from this curve, and it may be verified from Eq. (11.8c), that the pressure ratio increases without limit as the upstream Mach number increases.

The foregoing relations [Eqs. (11.8a), (11.8b), and (11.8c)] give each of the principal downstream quantities in terms of the upstream Mach number and the specific-heat ratio of the gas. The functional form of these results is shown qualitatively in Fig. 11.4, and quantitative data may be obtained from tables and figures which appear in the references at the end of this part of the book.

## 11.7 OBLIQUE SHOCK WAVES

Oblique shock waves are shock waves which are inclined to the free stream at an angle which is different from $\pi/2$. Such a shock wave is shown in Fig. 11.5, in which both the incoming and the outgoing velocity have been decomposed into components which are perpendicular to the shock wave and those which are parallel to the shock wave. The shock wave is inclined at an angle $\beta$ to the incoming flow direction, and the velocity vector is deflected through an angle $\delta$ by the shock wave.

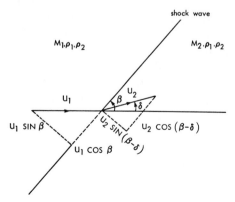

**FIGURE 11.5**
Configuration of oblique shock wave.

The components of the velocity vectors which are normal to the shock wave are $u_1 \sin \beta$ and $u_2 \sin (\beta - \delta)$ for the incoming and outgoing flow, respectively. These velocity components must obey the normal-shock-wave equations, so that

$$u_2 \sin (\beta - \delta) \leq u_1 \sin \beta$$

On the other hand, the tangential-velocity components must be equal, since there is no pressure differential or other force acting in the tangential direction. This shrinking of the normal-velocity component and preservation of the tangential-velocity component results in the downstream velocity vector $u_2$ being bent toward the shock wave as shown in Fig. 11.5.

The equations which determine the downstream values of the pressure, density, and Mach number need not be established from the governing equations in the same manner as was done in the previous section. Rather, the observations which have already been made regarding normal- and tangential-velocity components may be utilized in conjunction with the normal-shock-wave equations. Since the upstream normal-velocity component is now $u_1 \sin \beta$ rather than $u_1$, the upstream Mach number $M_1$ should be replaced by $M_1 \sin \beta$ in Eqs. (11.8a), (11.8b), and (11.8c). Likewise, the downstream Mach number $M_2$ should be replaced by $M_2 \sin (\beta - \delta)$. Thus the equations for the downstream Mach number, density, and pressure become

$$M_2^2 \sin^2 (\beta - \delta) = \frac{1 + [(\gamma - 1)/2] M_1^2 \sin^2 \beta}{\gamma M_1^2 \sin^2 \beta - (\gamma - 1)/2}$$

$$\frac{\rho_2}{\rho_1} = \frac{(\gamma + 1) M_1^2 \sin^2 \beta}{(\gamma - 1) M_1^2 \sin^2 \beta + 2}$$

$$\frac{p_2}{p_1} = 1 + \frac{2\gamma}{\gamma + 1} \left( M_1^2 \sin^2 \beta - 1 \right)$$

The foregoing equations express $M_2$, $\rho_2$, and $p_2$ in terms of $M_1$, $\beta$, and $\delta$. Although $M_1$ is usually known, only one of the angles $\beta$ and $\delta$ is typically known. If the shock wave is generated by the leading edge of a body, the angle $\delta$ will be known, since the downstream velocity vector must be tangent to the surface of the body. Then the angle $\beta$ is typically the unknown quantity. However, one more equation exists to close the system of equations. The conservation equations have been applied only to the normal components of the upstream and downstream velocity vectors. Since there are no forces acting along the shock wave, the conservation of mass and momentum in that direction are satisfied by equating the components of the velocity vectors in the tangential direction. This gives

$$u_1 \cos \beta = u_2 \cos (\beta - \delta)$$

hence

$$\frac{u_1}{u_2} = \frac{\cos (\beta - \delta)}{\cos \beta}$$

Since this equation is supposed to determine the shock-wave angle $\beta$, the velocity ratio should be eliminated in favor of known quantities. The continuity equation, which involves the normal-velocity components, gives

$$\frac{u_1}{u_2} = \frac{\rho_2}{\rho_1} \frac{\sin(\beta - \delta)}{\sin \beta}$$

Equating these two expressions for the velocity ratio results in the identity

$$\frac{\rho_2}{\rho_1} = \frac{\tan \beta}{\tan(\beta - \delta)}$$

Now the density ratio may be eliminated from the results which were deduced above from the normal-shock-wave equations. The result is the relation

$$\frac{(\gamma + 1)M_1^2 \sin^2 \beta}{(\gamma - 1)M_1^2 \sin^2 \beta + 2} = \frac{\tan \beta}{\tan(\beta - \delta)}$$

This equation is sufficient to determine the angle $\beta$, since both $M_1$ and $\delta$ are known. However, the result is an implicit expression for $\beta$ rather than an explicit expression. Although the equation is not readily rearranged to express $\beta$ in terms of $M_1$ and $\delta$, it is possible to solve for $M_1$ in terms of $\beta$ and $\delta$. Solving directly for $M_1$ gives

$$M_1^2 = \frac{2 \tan \beta}{\sin^2 \beta[(\gamma + 1)\tan(\beta - \delta) - (\gamma - 1)\tan \beta]}$$

This result may be simplified by first rearranging the numerator and denominator to give

$$M_1^2 = \frac{2 \cos(\beta - \delta)}{\sin \beta[(\gamma + 1)\sin(\beta - \delta)\cos \beta - (\gamma - 1)\sin \beta \cos(\beta - \delta)]}$$

Next, the trigonometric identities for multiple-angled functions may be employed to reduce the expression to the following form:

$$M_1^2 = \frac{2 \cos(\beta - \delta)}{\sin \beta[\sin(2\beta - \delta) - \gamma \sin \delta]} \tag{11.9a}$$

Equation (11.9a) connects three quantities, two of which will be known in any flow configuration. The form of the solution which is represented by Eq. (11.9a) is shown in Fig. 11.6a. These results show that, for given values of $M_1$ and the deflection angle $\delta$, two shock-wave angles $\beta$ are possible. The limiting values of $\beta$ may be established by recalling the condition for a normal shock wave, $M_1 \geq 1$, which here becomes $M_1 \sin \beta \geq 1$. Then $\beta$ must lie in the range

$$\sin^{-1}\left(\frac{1}{M_1}\right) \leq \beta \leq \frac{\pi}{2} \tag{11.9b}$$

where the upper limit corresponds to a normal shock wave. The lower limit will be recognized as the angle of a Mach wave which is the angle to the leading

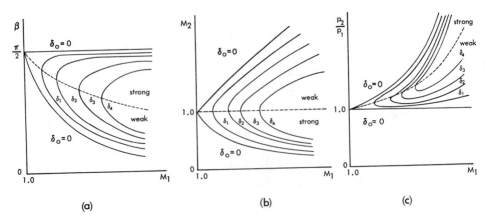

**FIGURE 11.6**
Oblique-shock-wave relations; (a) shock-wave inclination $\beta$, (b) downstream Mach number $M_2$, and (c) pressure ratio across the shock wave.

edge of a sound wave which is being continuously emitted by a source of sound in which the source is moving with Mach number $M_1$. Mach waves, of course, represent the sonic end of the shock-wave spectrum, so that the pressure ratio and the density ratio across Mach waves is unity. On the other hand, normal shock waves exhibit the maximum pressure and density ratio for a given approach Mach number. These observations lead to the classification of oblique shock waves as being either strong (if the value of $\beta$ is close to $\pi/2$) or weak [if the value of $\beta$ is close to $\sin^{-1}(1/M_1)$]. It will be shown shortly that the downstream flow is subsonic in the case of a strong shock wave and supersonic in the case of weak shock waves. The dotted line in Fig. 11.6a corresponds to $M_2 = 1$, which does not coincide with the minimum value of $M_1$ for fixed $\delta$, although these two values do not differ substantially.

The value of the downstream Mach number may be obtained from the equations which were already deduced from the normal-shock-wave equations. The expression is

$$M_2^2 = \frac{1 + [(\gamma - 1)/2] M_1^2 \sin^2 \beta}{\sin^2 (\beta - \delta)\left[\gamma M_1^2 \sin^2 \beta - (\gamma - 1)/2\right]} \qquad (11.9c)$$

Since $M_1$ and $\delta$ will be known from the problem definition and since $\beta$ will be known from Eq. (11.9a), the value of $M_2$ may be established from Eq. (11.9c). The results of such a solution are shown schematically in Fig. 11.6b. The figure clearly illustrates the possibilities of having either subsonic or supersonic flow downstream of the shock wave. In the case of normal shock waves it was found that the downstream flow had to be subsonic, but for oblique shock waves the unaffected tangential-velocity component, when added to the subsonic normal component, may again be supersonic, particularly for shallow angles $\beta$.

The expression for the pressure ratio across an oblique shock wave was also deduced from the normal-shock-wave equations and was shown to be

$$\frac{p_2}{p_1} = 1 + \frac{2\gamma}{\gamma + 1}(M_1^2 \sin^2 \beta - 1) \qquad (11.9d)$$

The form of the curves which are generated from this equation is shown in Fig. 11.6c. This diagram brings out the significance of the terminology "strong" and "weak" as applied to shock waves. The strength of a shock wave is defined by the nondimensional pressure difference $(p_2 - p_1)/p_1$, which is seen to be larger for the strong shock waves than for the weak shock waves.

The downstream Mach number and pressure ratio are two quantities of principal interest in shock-wave flows. However, the equation for the density ratio was also deduced from the normal-shock equations and was shown to be

$$\frac{\rho_2}{\rho_1} = \frac{(\gamma + 1)M_1^2 \sin^2 \beta}{(\gamma - 1)M_1^2 \sin^2 \beta + 2} \qquad (11.9e)$$

The foregoing equations are sufficient to completely determine the conditions downstream of an oblique shock wave, provided that the type of shock wave is known (that is, strong or weak). There is no mathematical criterion for determining whether the shock wave will belong to the strong family or the weak family. The configuration which will be adopted by nature depends upon the geometry of the projectile or boundary which is inducing the shock wave.

Figure 11.7 shows two different shapes of leading edge which are considered to be immersed in the same supersonic flow field. The boundary condition on the solid surface requires that the velocity vector be close to the vertical in

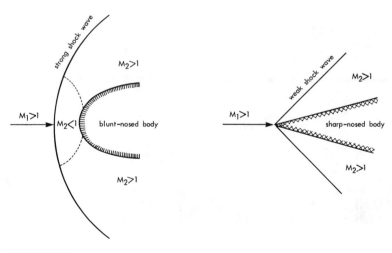

**FIGURE 11.7**
Supersonic flow approaching a blunt-nosed body and a sharp-nosed body.

the vicinity of the front stagnation point. This boundary condition may be realized only if a *detached* shock wave exists in front of the body as shown. Since the angle $\beta$ is close to $\pi/2$ for this shock wave, it will be of the strong variety, so that the downstream Mach number will be less than unity. The corresponding subsonic flow may then satisfy the required boundary condition in the usual way. Moving away from the front stagnation point along the surface of the body, the angle $\delta$ of the downstream velocity vector is continuously changing. Thus some point is eventually reached where the value of $\delta$ is such that matching the boundary condition by deflecting the flow through a weak shock wave is possible. The shock wave will therefore bend back with the flow far from the body so that the downstream flow becomes supersonic. Thus a region of subsonic flow will exist in the vicinity of the nose of the body and the rest of the flow field will be supersonic.

In the case of a sharp-nosed slender body an *attached* shock wave will exist as shown in Fig. 11.7. With this configuration the velocity vector will be deflected by the shock wave through just the correct angle to satisfy the boundary condition that the surface be a streamline. Since the shock wave will belong to the weak family in this case, the flow downstream of the shock wave will remain supersonic.

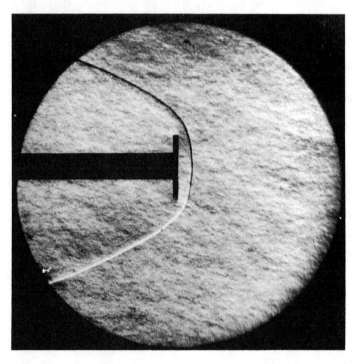

**PLATE 4**
Detached shock wave in front of a flat circular disk in flow at Mach number of 10.4. (*Photograph courtesy of National Research Council of Canada.*)

**PLATE 5**
Shock wave attached to cone of 15° half angle at 6° angle of attack and Mach number of 10.4.
(*Photograph courtesy of National Research Council of Canada.*)

## PROBLEMS

**11.1.** In general, the enthalpy $h$ depends on both the pressure and the temperature; that
is, $h(p, T)$. However, if $p = \rho RT$, it follows that $h = h(T)$ only. To show this,
obtain the first law of thermodynamics in the form

$$T\,ds = dh - v\,dp$$

where $v$ is the specific volume of the gas. Then, by considering $s = s(p, T)$ and
$h = h(p, T)$, show that

$$\frac{\partial s}{\partial T} = \frac{1}{T}\frac{\partial h}{\partial T}$$

$$\frac{\partial s}{\partial p} = \frac{1}{T}\left(\frac{\partial h}{\partial p} - v\right)$$

These are the reciprocity relations which are quoted in Appendix E. By eliminating
$s$ from these equations and utilizing the gas law $p = \rho RT$, show that $\partial h / \partial p = 0$ so
that $h = h(T)$ only.

**11.2.** In general, the internal energy $e$ depends on both the specific volume and the temperature so that $e = e(v, T)$. Show that, if $p = \rho R T$, it follows that $e = e(T)$ only.

**11.3.** Show that, for a calorically perfect gas, the entropy change involved in some event may be related to the temperature ratio and the pressure ratio or the temperature ratio and the density ratio by the following expressions:

$$s - s_0 = C_p \log \frac{T}{T_0} - R \log \frac{p}{p_0}$$

$$= C_v \log \frac{T}{T_0} + R \log \frac{p_0}{\rho}$$

**11.4.** The equation governing the fluid velocity induced by a finite-amplitude forward-running disturbance was shown to be

$$\frac{\partial u}{\partial t} + (u + a)\frac{\partial u}{\partial x} = 0$$

in which both $u$ and $a$ depend on both $x$ and $t$. Show, by direct substitution, that

$$u = f[x - (u + a)t]$$

is the general solution to this equation, where $f$ is any differentiable function.

**11.5.** The equation to be solved for $u$ in the above problem is

$$\frac{Du}{Dt} = 0$$

where

$$\frac{D}{Dt} = \frac{\partial}{\partial t} + (u + a)\frac{\partial}{\partial x}$$

and

$$a^2 = \frac{dp}{d\rho}$$

Show that the steepness of the wave front $\partial u / \partial x$ satisfies a relationship of the form

$$\frac{D}{Dt}\left(\frac{\partial u}{\partial x}\right) \sim \left(\frac{\partial u}{\partial x}\right)^2$$

and find the constant of proportionality. If the steepness of the wave front at time $t = 0$ is denoted by

$$\left.\frac{\partial u}{\partial x}\right|_0 = S$$

find the time required for $\partial u / \partial x$ to become infinite, and thus show that $S$ must be negative for a shock wave to form.

**11.6.** The entropy increase across a shock wave may be calculated from the expression

$$\frac{\Delta s}{C_v} = \log\left[\frac{p_2}{p_1}\left(\frac{\rho_1}{\rho_2}\right)^\gamma\right]$$

Using the normal-shock-wave equations, express $\Delta s / C_v$ as a function of $M_1$ and $\gamma$ only. Denoting $(M_1^2 - 1)$ by $\varepsilon$, express $\Delta s / C_v$ as the sum of three terms each of which has the form $(1 + \alpha\varepsilon)$ where $\alpha$ is a function of $\gamma$ only. Expand this result for small values of $\varepsilon$, and hence show that $\Delta s/C_v \sim \varepsilon^3$, which shows that weak shock waves are almost isentropic.

**11.7.** A normal shock wave occurs in a fluid which is not a perfect gas and for which the pressure and the density are related by the expression

$$\rho \frac{dp}{d\rho} = c$$

where $c$ is a constant.

(a) Using the continuity and momentum equations, together with the above relation and the general expression for the speed of sound, show that the upstream and downstream Mach numbers are related by the equation

$$\log \frac{M_1^2}{M_2^2} = M_1^2 - M_2^2$$

(b) The pressure rise across the shock wave $(p_2 - p_1)$ can, in principle, be expressed in terms of $M_1$ and $c$. This relation is implicit rather than explicit, but it can be solved for $M_1^2$ as a function of $(p_2 - p_1)$ and $c$. Find the expression.

**11.8.** The equation to be solved in the propagation of sound waves is the same as that which is to be solved for shallow-liquid waves. This leads to an analogy between sound waves in a gaseous medium and surface waves on a liquid. Find the corresponding physical quantities in this analogy, and find the value of $\gamma$ which makes the analogy complete.

**11.9.** The equations governing a wave which is approximately one-dimensional are as follows:

$$\frac{\partial u}{\partial x} + \frac{\partial v}{\partial y} = 0$$

$$\frac{\partial u}{\partial t} + u\frac{\partial u}{\partial x} + v\frac{\partial u}{\partial y} = -\frac{1}{\rho}\frac{\partial p}{\partial x}$$

$$0 = -\frac{1}{\rho}\frac{\partial p}{\partial y}$$

Look for a progressive-wave solution to these equations in which the pressure $p(x, t)$ is dependent on $x$ and $t$ only, and the velocity components follow the same $x$ and $t$ dependence as indicated below:

$$\frac{\partial p}{\partial t} + c\frac{\partial p}{\partial x} = 0$$

$$u(x, y, t) = U(p, y)$$

$$v(x, y, t) = V(p, y)\frac{\partial p}{\partial x}$$

In the above, the wave speed $c(p)$ is considered to be a function of the pressure $p$.

Without linearization, determine the equations to be satisfied by the functions $U(p, y)$ and $V(p, y)$.

Look for a similarity solution to the equations obtained above in the following form:

$$U(p, y) = p^{1/2} U^*(y)$$
$$V(p, y) = p^{-1/2} V^*(y)$$
$$c(p) = p^{1/2} C^*$$

where $C^*$ is a constant. Find the equations to be satisfied by $U^*(y)$ and $V^*(y)$.

# CHAPTER

# 12

# ONE-DIMENSIONAL FLOWS

This chapter deals with flow fields which are essentially one-dimensional and which are compressible, either subsonic or supersonic. Most of the topics considered involve sonic flow and so constitute a continuation of the topics treated in the previous section.

The topic of weak shock waves or sonic waves is treated from a general viewpoint by means of Riemann invariants. In this way the manner in which acoustic waves react in various situations is established. Particular situations which are treated include the release of waves in a shock tube, the reflection of waves at a solid boundary, reflection and refraction of waves at the interface of two gases, and waves generated by a moving piston. In order to show the quantitative differences due to finite-strength waves, the unlinearized shock-tube problem is also treated. Nonadiabatic flows are also treated through the technique of influence coefficients. This allows not only heat addition but also friction and area changes to be handled. Finally, the flow through convergent-divergent nozzles is treated.

## 12.1  WEAK WAVES

The topic of weak shock waves or acoustic waves will be further investigated in this section. The Riemann invariants for the governing equations will be established, which permits the treatment of general problems involving weak waves.

It was shown in Chap. 11 that weak waves are isentropic, so that $p$ may be considered to be a function of $\rho$ only. Thus

$$\frac{\partial p}{\partial x} = \frac{dp}{d\rho}\frac{\partial \rho}{\partial x} = a^2\frac{\partial \rho}{\partial x}$$

The continuity and momentum equations for a plane wave may therefore be written in the following form:

$$\frac{\partial \rho}{\partial t} + \rho\frac{\partial u}{\partial x} + u\frac{\partial \rho}{\partial x} = 0$$

$$\rho\frac{\partial u}{\partial t} + u\frac{\partial u}{\partial x} = -a^2\frac{\partial \rho}{\partial x}$$

For a fluid which is originally at rest before the wave passes through it, the density, pressure, and velocity may be written as their quiescent values plus a perturbation. That is,

$$\rho = \rho_0 + \rho'$$
$$p = p_0 + p'$$
$$u = 0 + u'$$

where, for a weak wave, $\rho'/\rho_0 \ll 1$, $p'/p_0 \ll 1$, and $u'/a_0 \ll 1$, where $a_0$ is the speed of sound in the undisturbed gas. Thus the linearized form of these equations, which will describe weak waves, is

$$\frac{\partial \rho'}{\partial t} + \rho_0\frac{\partial u'}{\partial x} = 0$$

$$\rho_0\frac{\partial u'}{\partial t} + a_0^2\frac{\partial \rho'}{\partial x} = 0$$

Since $\rho_0$ is a constant, it may be added to $\rho'$ when it appears inside a derivative. Thus the above equations may be rewritten in the following form:

$$\frac{\partial}{\partial t}(\rho_0 + \rho') + \rho_0\frac{\partial}{\partial x}(0 + u') = 0$$

$$\rho_0\frac{\partial}{\partial t}(0 + u') + a_0^2\frac{\partial}{\partial x}(\rho_0 + \rho') = 0$$

Using, again, the expansions for $\rho$ and $u$ shows that

$$\frac{\partial \rho}{\partial t} + \rho_0\frac{\partial u}{\partial x} = 0$$

$$\rho_0\frac{\partial u}{\partial t} + a_0^2\frac{\partial \rho}{\partial x} = 0$$

The quantities inside the differential operators will now be nondimensionalized by dividing the first equation by $\rho_0$ and the second equation by $\rho_0 a_0$. This gives

$$\frac{\partial}{\partial t}\left(\frac{\rho}{\rho_0}\right) + a_0\frac{\partial}{\partial x}\left(\frac{u}{a_0}\right) = 0$$

$$\frac{\partial}{\partial t}\left(\frac{u}{a_0}\right) + a_0\frac{\partial}{\partial x}\left(\frac{\rho}{\rho_0}\right) = 0$$

Finally, the desired form of the governing equations for weak waves is obtained by first adding, then subtracting, these two equations.

$$\frac{\partial}{\partial t}\left(\frac{u}{a_0} + \frac{\rho}{\rho_0}\right) + a_0\frac{\partial}{\partial x}\left(\frac{u}{a_0} + \frac{\rho}{\rho_0}\right) = 0$$

$$\frac{\partial}{\partial t}\left(\frac{u}{a_0} - \frac{\rho}{\rho_0}\right) - a_0\frac{\partial}{\partial x}\left(\frac{u}{a_0} - \frac{\rho}{\rho_0}\right) = 0$$

Both these equations are of the form of a material derivative of some quantity being zero. The material derivative is one in which the convection velocity is the speed of sound and in the first equation the convection is in the positive $x$ direction. Then, integrating these two equations gives

$$\frac{u}{a_0} + \frac{\rho}{\rho_0} = \text{constant along } x - a_0 t = \text{constant} \qquad (12.1a)$$

$$\frac{u}{a_0} - \frac{\rho}{\rho_0} = \text{constant along } x + a_0 t = \text{constant} \qquad (12.1b)$$

The lines $x - a_0 t$ and $x + a_0 t$ are called the *characteristics*, and the quantities $u/a_0 + \rho/\rho_0$ and $u/a_0 - \rho/\rho_0$, which are constant along the characteristic lines, are called *Riemann invariants*. Figure 12.1a shows the characteristics which pass through a typical $x$ location and the Riemann invariants for these characteristics. It will be noted that one of the characteristics is forward-running and the other is backward-running.

The Riemann invariants may be expressed in terms of the pressure and the velocity rather than the density and the velocity. Depending upon the problem being considered, this alternative formulation may be desirable. To obtain the alternative formulation, the density ratio is replaced by the pressure ratio through use of the isentropic gas law as follows:

$$\frac{p}{\rho^\gamma} = \frac{p_0}{\rho_0^\gamma}$$

$$\therefore \quad \frac{p}{p_0} = \left(\frac{\rho}{\rho_0}\right)^\gamma$$

$$= \left(1 + \frac{\rho'}{\rho_0}\right)^\gamma$$

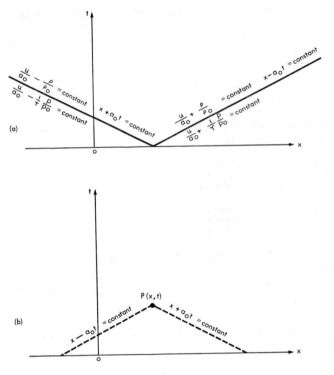

**FIGURE 12.1**
($a$) Characteristics and Riemann invariants in the $xt$ plane and ($b$) basis of evaluating field variables at an arbitrary point $P$.

Since $\rho'/\rho_0 \ll 1$, this expression may be linearized to give

$$\frac{p}{p_0} = 1 + \gamma \frac{\rho'}{\rho_0}$$

In order to eliminate the density ratio from the Riemann invariants, this expression must be rearranged to yield the density ratio in the form $\rho/\rho_0$. This may be done as follows:

$$\frac{\rho'}{\rho_0} = \frac{1}{\gamma}\left(\frac{p}{p_0} - 1\right)$$

but

$$\frac{\rho}{\rho_0} = 1 + \frac{\rho'}{\rho_0}$$

hence

$$\frac{\rho}{\rho_0} = 1 + \frac{1}{\gamma}\left(\frac{p}{p_0} - 1\right)$$

$$= \frac{1}{\gamma}\frac{p}{p_0} + \frac{\gamma - 1}{\gamma}$$

That is, the density ratio $\rho/\rho_0$ may be replaced by the pressure ratio as indicated so that, from Eqs. (12.1$a$) and (12.1$b$), the Riemann invariants may be written in the form

$$\frac{u}{a_0} + \frac{1}{\gamma}\frac{p}{p_0} = \text{constant along } x - a_0 t = \text{constant} \qquad (12.1c)$$

$$\frac{u}{a_0} - \frac{1}{\gamma}\frac{p}{p_0} = \text{constant along } x + a_0 t = \text{constant} \qquad (12.1d)$$

Figure 12.1$a$ shows the two characteristics which pass through a typical point $(x, 0)$ in the $xt$ plane and the alternative Riemann invariants along these characteristics.

Equations (12.1) may be used to evaluate the velocity, the density, and the pressure at any value of $x$ and any value of $t$ if the values of $u$, $\rho$, and $p$ are known as functions of $x$ at some time such as $t = 0$. The manner in which this is done may be explained with reference to Fig. 12.1$b$. A typical point $P(x, t)$ is shown in the $xt$ plane together with the two characteristics which originate along the $t = 0$ axis and which pass through the point $P$. Associated with these two characteristics are Riemann invariants whose constants may be evaluated from the known conditions at $t = 0$. Then at the point $P$ the Riemann invariants for $u$ and $\rho$ provide two algebraic equations for the two unknowns. Alternatively, the Riemann invariants for $u$ and $p$ provide two algebraic equations for these two unknowns. The following sections will utilize this approach to obtain solutions for particular flow situations.

## 12.2 WEAK SHOCK TUBES

The first application of the foregoing theory will be made to a shock tube in which a weak wave is released. A shock tube is a relatively long tube which is fitted with a diaphragm as shown in Fig. 12.2$a$. The gas on on side of the diaphragm is maintained at a pressure which is different from that on the other side. In general, the gases on either side of the diaphragm may be different and so may have different properties and states. In this instance, the gases will be considered to be the same, and only the states are assumed to differ. The initial pressure distribution, which is taken to be an equilibrium state, is shown in Fig. 12.2$b$. The diaphragm may be designed to burst at some predetermined value of the pressure $p_1$. A pressure wave is thus released from the vicinity of the diaphragm as the two regions tend to equalize their pressures. The problem is to determine the pressure and the velocity in the gas at any location and at any time.

The $xt$ diagram for the shock tube is shown in Fig. 12.2$c$. The time at which the diaphragm bursts is taken to be $t = 0$ for convenience, and the location of the diaphragm is chosen to be $x = 0$. Then a compression wave will emanate from the origin and will travel into the lower-pressure region while an expansion wave will emanate from the origin and travel into the region of higher

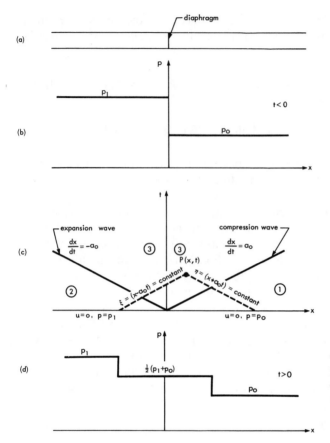

**FIGURE 12.2**
($a$) Shock tube, ($b$) initial pressure distribution, ($c$) $xt$ diagram, and ($d$) typical pressure distribution for $t > 0$.

pressure as indicated. Since weak waves are being considered, the two waves mentioned will travel at the speed of sound $a_0$. Then the slopes of the waves in the $xt$ plane are $a_0$ and $-a_0$ for the compression wave and the expansion wave, respectively. The $xt$ diagram in Fig. 12.2$c$ is divided into three regions which are defined by the waves which emanate from the origin of the diagram. Region 1 is that portion of the positive $x$ axis which has not yet been affected by the oncoming compression wave. In this region the velocity is zero and the pressure is $p_0$. Region 2 is defined as that portion of the negative $x$ axis which has not been influenced by the expansion wave. Here, $u = 0$ and $p = p_1$. The third region, denoted region 3, is that part of the $x$ axis which has been influenced by the compression wave and the expansion wave. Since the pressure and the velocity must be continuous across $x = 0$, both the positive and the negative portions of the $x$ axis in region 3 will experience the same pressure and velocity.

In order to determine the pressure and the velocity in region 3, an arbitrary point $P$ is considered as shown in Fig. 12.2$c$. The two characteristics which originate on the $x$ axis at $t = 0$ and which pass through the point $P$ are indicated and are denoted by $\xi = $ constant and $\eta = $ constant. The values of the

Riemann invariants along these characteristics may be determined from the known distributions along the $x$ axis at $t = 0$. Thus, along the characteristic $\xi = $ constant, Eq. $(12.1c)$ shows that

$$\frac{u}{a_0} + \frac{1}{\gamma}\frac{p}{p_0} = \frac{1}{\gamma}\frac{p_1}{p_0}$$

Here, the fact that $u = 0$ and $p = p_1$ at $t = 0$ for $x < 0$ has been used. From $\eta = $ constant.

$$\frac{u}{a_0} - \frac{1}{\gamma}\frac{p}{p_0} = -\frac{1}{\gamma}$$

The fact that $u = 0$ and $p = p_0$ at $t = 0$ along the positive $x$ axis has been used here. The solution to these two algebraic equations shows that the velocity and the pressure behind the compression and expansion waves are

$$\frac{u}{a_0} = \frac{1}{2\gamma}\left(\frac{p_1}{p_0} - 1\right) \tag{12.2a}$$

$$\frac{p}{p_0} = \frac{1}{2}\left(\frac{p_1}{p_0} + 1\right) \tag{12.2b}$$

Equation $(12.2a)$ shows that, for $p_1/p_0 > 1$, $u/a_0 > 0$, so that the fluid moves along the positive $x$ axis. This agrees with our previous finding that the fluid particles tend to follow compression waves and move away from expansion waves. Equation $(12.2b)$ shows that the pressure in region 3 is the arithmetic mean of the pressures in regions 1 and 3. The pressure distribution along the tube is shown for some time $t > 0$ in Fig. 12.2$d$. This figure illustrates that a compression wave of amplitude $(p_1 - p_0)/2$ moves along the positive $x$ axis with speed $a_0$ while an expansion wave of the same amplitude moves along the negative $x$ axis at the same speed. The expansion waves may be considered to be a discontinuity, since only weak waves are being considered. The analogous problem for finite-strength waves will be considered in a later section.

## 12.3   WALL REFLECTION OF WAVES

The behavior of a weak pressure wave when it strikes a solid boundary will be established in this section. This will be done by considering a shock tube which has a closed end so that the wave which travels along the tube will impinge upon it. In this way it will be shown that a compression wave is reflected by a wall as a compression wave of the same strength and an expansion wave is likewise reflected as an identical expansion wave.

Figure 12.3$a$ shows a shock tube similar to that which was considered in the previous section except that the tube is closed at one end. The $xt$ diagram for the gas conditions which result from bursting the diaphragm at time $t = 0$ is shown in Fig. 12.3$b$. As in the previous section, the outgoing waves divide this

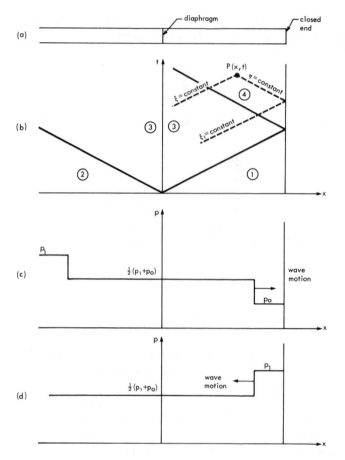

**FIGURE 12.3**
(*a*) Shock tube, (*b*) *xt* diagram, (*c*) pressure distribution at some time, and (*d*) at a later time.

diagram into distinct regions which are numbered 1, 2, and 3. Upon striking the closed end of the shock tube, the wave which was traveling in the positive $x$ direction will be reflected as a wave of some form. It is known that this reflected wave will travel at the speed of sound $a_0$, but it is not known a priori whether it will be an expansion wave or a compression wave and what the strength of this wave will be in relation to that of the incident wave. Thus the properties of the gas in region 4, which is that region which has been influenced by both the incident and the reflected waves, is not known.

Region 1 of Fig. 12.3*b* has not yet been influenced by the outgoing wave from the origin and so maintains its initial conditions of $u = 0$, $p = p_0$. Likewise, region 2 maintains its undisturbed condition of $u = 0$, $p = p_1$. Region 3 has been influenced by the outgoing waves, and so the velocity and the pressure there will be given by Eqs. (12.2$a$) and (12.2$b$). In order to determine the state of the gas in region 4, an arbitrary point $P(x, t)$ and its two characteristics are indicated in Fig. 12.3$b$. The $\xi = $ constant characteristic comes from region 3, where the velocity and the pressure are known. Hence this characteristic may be terminated at any point in region 3 where the value of the Riemann invariant may be established. The $\eta = $ constant characteristic runs parallel to the line of the reflected wave and eventually reaches the $x$ location of the closed tube end. Since the pressure here is unknown, the $\xi_1 = $ constant characteristic from this point is drawn into region 3, where the velocity and the pressure are known. This permits the Riemann invariant to be established for the $\eta = $ constant characteristics.

Denoting the pressure at the wall or closed end in region 4 by $p_w$, the Riemann invariant along the $\eta = $ constant characteristic gives

$$\frac{u}{a_0} - \frac{1}{\gamma}\frac{p}{p_0} = -\frac{1}{\gamma}\frac{p_w}{p_0}$$

Here the fact that $u = 0$ and $p = p_w$ on the wall has been used. To evaluate $p_w$, the Riemann invariant for the $\xi_1 = $ constant characteristic will be used. This gives

$$\frac{u}{a_0} + \frac{1}{\gamma}\frac{p}{p_0} = -\frac{1}{\gamma}\frac{p_w}{p_0} = \frac{1}{2\gamma}\left(\frac{p_1}{p_0} - 1\right) + \frac{1}{2\gamma}\left(\frac{p_1}{p_0} + 1\right)$$

where the values of the velocity and the pressure in region 3, as given by Eqs. (12.2$a$) and (12.2$b$), and at the wall in region 4 have been used. The last equality is satisfied by

$$p_w = p_1$$

Then the equation along the $\eta = $ constant characteristic becomes

$$\frac{u}{a_0} - \frac{1}{\gamma}\frac{p}{p_0} = -\frac{1}{\gamma}\frac{p_1}{p_0}$$

A second equation is obtained from the $\xi = $ constant characteristic.

$$\frac{u}{a_0} + \frac{1}{\gamma}\frac{p}{p_0} = \frac{1}{\gamma}\frac{p_1}{p_0}$$

Here the constant for the Riemann invariant has been evaluated in region 3 by again using Eqs. (12.2$a$) and (12.2$b$). The solution to the last two algebraic equations is

$$u = 0 \tag{12.3a}$$

$$p = p_1 \tag{12.3b}$$

Equation (12.3$a$) shows that the velocity of the gas in region 4 is zero. This result is due to the fact that region 4 includes the closed end of the tube and the boundary condition there requires zero velocity. Equation (12.3$b$) shows that the pressure in region 4 equals the pressure in region 2. This means that the pressure at any value of $x > 0$ varies as follows: Initially, the pressure is $p_0$, and as the first wave passes toward the closed end, the pressure jumps to $(p_1 + p_0)/2$. Finally, as the reflected wave passes, the pressure jumps to $p_1$ as shown in Figs. 12.3a and 12.3b. That is, the pressure differential across the incident wave is $(p_1 - p_0)/2$, which is also the pressure differential across the reflected wave. Since this result is valid for either $p_1 > p_0$ or $p_1 < p_0$, it follows that compression waves are reflected as compression waves of the same strength by solid boundaries and expansion waves are reflected as expansion waves of the same strength.

## 12.4 REFLECTION AND REFRACTION AT AN INTERFACE

When a wave encounters an interface between two different gases, part of the wave is transmitted through the interface and part of it is reflected by the interface. This conclusion may be reached by considering a shock tube in which an interface between two different gases exists part way down the tube. Such a configuration is shown in Fig. 12.4a. Initially, the velocity is zero everywhere while the pressure is $p_1$ for $x < 0$ and $p_0$ for $x > 0$. Partway down the positive $x$ axis the physical properties of the gas are assumed to change abruptly because there are two different gases in the tube, or the same gas may have two regions which are at different temperatures. In either case, the speed of sound is taken to be $a_{01}$ to the left of the interface and $a_{02}$ to the right of the interface. Likewise, the specific-heat ratio is denoted by $\gamma_1$ to the left of the interface and $\gamma_2$ to the right.

The $xt$ diagram which describes the sequence of events which results from bursting the diaphragm at time $t = 0$ is shown in Fig. 12.4b. It is assumed that the wave which emerges from the burst diaphragm and which travels in the

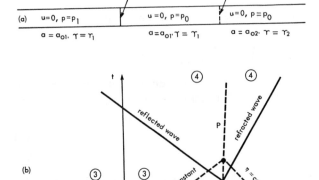

**FIGURE 12.4**
(a) Shock tube with gas interface and (b) $xt$ diagram subsequent to diaphragm's bursting.

positive $x$ direction toward the gaseous interface is partially transmitted and partially reflected at the interface. That is, in general it is assumed that part of the incident wave passes through the gaseous interface and is refracted while the other part of the wave is reflected by the interface. Thus Fig. 12.4$b$ is divided into four distinct regions as indicated. Region 1 represents the initial state of the gas which is located to the right of the diaphragm, and although the physical properties of the gas will be discontinuous at the gaseous interface, the velocity will be everywhere zero and the pressure will be everywhere $p_0$ in this region. In region 2 the velocity will be zero and the pressure will be $p_1$. In the regions marked 3, the gas will be influenced by passage of the waves which result from bursting the diaphragm, and the velocity and pressure there will be given by Eqs. (12.2$a$) and (12.2$b$).

In order to determine the velocity and the pressure in the regions marked 4, an arbitrary point $P(x, t)$ which lies on the interface between the two gases is considered. From this point the $\xi = $ constant and $\eta = $ constant characteristics are drawn, and by virtue of the fact that the point $P$ lies on the interface, each of these characteristics lies entirely in the domain of one gas only. The $\xi = $ constant characteristic may be terminated anywhere in region 3 where the velocity and pressure are known, while the $\eta = $ constant characteristic may be terminated anywhere in region 1. Since the velocity and the pressure must be continuous across the interface at all times, the two regions labeled 4 must have the same velocity and pressure. Since it is realized that, in general, the interface may move after being struck by the incident wave, the line which represents the interface in the regions 4 does not necessarily correspond to $x = $ constant.

Using the Riemann invariant along the $\xi = $ constant characteristic shows that the velocity and the pressure in region 4 must satisfy the condition

$$\frac{u}{a_{01}} + \frac{1}{\gamma_1}\frac{p}{p_0} = \frac{1}{2\gamma_1}\left(\frac{p_1}{p_0} - 1\right) + \frac{1}{2\gamma_1}\left(\frac{p_1}{p_0} + 1\right)$$

$$= \frac{1}{\gamma_1}\frac{p_1}{p_0}$$

where the known conditions for region 3 have been employed from Eqs. (12.2$a$) and (12.2$b$). Along the $\eta = $ constant characteristic we get

$$\frac{u}{a_{02}} - \frac{1}{\gamma_2}\frac{p}{p_0} = -\frac{1}{\gamma_2}$$

where the undisturbed conditions for region 1 have been used. The solution to these two equations is

$$\frac{u}{a_{01}} = \frac{p_1/p_0 - 1}{\gamma_1 + \gamma_2 a_{01}/a_{02}} \tag{12.4$a$}$$

$$\frac{p}{p_0} = \frac{p_1/p_0 + (\gamma_1/\gamma_2)(a_{02}/a_{01})}{1 + (\gamma_1/\gamma_2)(a_{02}/a_{01})} \tag{12.4$b$}$$

Equation (12.4a) shows that for $p_1/p_0 > 1$ the velocity $u$ in region 4 will be positive, so that the interface will move to the right in the positive $x$ direction. As was the case in the previous section, this confirms the result that the flow tends to follow compression waves, since for $p_1/p_0 > 1$ not only will the incident wave be a compression wave but so will the reflected wave. It was shown in the previous section that, for a solid interface, the reflected wave was of the same strength as the incident wave. In the present case it may be anticipated that the gaseous interface is not as efficient in reflecting waves as the solid interface. The actual strength of the reflected wave may be calculated from the solution given by Eq. (12.4b). If the pressure differential across the reflected wave is denoted by $\Delta p_r$, it follows that

$$\frac{\Delta p_r}{p_0} = \frac{p}{p_0} - \frac{1}{2}\left(\frac{p_1}{p_0} + 1\right)$$

where the solution for the pressure in region 3 has been used. Then, using the pressure given by Eq. (12.4b) for the value in region 4, the pressure differential across the reflected wave becomes

$$\frac{\Delta p_r}{p_0} = \frac{[1 - (\gamma_1/\gamma_2)(a_{02}/a_{01})](p_1/p_0 - 1)}{2[1 + (\gamma_1/\gamma_2)(a_{02}/a_{01})]} \tag{12.5a}$$

If the speed of sound $a_{02}$ becomes very small compared with $a_{01}$, which corresponds to a high-density gas beyond the interface, this result reduces to Eq. (12.3b) for a solid boundary. That is, as the density difference at the interface increases, the foregoing result reduces to that for an impermeable boundary corresponding to perfect reflection.

If the pressure differential across the transmitted or refracted wave is denoted by $\Delta p_t$, it follows that

$$\frac{\Delta p_t}{p_0} = \frac{p}{p_0} - 1$$

where the fact that $p = p_0$ in region 1 has been used. Then, from Eq. (12.4b),

$$\frac{\Delta p_t}{p_0} = \frac{(p_1/p_0 - 1)}{1 + (\gamma_1/\gamma_2)(a_{02}/a_{01})} \tag{12.5b}$$

Equations (12.5a) and (12.5b) show that the strength of the reflected wave and that of the refracted wave depend upon the nature of the interface between the two gases. For $\gamma_2 = \gamma_1 = \gamma$ and $a_{02} = a_{01} = a_0$. Eqs. (12.5a) and (12.5b) show that there is no reflected wave and that the transmitted wave is identical to the incident wave. For $a_{02}/a_{01} \to 0$ the reflection becomes total, as was discussed earlier. Intermediate to these two limiting cases both a reflected wave and a transmitted wave will exist.

## 12.5  PISTON PROBLEM

The so-called piston problem is a classical one and may be stated as follows: Figure 12.5a shows a cylinder or a tube inside which a piston slides. Initially the piston and the gas ahead of it are stationary, when the piston is suddenly jerked into motion at some constant velocity. The problem is to find the velocity and the pressure ahead of the piston after the motion has started.

The $xt$ diagram for such a situation is shown in Fig. 12.5b. One of the two lines on this diagram corresponds to a wave front which is generated by the impulsive acceleration of the piston and which travels down the cylinder ahead of the piston with velocity $a_0$. The second line represents the instantaneous location of the piston, which is moving with constant velocity $U$ for $t > 0$. Since, according to our linear theory. $U/a_0 \ll 1$ so that the piston will always be close to $x = 0$, compared with the location of the wave front, the boundary condition that $u = U$ on the piston face may be imposed on $x = 0$ rather than $x = Ut$. This yields the modified $xt$ diagram shown in Fig. 12.5c.

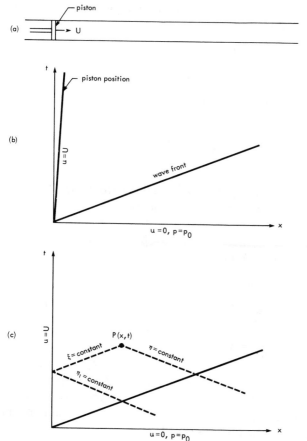

**FIGURE 12.5**

(a) Piston and cylinder, (b) actual $xt$ diagram, and (c) linearized $xt$ diagram.

The $xt$ diagram shown in Fig. 12.5c is divided into two distinct regions by the wave which leaves the piston face. Region 1 contains the undisturbed gas, which is stationary and whose pressure is $p_0$. In order to determine the velocity and pressure in region 2, a typical point $P(x, t)$ is considered. The $\eta =$ constant characteristic from this point enters region 1, where the values of $u$ and $p$ are known. The $\xi =$ constant characteristic runs parallel to the wave front and eventually encounters the position of the piston face at $x = 0$. Although the velocity there is known, the pressure is not; hence the $\eta_1 =$ constant characteristic is drawn from the point where the $\xi =$ constant characteristic terminates. These three characteristics permit the values of the velocity and the pressure at the point $P$ to be evaluated.

Denoting the value of the pressure at the piston face by $p_p$, the Riemann invariant along $\xi =$ constant gives

$$\frac{u}{a_0} + \frac{1}{\gamma}\frac{p}{p_0} = \frac{U}{a_0} + \frac{1}{\gamma}\frac{p_p}{p_0}$$

The pressure at the piston may be evaluated from the Riemann invariant for the $\eta_1 =$ constant characteristic, which is evaluated first on the piston face, then in region 1. The resulting equation is

$$\frac{u}{a_0} - \frac{1}{\gamma}\frac{p}{p_0} = \frac{U}{a_0} - \frac{1}{\gamma}\frac{p_p}{p_0} = \frac{1}{\gamma}$$

Using the last equality to evaluate $p_p$, the equation for the $\xi =$ constant characteristic becomes

$$\frac{u}{a_0} + \frac{1}{\gamma}\frac{p}{p_0} = 2\frac{U}{a_0} + \frac{1}{\gamma}$$

Another equation connecting $u$ and $p$ may be obtained from the Riemann invariant for the $\eta =$ constant characteristic which yields

$$\frac{u}{a_0} - \frac{1}{\gamma}\frac{p}{p_0} = -\frac{1}{\gamma}$$

The solution to the last two equations is

$$u = U \qquad\qquad (12.6a)$$

$$\frac{p}{p_0} = \gamma\frac{U}{a_0} + 1 \qquad\qquad (12.6b)$$

Equation (12.6a) shows that the gas velocity in region 2 is everywhere the same as that of the piston. Equation (12.6b) shows that the pressure ahead of the piston but behind the outgoing wave is greater than the initial value by an amount which is proportional to the piston speed $U$.

## 12.6  FINITE-STRENGTH SHOCK TUBES

In the previous four sections some properties of internal waves have been exposed through reference to weak shock tubes. In reality finite-strength waves exist and their properties may be established through reference to finite-strength shock tubes. Although the qualitative behavior of strong waves is the same as that for weak waves, the quantitative results are different. The nature of these differences will be established in this section by carrying out an analysis for the problem which is analogous to that treated in Sec. 12.2 for a weak shock tube.

Figure 12.6$a$ shows a shock tube in which the initial velocity is everywhere zero and in which the initial pressure distribution is $p_1$ to the right of the diaphragm and $p_4$ to the left of the diaphragm. The initial pressure distribution is shown in Fig. 12.6$b$, and it is assumed that the pressure differential $p_4 - p_1$ is substantial so that a linear theory is no longer valid. Because of the substantial pressure differential and/or the fact that the gases may differ, the specific-heat ratio and the speed of sound will be different on either side of the diaphragm and are denoted by $\gamma_1, a_1$ and $\gamma_4, a_4$, as indicated in Fig. 12.6$a$.

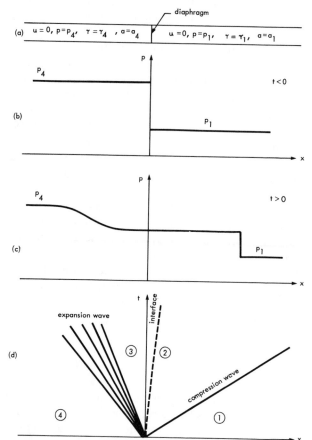

**FIGURE 12.6**
($a$) Shock-tube configuration, ($b$) initial pressure distribution, ($c$) typical pressure distribution for $t > 0$, and ($d$) $xt$ diagram.

The form of the pressure distribution for some $t > 0$, corresponding to the diaphragm bursting at $t = 0$, is shown in Fig. 12.6c. A compression wave of finite strength will travel down the tube in the positive $x$ direction. It is known from Sec. 11.2 that this wave will steepen as it travels and will develop into a shock wave as shown in Fig. 12.6c. An expansion wave will travel in the negative $x$ direction, and it is known that such waves tend to smooth out as they propagate. The $xt$ diagram for this situation is shown in Fig. 12.6d. The shock wave may be represented by a single-line discontinuity, but the expansion wave will extend over a substantial portion of the $x$ axis and is thus represented by an expansion fan. This consists of a series of lines which emanate from the origin of the $xt$ diagram and may be thought of as a very large number of weak waves.

The location of the interface between the two bodies of gas is also shown in Fig. 12.6d so that the possibility of two different gases may be covered. Thus the $xt$ diagram is seen to be divided into four distinct regions. Region 1 consists of gas 1 and represents those locations which have been affected by the passage of the compression wave. Region 3 consists of gas 4 in those locations which have been affected by the expansion wave. The principal quantities of interest are the strength of the shock wave which results from bursting the diaphragm, for given values of $p_4$ and $p_1$, and the speed with which the shock wave moves along the tube.

The boundary conditions at the interface between regions 3 and 2 are $u_3 = u_2$ and $p_3 = p_2$. These conditions guarantee continuity of the velocity and the pressure, and they may be used to determine the strength of the shock wave as follows. By employing a galilean transformation to a stationary normal shock wave, the results obtained in Chap. 11 may be used to obtain an expression for $u_2$ in terms of $p_2/p_1$. By an analogous procedure, the velocity $u_3$ may be expressed in terms of the pressure ratio across the expansion wave $p_4/p_3$. The matching conditions at the interface will then give an equation which relates the pressure ratio $p_2/p_1$ across the shock wave to the initial pressure ratio $p_4/p_1$ across the diaphragm.

Considering first the compression wave, let $u_{1n}$ and $u_{2n}$ be the gas velocities $u_2$ and $u_1$, respectively, expressed in a frame of reference in which the shock wave is stationary. Then, in order that $u_1$ may be zero here, a galilean transformation of magnitude $u_{2n}$ must be made on the velocities. The relationships between the normal shock velocities $u_{2n}$ and $u_{1n}$, which refer to Fig. 11.3a, and the present velocities $u_1$ and $u_2$, which refer to Fig. 12.6d, are

$$u_1 = u_{2n} - u_{2n} = 0$$

$$u_2 = u_{1n} - u_{2n} = u_{1n}\left(1 - \frac{u_{2n}}{u_{1n}}\right)$$

This now represents a shock wave which is moving with velocity $u_{2n}$ through a stationary gas in which the velocity of the gas behind the shock wave is $u_2$. But $u_{2n}/u_{1n}$ may be evaluated from the Rankine-Hugoniot relation [Eq. (11.4)], which gives

$$u_2 = u_{1n}\left[1 - \frac{(\gamma_1 + 1)/(\gamma_1 - 1) + p_1/p_2}{1 + (\gamma_1 + 1)/(\gamma_1 - 1)(p_1/p_2)}\right]$$

where $p_1$ and $p_2$ now refer to Fig. 12.6d. But $u_{1n} = a_1 M_{1n}$ where $M_{1n}$ is the Mach number of the flow approaching a stationary shock wave. From Eq. (11.8c)

$$M_{1n}^2 = \frac{\gamma_1 + 1}{2\gamma_1}\left(\frac{p_1}{p_2} - 1\right) + 1$$

where, again, the pressures now refer to Fig. 12.6c. The expression for the velocity in region 2 then becomes

$$u_2 = a_1\left[\frac{\gamma_1 + 1}{2\gamma_1}\left(\frac{p_1}{p_2} - 1\right) + 1\right]^{1/2}\left[1 - \frac{(\gamma_1 + 1)/(\gamma_1 - 1) + p_1/p_2}{1 + (\gamma_1 + 1)/(\gamma_1 - 1)(p_1/p_2)}\right]$$

This result may be simplified to give

$$u_2 = a_1\left\{\frac{2(p_1/p_2 - 1)^2}{\gamma_1[(\gamma_1 - 1) + (\gamma_1 + 1)(p_1/p_2)]}\right\}^{1/2}$$

Consider next, the expansion wave. It was established in Chap. 11 that expansion waves, contrary to compression waves, tend to smooth out and spread themselves over substantial distances. Thus the expansion from $p_4$ to $p_3$ takes place in a continuous manner which may be thought of as consisting of a very large number of weak expansion waves, each of which is isentropic. Thus, from Eq. (11.2a), it follows that at any point in the expansion wave

$$\frac{du}{a} = -\frac{d\rho}{\rho}$$

where the minus sign denotes that the wave is traveling in the negative $x$ direction. But

$$a^2 = \gamma_4\frac{p}{\rho} = \frac{\gamma_4}{\rho}\frac{p_4\rho^{\gamma_4}}{\rho_4^{\gamma_4}}$$

$$= a_4^2\left(\frac{\rho}{\rho_4}\right)^{\gamma_4 - 1}$$

where the isentropic law has been used to relate $p$ to $\rho$. Thus the expression for $du$ becomes

$$du = -\frac{a_4}{\rho_4^{(\gamma_4 - 1)/2}}\rho^{(\gamma_4 - 3)/2}\,d\rho$$

Integrating this expression and noting that $u = 0$ when $\rho = \rho_4$ yields the following expression for the local value of the velocity $u$ in the expansion wave:

$$u = -\frac{2a_4}{\gamma_4 - 1}\left[\left(\frac{\rho}{\rho_4}\right)^{(\gamma_4-1)/2} - 1\right]$$

The local density $\rho$ may be replaced by the local pressure $p$ through use of the isentropic law. This gives

$$u = \frac{2a_4}{\gamma_4 - 1}\left[1 - \left(\frac{p}{p_4}\right)^{(\gamma_4-1)/2\gamma_4}\right]$$

In particular, at the trailing edge of the expansion wave $p = p_3$ and $u = u_3$ so that

$$u_3 = \frac{2a_4}{\gamma_4 - 1}\left[1 - \left(\frac{p_3}{p_4}\right)^{(\gamma_4-1)/2\gamma_4}\right]$$

The expressions obtained above for $u_2$, from the compression wave, and $u_3$, from the expansion wave, will now be used in conjunction with the interface matching conditions. Thus setting $u_3 = u_2$ and at the same time replacing $p_3$ by $p_2$ yields the following identity:

$$\frac{2a_4}{\gamma_4 - 1}\left[1 - \left(\frac{p_2}{p_4}\right)^{(\gamma_4-1)/2\gamma_4}\right] = a_1\left\{\frac{2(p_1/p_2 - 1)^2}{\gamma_1[(\gamma_1 - 1) + (\gamma_1 + 1)(p_1/p_2)]}\right\}^{1/2}$$

Although this equation cannot be solved to yield an explicit expression for the shock-wave pressure ratio $p_2/p_1$ in terms of the initial diaphragm pressure ratio $p_4/p_1$, the converse is not true. Thus, solving this equation for $p_4$ gives

$$\frac{p_4}{p_1} = \frac{p_2}{p_1}\left\{1 - \frac{(\gamma_4 - 1)(a_1/a_4)(p_1/p_2 - 1)}{\sqrt{2\gamma_1[(\gamma_1 - 1) + (\gamma_1 + 1)(p_1/p_2)]}}\right\}^{-2\gamma_4(\gamma_4-1)} \tag{12.7a}$$

If $p_1/p_2$ is replaced by $1 - \varepsilon$, Eq. (12.7a) shows that $p_4/p_1 = 1 + 2\varepsilon$ to the first order in $\varepsilon$. That is, for weak waves the result obtained in Sec. 12.2 is recovered.

It was shown earlier in this section that the results for normal shock waves could be related to those of a shock wave which is moving with velocity $u_{2n}$ into a stationary gas in which the gas velocity behind the shock wave is $u_{1n} - u_{2n}$. Thus if $M_s$ denotes the Mach number with which the shock wave propagates through fluid 1, it follows that

$$M_s = M_{1n}$$

Then, from Eq. (11.8a),

$$M_s = \left\{\frac{1 + [(\gamma_1 - 1)/2]M_{1n}^2}{\gamma_1 M_{1n}^2 - (\gamma_1 - 1)/2}\right\}^{1/2}$$

But, from Eq. (11.8c),

$$M_{1n}^2 = 1 + \frac{\gamma_1 + 1}{2\gamma_1}\left(\frac{p_1}{p_2} - 1\right)$$

where $p_1$ and $p_2$ refer to the problem at hand. Thus the expression for the Mach number of the compression wave becomes

$$M_s = \left\{\frac{1 + (\gamma_1 - 1)/2 + [(\gamma_1 - 1)(\gamma_1 + 1)/4\gamma_1](p_1/p_2 - 1)}{\gamma_1 + [(\gamma_1 + 1)/2](p_1/p_2 - 1) - (\gamma_1 - 1)/2}\right\}^{1/2}$$

This result may be simplified to yield the following equation:

$$M_s = \left[\frac{(\gamma_1 - 1) + (\gamma_1 + 1)(p_2/p_1)}{2\gamma_1}\right]^{1/2} \tag{12.7b}$$

As $p_2/p_1$ approaches unity, $M_s$ also approaches unity. That is, for weak shock waves the front travels at the speed of sound, which confirms the results of Chap. 11. Equation (12.7b) also shows that the Mach number of the shock wave can be considerably greater than unity for strong shock waves.

## 12.7 NONADIABATIC FLOWS

The physical situations which will be treated here differ from those of the previous sections in several ways. As the heading suggests, the most significant difference is that flows in which heat is being added to the gas, or removed from it, will be covered. In addition, external body forces such as frictional forces may be included. Another difference from previous treatments is that variations in the flow area may be included, provided that the flow may be considered to be essentially one-dimensional. In the previous sections of this chapter the flow configurations have been exactly one-dimensional. Here, converging and diverging boundaries will be permitted, and the velocity which exists at any streamwise location will be considered to be the average value at that location. Finally, the flow situations considered here differ from those of the previous sections in the sense that the flow varies continuously rather than abruptly through either Mach waves or shock waves. That is, the heat addition, external forces, and area changes will be assumed to be such that they vary the flow properties continuously rather than abruptly so that the use of derivatives will be valid.

Figure 12.7a shows a typical flow configuration of the type to be considered here. At the location defined by $x$ the flow area is $A$ and at the location $x + dx$ the flow area is $A + dA$. The element of length $dx$ is subjected to an external force $\delta f(x)$, and an amount of heat $\delta q(x)$ is added to it.

The equations of motion for the gas may be readily derived in differential form as follows: The continuity equation requires that $\rho u A = $ constant where $\rho$ and $u$ are, respectively, the average density and velocity of the gas at the

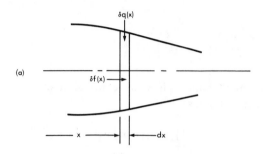

(a)

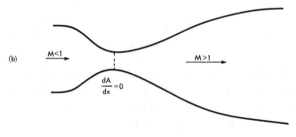

(b)

**FIGURE 12.7**
(a) Element of one-dimensional flow field and (b) flow through a typical nozzle.

location $x$. Then

$$d(\rho uA) = 0$$

so that performing the indicated differentiation and dividing by $\rho uA$ gives

$$\frac{d\rho}{\rho} + \frac{du}{u} = -\frac{dA}{A} \tag{12.8a}$$

Equation (12.8a) supplies one equation connecting three of the variables. The momentum balance for the element shown in Fig. 12.7a is

$$\rho u \, du = -dp + \delta f$$

Dividing this equation by $p$ and using the fact that $a^2 = \gamma p/\rho$ reduces it to the form

$$\gamma M^2 \frac{du}{u} + \frac{dp}{p} = \frac{\delta f}{p} \tag{12.8b}$$

where the local Mach number $M = u/a$ has been introduced. The thermal-energy balance for the case of a perfect gas may be written in the form

$$C_p \, dT + u \, du = \delta q$$

where $T$ is the local temperature of the gas. Dividing this equation by $C_pT$ gives

$$\frac{dT}{T} + \frac{u \, du}{C_pT} = \frac{\delta q}{C_pT}$$

But $T$ may be replaced by $p/\rho R$ for a perfect gas and $RC_p = (\gamma - 1)/\gamma$, so that the energy equation may be written in the form

$$\frac{dT}{T} + (\gamma - 1)M^2\frac{du}{u} = \frac{\delta q}{C_pT} \tag{12.8c}$$

Here, again, the expression for the speed of sound and the definition of the Mach number have been employed. Finally, the equation of state for an ideal gas gives

$$\frac{p}{\rho T} = R$$

hence

$$\frac{dp}{p} - \frac{d\rho}{\rho} - \frac{dT}{T} = 0 \tag{12.8d}$$

Equations (12.8a), (12.8b), (12.8c), and (12.8d) represent four algebraic equations for the differentials $du$, $d\rho$, $dp$, and $dT$ in terms of the local values of the variables $u$, $\rho$, $p$, $T$, $M$, $f$, $q$, and $A$. Then these equations may be solved to yield expressions for each of the differentials separately. To obtain the expression for $du$, for example, use Eq. (12.8d) and eliminate $dp/p$ by using Eq. (12.8b) and $dT/T$ by using Eq. (12.8c). This gives

$$\left(-\gamma M^2\frac{du}{u} + \frac{\delta f}{p}\right) - \frac{d\rho}{\rho} + \left[(\gamma - 1)M^2\frac{du}{u} - \frac{\delta q}{C_pT}\right] = 0$$

Adding Eq. (12.8a) to this result to eliminate $d\rho/\rho$ results in the following expression:

$$\frac{du}{u} = \frac{1}{M^2 - 1}\left(\frac{dA}{A} + \frac{\delta f}{p} - \frac{\delta q}{C_pT}\right) \tag{12.9a}$$

Equation (12.9a) gives the change in speed $du$ which is associated with a change in area $dA$, application of an external force $\delta f$, and the addition of an amount of heat $\delta q$. The coefficients of the quantities $dA/A$, $\delta f/p$, and $\delta q/C_pT$ are called *influence coefficients*, since they represent the influence of some external process, such as heat addition, on some flow variable, such as the gas velocity. Thus with respect to the normalized velocity $du/u$, the influence coefficient for $dA/A$ and that for $\delta f/p$ is $1/(M^2 - 1)$, while the influence coefficient for $\delta q/C_pT$ is $-1/(M^2 - 1)$. As a special case, consider adiabatic flow without external forces. Then Eq. (12.9a) becomes

$$\frac{du}{u} = \frac{1}{M^2 - 1}\frac{dA}{A}$$

This equation expresses the familiar result that in order to accelerate ($du > 0$) subsonic ($M < 1$) flow, the flow area should be decreased ($dA < 0$). On the other hand, for supersonic flow the area should be increased to accelerate the flow. This leads to nozzle shapes of the form shown in Fig. 12.7b where the throat, which corresponds to $dA = 0$, has sonic conditions.

An expression similar to Eq. (12.9a) may be obtained for $dp/p$ by using Eq. (12.8a) and by eliminating $du/u$ from it through use of Eq. (12.9a). This gives

$$\frac{dp}{p} = -\frac{\gamma M^2}{M^2 - 1}\frac{dA}{A} - \frac{1 + (\gamma - 1)M^2}{M^2 - 1}\frac{\delta f}{p} + \frac{\gamma M^2}{M^2 - 1}\frac{\delta q}{C_p T} \quad (12.9b)$$

For adiabatic flow without external forces, Eq. (12.9b) becomes

$$\frac{dp}{p} = -\frac{\gamma M^2}{M^2 - 1}\frac{dA}{A}$$

This result shows that in order to expand a gas in a nozzle continuously, the flow area should decrease when the flow is subsonic and increase when the flow is supersonic. This agrees with our conclusion regarding accelerating gases in nozzles. For flow in a constant-area channel in which there are no external forces acting on the gas, Eq. (12.9b) becomes

$$\frac{dp}{p} = \frac{\gamma M^2}{M^2 - 1}\frac{\delta q}{C_p T}$$

This equation shows that in order to expand a gas in a pipe by thermal means, heat should be added if the flow is subsonic, whereas heat should be removed if the flow is supersonic.

The variation of the temperature may be obtained by using Eq. (12.8a) to eliminate $du/u$ from Eq. (12.8c). This gives

$$\frac{dT}{T} = -\frac{(\gamma - 1)M^2}{M^2 - 1}\frac{dA}{A} - \frac{(\gamma - 1)M^2}{M^2 - 1}\frac{\delta f}{p} + \frac{\gamma M^2 - 1}{M^2 - 1}\frac{\delta q}{C_p T} \quad (12.9c)$$

For adiabatic flow without external forces Eq. (12.9c) reduces to

$$\frac{dT}{T} = -\frac{(\gamma - 1)M^2}{M^2 - 1}\frac{dA}{A}$$

Hence, for $\gamma > 1$, the temperature will drop in a converging-flow area if the flow is subsonic or in a diverging-flow area if the flow is supersonic. For flow in a constant-area channel in which there are not external forces, Eq. (12.9c) becomes

$$\frac{dT}{T} = \frac{\gamma M^2 - 1}{M^2 - 1}\frac{\delta q}{C_p T}$$

The influence coefficient here changes sign at $M = 1/\sqrt{\gamma}$ and at $M = 1$. Then, for $\gamma > 1$, the effect of heat addition will be to increase the temperature for $M < 1/\sqrt{\gamma}$ and for $M > 1$, but the temperature will decrease in the range

$1/\sqrt{\gamma} < M < 1$. For adiabatic flow in a constant-area duct, Eq. (12.9c) becomes

$$\frac{dT}{T} = -\frac{(\gamma - 1)M^2}{M^2 - 1} \frac{\delta f}{p}$$

Hence for $\gamma > 1$ the effect of an external force such as wall friction is to increase the temperature of subsonic flow and to decrease the temperature of supersonic flow.

Two well-known results which may be established in this way are the *Fanno line* and the *Rayleigh line*. In each case, the variation of temperature or enthalpy is considered as a function of the entropy. The resulting curve for the case of adiabatic flow in a constant-area duct is called a Fanno line, while the curve for the case of a constant-area duct without external forces is called a Rayleigh line. That is, the Fanno line shows the effect of friction in a constant-area duct while the Rayleigh line shows the effect of heat addition. One of the most practical and significant results which may be deduced from these diagrams, or from the equations established above, is that choking takes place at $M = 1$. Thus, adding heat to a constant-area flow will accelerate it until $M = 1$, and no more heat can be added beyond this point. For long pipelines the effect of friction is similar, so that a point may be reached beyond which no more gas can be pumped through the pipe without removing some heat.

## 12.8  ISENTROPIC-FLOW RELATIONS

For isentropic flows, simple and useful relations exist between the local value of some variable, such as the temperature or pressure, and the local Mach number. These relations may be obtained from the thermal-energy equation as follows.

For steady, isentropic flow the thermal-energy equation (IV.3b) becomes

$$\rho(\mathbf{u} \cdot \boldsymbol{\nabla})h = (\mathbf{u} \cdot \boldsymbol{\nabla})p$$

But, forming the scalar product of $\mathbf{u}$ and the Euler equation [Eq. (IV.2)] shows that

$$(\mathbf{u} \cdot \boldsymbol{\nabla})p = -\mathbf{u} \cdot [\rho(\mathbf{u} \cdot \boldsymbol{\nabla})\mathbf{u}]$$

$$= -\rho(\mathbf{u} \cdot \boldsymbol{\nabla})(\tfrac{1}{2}\mathbf{u} \cdot \mathbf{u})$$

Hence, the energy equation may be written in the form

$$\rho(\mathbf{u} \cdot \boldsymbol{\nabla})(h + \tfrac{1}{2}\mathbf{u} \cdot \mathbf{u}) = 0$$

This means that the quantity $h + \mathbf{u} \cdot \mathbf{u}/2$ is constant along each streamline so that

$$h + \tfrac{1}{2}\mathbf{u} \cdot \mathbf{u} = h_0$$

along each streamline. The quantity $h_0$ is called the *stagnation enthalpy*, and it corresponds to the enthalpy which the fluid would have at zero velocity. Of course, it may be known that in some part of the flow field the stagnation enthalpy is constant, which is usually the case, so that $h_0$ will become the constant for every streamline.

The foregoing result for the enthalpy may be recast in terms of the temperature, the pressure, or the density. For a perfect gas whose physical properties are constant, $h = C_p T$, so that

$$C_p T + \tfrac{1}{2} u^2 = C_p T_0$$

Here, the fact that we are dealing with one-dimensional flows only has been used, so that $\mathbf{u} \cdot \mathbf{u} = u^2$. The quantity $T_0$ which has been introduced here is the stagnation temperature and corresponds to the temperature which the fluid would have if it were brought to rest. Solving this equation for the temperature ratio shows that

$$\frac{T_0}{T} = 1 + \frac{u^2}{2 C_p T}$$

The quantity $u^2/T$ may be rewritten as $\gamma R u^2/(\gamma R T) = \gamma R M^2$, since $\gamma R T$ is the square of the local value of the speed of sound. Then, since $R/C_p = (\gamma - 1)/\gamma$, the expression for the temperature may be written in the following form:

$$\frac{T_0}{T} = 1 + \frac{\gamma - 1}{2} M^2 \qquad (12.10a)$$

The following relations are known from thermodynamics to be valid for isentropic flows:

$$\frac{T_0}{T} = \left( \frac{p_0}{p} \right)^{(\gamma - 1)/\gamma} = \left( \frac{\rho_0}{\rho} \right)^{\gamma - 1}$$

Using these identities and Eq. (12.10a) shows that the following relations hold:

$$\frac{p_0}{p} = \left( 1 + \frac{\gamma - 1}{2} M^2 \right)^{\gamma/(\gamma - 1)} \qquad (12.10b)$$

$$\frac{\rho_0}{\rho} = \left( 1 + \frac{\gamma - 1}{2} M^2 \right)^{1/(\gamma - 1)} \qquad (12.10c)$$

Here, $p_0$ and $\rho_0$ are, respectively, the stagnation pressure and the stagnation density, and $M$ is the local Mach number.

## 12.9 FLOW THROUGH NOZZLES

It was shown in Sec. 12.7 that if a nozzle is required to expand a subsonic flow to supersonic speeds, its shape should be of the form shown in Fig. 12.7b. Such a flow configuration will be considered here in which the flow reaches sonic conditions at the throat and is supersonic downstream of the throat. Since the flow is adiabatic and frictional losses may be considered to be negligible, the flow will be isentropic. This means that the results of the previous section may be employed.

The notation which was introduced in Sec. 11.4 to indicate sonic conditions will again be used here. Thus the temperature, pressure, and density corresponding to $M = 1$ will be denoted by $T_*$, $p_*$, and $\rho_*$, respectively. Then, from Eqs. (12.10$a$), (12.10$b$), and (12.10$c$),

$$\frac{T_0}{T_*} = \frac{\gamma + 1}{2} \tag{12.11$a$}$$

$$\frac{p_0}{p_*} = \left(\frac{\gamma + 1}{2}\right)^{\gamma/(\gamma - 1)} \tag{12.11$b$}$$

$$\frac{\rho_0}{\rho_*} = \left(\frac{\gamma + 1}{2}\right)^{1/(\gamma - 1)} \tag{12.11$c$}$$

That is, the temperature, pressure, and density at the throat of the nozzle may be determined if the stagnation values are known. The stagnation conditions will be know directly if the flow originates in a large reservoir where the fluid speed is zero and its other properties are known. If the fluid properties and speed are given at the inlet to the nozzle, then the stagnation properties may be calculated from Eqs. (12.10).

The variation of the Mach number of the flow with the flow area of the nozzle may be established as follows. The continuity equation written for an arbitrary section and for the throat of the nozzle gives $\rho u A = \rho_* u_* A_*$; hence

$$\frac{A}{A_*} = \frac{\rho_*}{\rho}\frac{M_* a_*}{Ma}$$

But $M_* = 1$ by definition and $a_*/a = \sqrt{T_*/T}$, so that

$$\frac{A}{A_*} = \frac{\rho_* \rho_0}{\rho_0 \rho}\frac{1}{M}\sqrt{\frac{T_* T_0}{T_0 T}}$$

Now $\rho_*/\rho_0$ is given by Eq. (12.11$c$), while $\rho_0/\rho$ is given in terms of the local Mach number by Eq. (12.10$c$). Likewise, $T_*/T_0$ is given by Eq. (12.11$a$) and $T_0/T$ is given by Eq. (12.10$a$). Thus the expression for the area ratio may be written in the form

$$\frac{A}{A_*} = \left(\frac{2}{\gamma + 1}\right)^{1/(\gamma - 1)}\left(1 + \frac{\gamma - 1}{2}M^2\right)^{1/(\gamma - 1)}\frac{1}{M}\left(\frac{2}{\gamma + 1}\right)^{1/2}$$

$$\times \left(1 + \frac{\gamma - 1}{2}M^2\right)^{1/2}$$

This result may be simplified to yield the following expression for the ratio of the local flow area, for which the Mach number is $M$, to the area of the nozzle throat:

$$\frac{A}{A_*} = \frac{1}{M}\left[\frac{2}{\gamma + 1}\left(1 + \frac{\gamma - 1}{2}M^2\right)\right]^{(\gamma + 1)/2(\gamma - 1)} \qquad (12.12a)$$

Equation (12.12 a) relates the local flow area to that of the throat, so that it is of interest to obtain an expression which relates the throat area $A_*$ to the mass flow rate through the nozzle. Denoting this quantity by $\dot{m}$, it follows that

$$\dot{m} = \rho_* u_* A_*$$

$$= \frac{\rho_* \rho_0}{\rho_0}(M_* a_*) A_*$$

Using Eq. (12.11c) again together with the facts that $M_* = 1$ and $a_* = \sqrt{\gamma R T_*} = \sqrt{\gamma R T_0 T_* / T_0}$, this equation becomes

$$\dot{m} = \left(\frac{2}{\gamma + 1}\right)^{1/(\gamma - 1)} \rho_0 \sqrt{\gamma R T_0}\left(\frac{2}{\gamma + 1}\right)^{1/2} A_*$$

where Eq. (12.11a) has been used for $T_*/T_0$. Using the ideal-gas law to write $\rho_0 = p_0/(R T_0)$, the final form of the expression for the mass flow through the nozzle becomes

$$\dot{m} = \frac{p_0 A_*}{\sqrt{T_0}}\sqrt{\frac{\gamma}{R}\left(\frac{2}{\gamma + 1}\right)^{(\gamma + 1)/(\gamma - 1)}} \qquad (12.12b)$$

As might be expected, the mass flow rate through the nozzle is proportional to the throat area $A_*$. Equation (12.12b) further shows that the mas flow rate is proportional to the stagnation pressure of the gas and inversely proportional to the stagnation temperature of the gas.

The foregoing expressions are sufficient to design convergent-divergent nozzles. Typically, the conditions in the gas at the entrance to the nozzle are given together with the mass flow rate (or inlet area) and the exit pressure to which the gas must be expanded. From the inlet conditions the stagnation properties may be evaluated from Eqs. (12.10). The required throat area for the nozzle may then be calculated from Eq. (12.12b). Since the stagnation properties of the gas are constant through the nozzle, Eq. (12.10b) permits the exit. Mach number to be determined. Equation (12.12a) then determines the exit flow area of the nozzle.

## PROBLEMS

**12.1.** Consider a weak shock tube as shown in Fig. 12.3a in which the tube is finite in length in the positive x direction. Rather than having a closed end as shown in Fig. 12.3a, suppose that the end of the tube is open. Draw the xt diagram for the wave which results from bursting the diaphragm. Assuming that the pressure at the end of the tube is at all times $p_0$, obtain expressions for the velocity and the pressure behind the wave which is reflected from the open end.

**12.2.** Figure 12.8 shows a piston located at $x = 0$ in a cylinder of length $L$ which encloses two temperature regions whose interface is at $x = \alpha L$. At time $t = 0$ the piston is impulsively set into motion with constant velocity $U$, which may be assumed to be small compared with the acoustic velocities $a_{01}$ and $a_{02}$. At the interface, part of the resulting wave is reflected and part of it is refracted. If the reflected wave reaches the piston at time $t = \tau$ and the transmitted wave reaches the end of the cylinder at the same instant, draw the $xt$ diagram and use linear theory to find the following:

(a) The temperature ratio $T_{02}/T_{01}$ in terms of $\alpha$.

(b) The velocity and the pressure for $0 < x < L$ and for $0 < t < \tau$ in terms of $U$, $p_0$, $\gamma$, $a_{01}$, and $\alpha$.

(c) The ratio of the strength of the reflected wave to that of the transmitted wave.

piston  interface  closed end

$u = o, \; p = p_0$    $u = o, \; p = p_0$

$\gamma, \; T_{01}, \; a_{01}$    $\gamma, \; T_{02}, \; a_{02}$

$x = 0$    $x = \alpha L$    $x = L$

**FIGURE 12.8**
Piston in a cylinder containing a gas which has two isothermal regions.

**12.3.** Show that, for a finite-strength shock tube, the Mach number of the flow behind the shock wave is given by

$$M_2 = \frac{1}{\gamma_1}\left(\frac{p_2}{p_1} - 1\right)\left\{\left[1 + \frac{\gamma_1 - 1}{2\gamma_1}\left(\frac{p_2}{p_1} - 1\right)\right]\frac{p_2}{p_1}\right\}^{-1/2}$$

The subscripts correspond to the regions indicated in Fig. 12.6.

**12.4.** Heat is being added to a perfect gas which is flowing in a constant-area channel. Neglecting all external forces, determine the influence coefficient $\beta$ in the equation

$$\frac{dM}{M} = \beta\frac{\delta q}{C_p T}$$

Use the resulting equation and the equation

$$\frac{dp}{p} = \frac{\gamma M^2}{M^2 - 1}\frac{\delta q}{C_p T}$$

to establish a differential relation between $p$ and $M$. Integrate this relation to obtain an expression for the pressure ratio $p_2/p_1$ between two sections in the flow whose Mach numbers are $M_2$ and $M_1$.

**12.5.** Use the results of Prob. 12.4 and the isentropic-flow relations to obtain the temperature ratio $T_2/T_1$ and the density ratio $\rho_2/\rho_1$ in terms of the Mach numbers $M_2$ and $M_1$.

**12.6.** The entropy change between two flow conditions of a perfect gas may be expressed in the form

$$s - s_1 = C_v \log\left[\frac{p}{p_1}\left(\frac{\rho_1}{\rho}\right)^\gamma\right]$$

where the subscript 1 denotes some inlet conditions. Apply this to the case of steady, adiabatic flow in a channel of constant cross-sectional area to establish the equation of the Fanno line. To do this, use the following forms of the continuity and energy equations, and the equation of state:

$$\rho u A = \dot{m}$$

$$h + \frac{u^2}{2} = h_0$$

$$p = \rho R T$$

$$h = C_p T$$

Here $\dot{m}$ is the mass flow rate and $h_0$ is the stagnation enthalpy. Hence show that the equation of the Fanno line is

$$\frac{s - s_1}{C_p} = \log\left[h(h_0 - h)^{(\gamma - 1)/2}\right] + \log\left[\frac{\rho_1^{\gamma}}{p_1}\frac{R}{C_p}\left(\frac{2A^2}{\dot{m}^2}\right)^{(\gamma - 1)/2}\right]$$

From this result show that the entropy reaches a maximum when $M = 1$.

12.7. To obtain the equation of the Raleigh line, use the same equations which were used in the previous problem except, since heat addition is involved here, replace the energy equation by the following form of the momentum equation:

$$\rho u^2 + p = p_0$$

Hence obtain the equation of the Rayleigh line and show that the entropy reaches a maximum when $M = 1$ and that the enthalpy reaches a maximum when $M = 1/\sqrt{\gamma}$.

# CHAPTER

# 13

# MULTI-DIMENSIONAL
# FLOWS

This chapter deals with some steady two-dimensional and three-dimensional flow problems, both supersonic and subsonic. The governing equations are first established for irrotational motion, and then solutions to these equations are sought. The Janzen-Rayleigh expansion is discussed first. This expansion is a parameter expansion in powers of the Mach number squared, and so it is valid only for Mach numbers somewhat less than unity. Small-perturbation theory is next discussed. This approximation assumes that the body about which the flow is sought disturbs the free stream in a minor way only. Since the many real flow situations satisfy this condition, a small-perturbation theory is widely used.

Small-perturbation theory is used to study some specific subsonic and supersonic flows. The Prandtl-Glauert rule for subsonic flows is then covered. This rule relates subsonic compressible flows to the corresponding incompressible flows. Ackeret's theory for supersonic flows, which is also based on small-perturbation theory, is then discussed.

Leaving the topic of small-perturbation theory, the chapter ends with a discussion of an exact solution. The flow treated is that of supersonic flow turning around a corner which bends away from the free stream. This flow is known as *Prandtl-Meyer flow*.

## 13.1 IRROTATIONAL MOTION

As was the case for incompressible flows, many of the flow fields of interest are irrotational by virtue of the fact that they originate in a uniform flow. Then, according to Crocco's equation, the flow will also be isentropic. Then the pressure term in the momentum equation may be rewritten as follows:

$$\nabla p = \frac{dp}{d\rho}\,\nabla\rho = a^2\,\nabla\rho$$

where the fact that $p$ is a function of $\rho$ only, owing to the isentropic nature of the flow, has been used. Then the momentum equation (IV.2) becomes

$$\frac{\partial \mathbf{u}}{\partial t} + (\mathbf{u}\cdot\nabla)\mathbf{u} = -\frac{a^2}{\rho}\,\nabla\rho$$

Forming this scalar product of $\mathbf{u}$ with this vector equation gives

$$\frac{1}{2}\frac{\partial}{\partial t}(\mathbf{u}\cdot\mathbf{u}) + \mathbf{u}\cdot\left[(\mathbf{u}\cdot\nabla)\mathbf{u}\right] = -\frac{a^2}{\rho}\,\mathbf{u}\cdot\nabla\rho$$

The term on the right-hand side of this equation may be recast by use of the continuity equation in the form

$$\mathbf{u}\cdot\nabla\rho = -\frac{\partial\rho}{\partial t} - \rho\nabla\cdot\mathbf{u}$$

Thus the foregoing form of the momentum equation becomes

$$\frac{1}{2}\frac{\partial}{\partial t}(\mathbf{u}\cdot\mathbf{u}) + \mathbf{u}\cdot\left[(\mathbf{u}\cdot\nabla)\mathbf{u}\right] = \frac{a^2}{\rho}\frac{\partial\rho}{\partial t} + a^2\,\nabla\cdot\mathbf{u}$$

The density may be completely eliminated from this equation by taking the time derivative of the Bernoulli equation. Thus

$$\frac{\partial^2\phi}{\partial t^2} + \frac{1}{2}\frac{\partial}{\partial t}(\mathbf{u}\cdot\mathbf{u}) = -\frac{\partial}{\partial t}\left(\int\frac{dp}{\rho}\right)$$

$$= -\frac{\partial}{\partial t}\left(\int\frac{dp}{d\rho}\frac{d\rho}{\rho}\right)$$

$$= -\frac{\partial}{\partial t}\left(\int\frac{a^2}{\rho}\,d\rho\right)$$

$$= -\frac{d}{d\rho}\left(\int\frac{a^2}{\rho}\,d\rho\right)\frac{\partial\rho}{\partial t}$$

Now the two inverse operations, differentiation with respect to $\rho$ and integration with respect to $\rho$, cancel one another, so that the Bernoulli equation becomes

$$\frac{\partial^2\phi}{\partial t^2} + \frac{1}{2}\frac{\partial}{\partial t}(\mathbf{u}\cdot\mathbf{u}) = -\frac{a^2}{\rho}\frac{\partial\rho}{\partial t}$$

Using this equation to rewrite the nonsteady term on the right-hand side of the momentum equation gives

$$\frac{1}{2}\frac{\partial}{\partial t}(\mathbf{u}\cdot\mathbf{u}) + \mathbf{u}\cdot[(\mathbf{u}\cdot\nabla)\mathbf{u}] = -\frac{\partial^2\phi}{\partial t^2} - \frac{1}{2}\frac{\partial}{\partial t}(\mathbf{u}\cdot\mathbf{u}) + a^2\nabla\cdot\mathbf{u}$$

Solving this equation for $\nabla\cdot\mathbf{u}$ yields the following equation governing the irrotational motion of a compressible fluid:

$$\nabla\cdot\mathbf{u} = \frac{1}{a^2}\left\{\mathbf{u}\cdot[(\mathbf{u}\cdot\nabla)\mathbf{u}] + \frac{\partial}{\partial t}\left(\frac{\partial\phi}{\partial t} + \mathbf{u}\cdot\mathbf{u}\right)\right\}$$

In terms of the velocity potential $\phi$ this equation becomes

$$\nabla^2\phi = \frac{1}{a^2}\left\{\nabla\phi\cdot[(\nabla\phi\cdot\nabla)\nabla\phi] + \frac{\partial}{\partial t}\left(\frac{\partial\phi}{\partial t} + \nabla\phi\cdot\nabla\phi\right)\right\} \qquad (13.1)$$

Equation (13.1) is the differential equation to be satisfied by the velocity potential $\phi$ for irrotational motion of a compressible fluid. The equation differs drastically from the Laplace equation, which was shown to be the governing equation for incompressible flow. In fact, Eq. (13.1) becomes $\nabla^2\phi = 0$ as $a^2 \to \infty$, which corresponds to $\rho = $ constant. This may be verified by noting that

$$a^2 = \frac{dp}{d\rho}$$

where the derivative is evaluated at constant entropy. But for $\rho = $ constant, $d\rho = 0$, so that $a^2 \to \infty$. Thus it may be concluded that for constant density the governing equation for irrotational motion is linear but for variable density the governing equation becomes nonlinear. It should also be noted that the nonlinearity must be dealt with directly here, since the equations of kinematics and dynamics are no longer separable.

Clearly, Eq. (13.1) represents a formidable analytic problem for any specific flow problem which is to be solved. The difficulty of obtaining exact solutions has led to the development of approximate methods, and two of these will be discussed in the following sections.

## 13.2 JANZEN-RAYLEIGH EXPANSION

The Janzen-Rayleigh expansion is an expansion of the steady-state form of Eq. (13.1) which is valid for any shape of body but only for Mach numbers less than about 0.5. For steady flow, Eq. (13.1) becomes

$$\nabla^2\phi = \frac{1}{a^2}\nabla\phi\cdot[(\nabla\phi\cdot\nabla)\nabla\phi]$$

In tensor notation this equation is

$$\frac{\partial^2\phi}{\partial x_i\partial x_i} = \frac{1}{a^2}\frac{\partial\phi}{\partial x_i}\frac{\partial\phi}{\partial x_j}\frac{\partial^2\phi}{\partial x_j\partial x_i} \qquad (13.2a)$$

As the speed of sound becomes infinite, that is, as the Mach number tends to zero, this equation reduces to the Laplace equation. That is, the right-hand side of the foregoing equation represents compressible effects, so that these effects vary as $a^{-2}$. It seems reasonable, then, that an approximate solution for slightly compressible flows could be sought in which the first correction due to compressibility varies as the Mach number squared.

The foregoing remarks form the basis of an expansion for $\phi$ of the following form:

$$\phi(x, y, z) = U \sum_{n=0}^{\infty} M_{\infty}^{2n} \phi_n(x, y, z) \tag{13.2b}$$

It is assumed here that a uniform flow of magnitude $U$ approaches the body under consideration and $M_{\infty} = U/a_{\infty}$ is the Mach number far from the body where $a_{\infty}$ is the speed of sound there. Substituting Eq. (13.2b) into Eq. (13.2a) gives

$$U \sum_n M_{\infty}^{2n} \frac{\partial^2 \phi_n}{\partial x_i \partial x_i} = \frac{U^3}{a^2} \sum_n M_{\infty}^{2n} \frac{\partial \phi_n}{\partial x_i} \sum_n M_{\infty}^{2n} \frac{\partial \phi_n}{\partial x_j} \sum_n M_{\infty}^{2n} \frac{\partial^2 \phi_n}{\partial x_i \partial x_j}$$

The significance of the coefficient $U$ in Eq. (13.2b) is now apparent; the coefficients of the series may be made dimensionless, yielding the following equation:

$$\sum_n M_{\infty}^{2n} \frac{\partial^2 \phi_n}{\partial x_i \partial x_i} = \frac{a_{\infty}^2}{a^2} M_{\infty}^2 \sum_n M_{\infty}^{2n} \frac{\partial \phi_n}{\partial x_i} \sum_n M_{\infty}^{2n} \frac{\partial \phi_n}{\partial x_j} \sum_n M_{\infty}^{2n} \frac{\partial^2 \phi_n}{\partial x_i \partial x_j}$$

Although the coefficients are all dimensionless, the quantity $a_{\infty}^2/a^2$ is not constant and should also be expressed as a series in $M_{\infty}^2$. This may be done by using the energy equation in the form

$$\frac{1}{2} \mathbf{u} \cdot \mathbf{u} + \frac{a^2}{\gamma - 1} = \frac{1}{2} U^2 + \frac{a_{\infty}^2}{\gamma - 1}$$

where the constant has been evaluated far from the body. Solving this equation for the ratio of the local value of the speed of sound to that far from the origin gives

$$\frac{a^2}{a_{\infty}^2} = 1 + \frac{\gamma - 1}{2} \left( M_{\infty}^2 - \frac{\mathbf{u} \cdot \mathbf{u}}{a_{\infty}^2} \right)$$

Substituting $\mathbf{u} = \nabla \phi$ and inserting the expansion for $\phi$ yields the following expansion for $a^2$:

$$\frac{a^2}{a_{\infty}^2} = 1 + \frac{\gamma - 1}{2} M_{\infty}^2 \left[ 1 - \left( \sum_n M_{\infty}^{2n} \frac{\partial \phi_n}{\partial x_i} \right)^2 \right]$$

$$= 1 + \frac{\gamma - 1}{2} M_{\infty}^2 \left[ 1 - \left( \frac{\partial \phi_0}{\partial x_i} \right)^2 \right] + O(M_{\infty}^4)$$

Inverting this expression defines the quantity which appears in the governing equation:

$$\frac{a_\infty^2}{a^2} = 1 - \frac{\gamma - 1}{2} M_\infty^2 \left[ 1 - \left( \frac{\partial \phi_0}{\partial x_i} \right)^2 \right] + O(M_\infty^4)$$

Substituting this result into the expanded form of the governing equation for $\phi$ yields

$$\sum_n M^{2n} \frac{\partial^2 \phi_n}{\partial x_i \partial x_i} = M_\infty^2 \left\{ 1 - \frac{\gamma - 1}{2} M_\infty^2 \left[ 1 - \left( \frac{\partial \phi_0}{\partial x_i} \right)^2 \right] + O(M_\infty^4) \right\}$$

$$\times \sum_n M_\infty^{2n} \frac{\partial \phi_n}{\partial x_i} \sum_n M_\infty^{2n} \frac{\partial \phi_n}{\partial x_j} \sum_n M_\infty^{2n} \frac{\partial^2 \phi_n}{\partial x_i \partial x_j}$$

The expansion (13.2b) is assumed to be uniformly valid in $M_\infty^2$. This means that the coefficients of like powers of this quantity must balance on each side of the foregoing equation. This gives the following sequence of equations which represents the coefficients of $M_\infty^0$, $M_\infty^2$, $M_\infty^4$, etc.:

$$\frac{\partial^2 \phi_n}{\partial x_i \partial x_i} = 0 \tag{13.2c}$$

$$\frac{\partial^2 \phi_1}{\partial x_i \partial x_i} = \frac{\partial \phi_0}{\partial x_i} \frac{\partial \phi_0}{\partial x_j} \frac{\partial^2 \phi_0}{\partial x_i \partial x_j} \tag{13.2d}$$

$$\frac{\partial^2 \phi_2}{\partial x_i \partial x_i} = -\frac{\gamma - 1}{2} \left[ 1 - \left( \frac{\partial \phi_0}{\partial x_i} \right)^2 \right] \frac{\partial \phi_0}{\partial x_i} \frac{\partial \phi_0}{\partial x_j} \frac{\partial^2 \phi_0}{\partial x_i \partial x_j} + \frac{\partial \phi_1}{\partial x_i} \frac{\partial \phi_0}{\partial x_j} \frac{\partial^2 \phi_0}{\partial x_i \partial x_j}$$

$$+ \frac{\partial \phi_0}{\partial x_i} \frac{\partial \phi_1}{\partial x_j} \frac{\partial^2 \phi_0}{\partial x_i \partial x_j} + \frac{\partial \phi_0}{\partial x_i} \frac{\partial \phi_0}{\partial x_j} \frac{\partial^2 \phi_1}{\partial x_i \partial x_j} \tag{13.2e}$$

etc.

The equation to be solved for $\phi_0$ [Eq. (13.2c)] represents the incompressible-flow problem corresponding to $M_\infty \to 0$. The problem for $\phi_1$, represented by Eq. (13.2d), is a linear one, although the differential equation is nonhomogeneous. Having solved the problem for $\phi_0$, the right-hand side of Eq. (13.2d) will become an explicit function of the spatial coordinates. Likewise, having obtained expressions for $\phi_0$ and $\phi_1$, Eq. (13.2e) represents a linear, nonhomogeneous equation for $\phi_2$. In this way solutions for $\phi_0$, $\phi_1$, $\phi_2$, etc., may be obtained sequentially, and each of these solutions represents a term in the perturbation solution (13.2b). It may be seen from the equations for $\phi_1$ and $\phi_2$ that the differential equation to be solved becomes complicated rapidly, and it is not practical to carry out the solution beyond the first two or three terms. This means that the solution so obtained will be valid only for Mach numbers which are of the order of 0.5 or smaller. The advantage of the Janzen-Rayleigh expansion, on the other hand, it that is is valid for any shape of body, not just slender bodies.

## 13.3   SMALL-PERTURBATION THEORY

An alternative approximate method of solution to the equation for compressible potential flows is small-perturbation theory. This approximation is valid for supersonic flows as well as subsonic flows, but it is restricted to relatively slender bodies.

Suppose that a uniform flow approaches a body which is sufficiently slender that it causes a small perturbation to the free stream. Then the velocity potential may be written in the form

$$\phi = Ux + \Phi \tag{13.3a}$$

where

$$\frac{1}{U}\left|\frac{\partial \Phi}{\partial x_i}\right| \ll 1$$

Then, for steady flows, Eq. (13.1) becomes

$$\nabla^2 \Phi = \frac{1}{a^2}(U\mathbf{e}_x + \nabla\Phi) \cdot [(U\mathbf{e}_x + \nabla\Phi) \cdot \nabla](U\mathbf{e}_x + \nabla\Phi)$$

This expression is still exact within the inviscid theory, but now that fact that the perturbation velocity potential $\Phi$ leads to small-velocity components will be used to eliminate quadratic and higher terms. Thus the linearized form of the governing equation is

$$\nabla^2 \Phi = \frac{U^2}{a^2}\frac{\partial^2 \Phi}{\partial x^2}$$

This simplified equation retains only one term out of the compressible correction terms, and the retained term corresponds to the direction of the free stream. In its present form the compressible correction term contains a variable coefficient $a$, which should also be linearized for consistency. This may be done by appealing to the energy equation in the form

$$\tfrac{1}{2}\mathbf{u} \cdot \mathbf{u} + \frac{a^2}{\gamma - 1} = \tfrac{1}{2}U^2 + \frac{a_\infty^2}{\gamma - 1}$$

but

$$\mathbf{u} \cdot \mathbf{u} = (U + u')^2 + (v')^2 + (w')^2$$

where $u'$, $v'$, and $w'$ are the velocity perturbations to the free-stream velocity $U$. Thus the linearized form of the kinetic-energy term is

$$\mathbf{u} \cdot \mathbf{u} = U^2 + 2Uu'$$

Using this form, the energy equation becomes

$$Uu' + \frac{a^2}{\gamma - 1} = \frac{a_\infty^2}{\gamma - 1}$$

or
$$a^2 = a_\infty^2 \left[ 1 - (\gamma - 1) \frac{Uu'}{a_\infty^2} \right]$$

Substituting this expression into the simplified equation for the perturbation velocity potential gives

$$\nabla^2\Phi = \frac{U^2}{a_\infty^2} \left[ 1 - (\gamma - 1) \frac{U}{a_\infty^2} \frac{\partial\Phi}{\partial x} \right]^{-1} \frac{\partial^2\Phi}{\partial x^2}$$

This equation, in turn, may now be linearized to yield the following linear equation with constant coefficients:

$$\nabla^2\Phi = \frac{U^2}{a_\infty^2} \frac{\partial^2\Phi}{\partial x^2}$$

In cartesian coordinates this equation is

$$\left(1 - M_\infty^2\right) \frac{\partial^2\Phi}{\partial x^2} + \frac{\partial^2\Phi}{\partial y^2} + \frac{\partial^2\Phi}{\partial z^2} = 0 \qquad (13.3b)$$

Equation (13.3b) shows that for subsonic flow the governing equation is elliptic and so will have no real characteristics. On the other hand, for supersonic flow the governing equation is hyperbolic and so will have real characteristics. This observation is compatible with our previous result that shock waves can occur only in supersonic flow. Equation (13.3b) is valid for supersonic flows as well as subsonic flows, but as will be demonstrated later, it is invalid near $M_\infty = 1$. Also, by virtue of the linearization, the equation is valid only for flows which involve relatively slender bodies.

## 13.4  PRESSURE COEFFICIENT

The principal quantity which will be of interest in the solution to specific problems is the pressure in the fluid, since the integral of the pressure around the surface of a body defines the lift and drag forces acting on the body. The usual way of expressing the pressure is by means of the dimensionless pressure coefficient. It is therefore of interest to obtain a linear expression for the pressure coefficient which will be compatible with the linearized equation which was derived in the previous section. Such an expression will be derived in this section.

The pressure coefficient $C_p$ is defined in the following way:

$$C_p = \frac{p - p_\infty}{\frac{1}{2}\rho_\infty U^2}$$

Here $p_\infty$, $\rho_\infty$, and $U$ are, respectively, the pressure, density, and fluid velocity far from the body around which the flow is being studied. The pressure coefficient may be readily expressed in terms of the pressure ratio as follows:

$$C_p = 2\frac{p_\infty}{\rho_\infty U^2}\left(\frac{p}{p_\infty} - 1\right)$$

$$= \frac{2}{\gamma M_\infty^2}\left(\frac{p}{p_\infty} - 1\right)$$

In order to relate the pressure ratio to the velocity, the energy equation will be employed.

$$\tfrac{1}{2}\mathbf{u}\cdot\mathbf{u} + \frac{\gamma}{\gamma - 1}\frac{p}{\rho} = \tfrac{1}{2}U^2 + \frac{\gamma}{\gamma - 1}\frac{p_\infty}{\rho_\infty}$$

Since the flow is irrotational, it is also isentropic, so that the quantity $p/\rho$ may be expressed in terms of $p$ only by use of the isentropic law. Thus

$$\rho = \rho_\infty\left(\frac{p}{p_\infty}\right)^{1/\gamma}$$

$$\therefore \quad \frac{p}{\rho} = \frac{p}{\rho_\infty}\left(\frac{p}{p_\infty}\right)^{-1/\gamma}$$

$$= \frac{a_\infty^2}{\gamma}\left(\frac{p}{p_\infty}\right)^{1-1/\gamma}$$

Substituting this expression into the energy equation gives

$$\tfrac{1}{2}\mathbf{u}\cdot\mathbf{u} + \frac{a_\infty^2}{\gamma - 1}\left(\frac{p}{p_\infty}\right)^{(\gamma-1)/\gamma} = \tfrac{1}{2}U^2 + \frac{a_\infty^2}{\gamma - 1}$$

From this equation the following expression is obtained for the pressure ratio:

$$\frac{p}{p_\infty} = \left[1 + \frac{\gamma - 1}{2a_\infty^2}(U^2 - \mathbf{u}\cdot\mathbf{u})\right]^{\gamma/(\gamma-1)}$$

Using this result, our expression for the pressure coefficient becomes

$$C_p = \frac{2}{\gamma M_\infty^2}\left\{\left[1 + \frac{\gamma - 1}{2a_\infty^2}(U^2 - \mathbf{u}\cdot\mathbf{u})\right]^{\gamma/(\gamma-1)} - 1\right\} \qquad (13.4a)$$

Equation (13.4a), which expresses the local value of the pressure coefficient in terms of the local velocity, is still exact within the inviscid, adiabatic assumptions. In order to obtain an expression for the pressure coefficient which is compatible with small-perturbation theory, Eq. (13.4a) will now be linearized in the perturbation velocity. The velocity term in the foregoing equation may thus be rewritten as follows:

$$U^2 - \mathbf{u}\cdot\mathbf{u} = U^2 - \left[(U + u')^2 + (v')^2 + (w')^2\right]$$

$$= -2Uu'$$

Substituting this linearized expression into Eq. (13.4a) yields the following simplified expression for the pressure coefficient:

$$C_p = \frac{2}{\gamma M_\infty^2} \left\{ \left[ 1 - (\gamma - 1) \frac{U u'}{a_\infty^2} \right]^{\gamma/(\gamma - 1)} - 1 \right\}$$

But, to the first order in the perturbation velocity,

$$\left[ 1 - (\gamma - 1) \frac{U u'}{a_\infty^2} \right]^{\gamma/(\gamma - 1)} = 1 - \gamma \frac{U u'}{a_\infty^2}$$

Thus the linearized form of Eq. (13.4a) is

$$C_p = -2 \frac{u'}{U} \tag{13.4b}$$

This simple result will be used in conjunction with approximate solutions to the compressible-flow equations which are established through use of Eq. (13.3b).

## 13.5  FLOW OVER A WAVE-SHAPED WALL

The first application of small-perturbation theory will be made to flow over a sinuous wall. This flow is relatively simple, yet it has the property of illustrating clearly the distinctions between subsonic and supersonic flows.

Figure 13.1a shows a sinusoidal surface over which a uniform flow of magnitude $U$ is assumed to flow such that compressible effects are not negligible. The equation of the wavy surface is taken to be

$$y = \eta(x) = \varepsilon \sin \frac{2\pi x}{\lambda}$$

where $\varepsilon/\lambda$ is assumed to be small compared with unity, so that the linear theory will be valid. The differential equation to be solved is the two-dimensional form of Eq. (13.3b). The boundary condition to be satisfied on $y = \eta(x)$ is

$$\frac{v'}{U + u'} = \frac{dy}{dx} = \frac{2\pi}{\lambda} \varepsilon \cos \frac{2\pi x}{\lambda}$$

In view of our linear theory the quantity $v'/(U + u')$ may be reduced to $v'/U$, so that this boundary condition becomes

$$\frac{\partial \Phi}{\partial y}(x, \eta) = U \frac{2\pi}{\lambda} \varepsilon \cos \frac{2\pi x}{\lambda}$$

The left-hand side of this equation may be expanded in a Taylor series and linearized so that the linear form of this boundary condition is

$$\frac{\partial \Phi}{\partial y}(x, 0) = U \frac{2\pi}{\lambda} \varepsilon \cos \frac{2\pi x}{\lambda}$$

From the foregoing discussion it is evident that, within the small-perturbation theory, the problem to be solved for the perturbation velocity potential

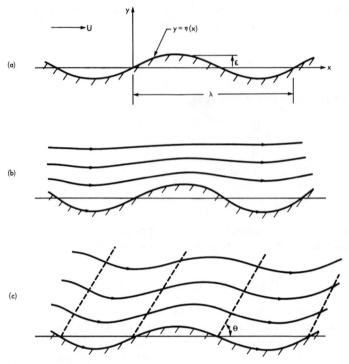

**FIGURE 13.1**
(a) Wave-shaped wall, (b) flow for $M_\infty < 1$, and (c) flow for $M_\infty > 1$.

$\Phi(x, y)$ is the following:

$$\left(1 - M_\infty^2\right)\frac{\partial^2 \Phi}{\partial x^2} + \frac{\partial^2 \Phi}{\partial y^2} = 0 \tag{13.5a}$$

$$\frac{\partial \Phi}{\partial y}(x, 0) = U\frac{2\pi}{\lambda}\varepsilon \cos \frac{2\pi x}{\lambda} \tag{13.5b}$$

$$\frac{\partial \Phi}{\partial x}(x, y) = \text{finite} \qquad \text{as } y \to \infty \tag{13.5c}$$

Since Eq. (13.5a) may be either elliptic or hyperbolic, depending upon whether $M_\infty$ is less than unity or greater than unity, it is convenient to discuss the solution to Eqs. (13.5) for subsonic flow and supersonic flow separately.

Consider, first, the case of subsonic flow. It will be found convenient to replace $x$ by a new coordinate $\xi$ which is defined by

$$\xi = \frac{x}{\sqrt{1 - M_\infty^2}}$$

Then, in terms of the coordinates $\xi$ and $y$, Eqs. (13.5) become

$$\frac{\partial^2 \Phi}{\partial \xi^2} + \frac{\partial^2 \Phi}{\partial y^2} = 0$$

$$\frac{\partial \Phi}{\partial y}(\xi, 0) = U\frac{2\pi}{\lambda}\varepsilon \cos\left(\frac{2\pi}{\lambda}\sqrt{1 - M_\infty^2}\,\xi\right)$$

$$\frac{\partial \Phi}{\partial \xi}(\xi, y) = \text{finite} \qquad \text{as } y \to \infty$$

The governing equation is now seen to be Laplace's equation in this new coordinate system. Solving this equation by separation of variables, the solution should be trigonometric in $\xi$ in view of the first boundary condition. Then, the $y$ dependence will be either exponential or hyperbolic, and in view of the semi-infinite domain in the $y$ direction, the exponential form will be used. However, the second boundary condition rules out the possibility of a positive exponential, so that the required solution will be of the form

$$\Phi(\xi, y) = (A \cos \alpha\xi + B \sin \alpha\xi)e^{-\alpha y}$$

But the first boundary condition has been used only to obtain the function form of $\Phi$. Thus, imposing this boundary condition completely gives

$$\frac{\partial \Phi}{\partial y}(\xi, 0) = -\alpha(A \cos \alpha\xi + B \sin \alpha\xi) = U\frac{2\pi}{\lambda}\varepsilon \cos\left(\frac{2\pi}{\lambda}\sqrt{1 - M_\infty^2}\,\xi\right)$$

Equating coefficients and arguments of the two trigonometric terms involved in the last equality gives

$$-\alpha A = U\frac{2\pi}{\lambda}\varepsilon$$

$$B = 0$$

$$\alpha = \frac{2\pi}{\lambda}\sqrt{1 - M_\infty^2}$$

Thus the complete solution for $\Phi$ is

$$\Phi(\xi, y) = -\frac{U\varepsilon}{\sqrt{1 - M_\infty^2}} \cos\left(\frac{2\pi}{\lambda}\sqrt{1 - M_\infty^2}\,\xi\right) \exp\left(-\frac{2\pi}{\lambda}\sqrt{1 - M_\infty^2}\,y\right)$$

Returning now to the original coordinate system gives the following solution for the perturbation velocity potential:

$$\Phi(x, y) = -\frac{U\varepsilon}{\sqrt{1 - M_\infty^2}} \cos\frac{2\pi x}{\lambda} \exp\left(-\frac{2\pi}{\lambda}\sqrt{1 - M_\infty^2}\,y\right) \qquad (13.6a)$$

Equation (13.6a) shows that the perturbation to the free stream is in phase with the wall, and it leads to a flow pattern as shown in Fig. 13.1b. It is also evident that the perturbation dies exponentially with distance from the surface. Since it was assumed that the perturbation velocity should be small compared with the free-stream velocity, Eq. (13.6a) shows that

$$\text{Max}\left(\frac{u'}{U}\right) = \frac{(2\pi/\lambda)\varepsilon}{\sqrt{1 - M_\infty^2}} \ll 1$$

Using the expression (13.4b) and the solution (13.6a) yields the following expression for the pressure coefficient in the fluid:

$$C_p(x, y) = -\frac{(4\pi/\lambda)\varepsilon}{\sqrt{1 - M_\infty^2}} \sin\frac{2\pi x}{\lambda} \exp\left(-\frac{2\pi}{\lambda}\sqrt{1 - M_\infty^2}\, y\right) \quad (13.6b)$$

This result shows that the maximum pressure on the wall corresponds to the bottom of the troughs and the minimum pressure corresponds to the top of the humps. That is, the pressure is symmetrically distributed about the humps on the wall, so that there will be no induced drag on the wall. This result will be further discussed later in this section.

Considering now the case of supersonic flow, the governing partial differential equation is

$$\frac{\partial^2 \Phi}{\partial x^2} - \frac{1}{(M_\infty^2 - 1)}\frac{\partial^2 \Phi}{\partial y^2} = 0$$

This is the so-called "one-dimensional wave equation" whose general solutions will be of the form

$$\Phi(x, y) = f\left(x - \sqrt{M_\infty^2 - 1}\, y\right) + g\left(x + \sqrt{M_\infty^2 - 1}\, y\right)$$

where $f$ and $g$ are any differentiable functions. The first solution represents a wave which slopes downstream and away from the wall, so that perturbations which are generated by the wall will travel downstream only according to this solution. On the other hand, the second function in the above solution represents signals which travel upstream as they move away from the wall. Since such a solution has no physical meaning in supersonic flow, it must be rejected here, so that the general solution becomes

$$\Phi(x, y) = f\left(x - \sqrt{M_\infty^2 - 1}\, y\right)$$

The function $f$ may be evaluated by imposing the surface boundary condition (13.5b). This gives

$$\frac{\partial \Phi}{\partial y}(x, 0) = -\sqrt{M_\infty^2 - 1}\, f'(x) = U\frac{2\pi}{\lambda}\varepsilon \cos\frac{2\pi x}{\lambda}$$

hence

$$f(x) = -\frac{U\varepsilon}{\sqrt{M_\infty^2 - 1}} \sin\frac{2\pi x}{\lambda}$$

Thus the perturbation velocity becomes

$$\Phi(x, y) = -\frac{U\varepsilon}{\sqrt{M_\infty^2 - 1}} \sin\left[\frac{2\pi}{\lambda}\left(x - \sqrt{M_\infty^2 - 1}\,y\right)\right] \qquad (13.7a)$$

This solution satisfies the remaining boundary condition [Eq. (13.5c)]. From this solution and Eq. (13.4b) the value of the pressure coefficient is found to be

$$C_p = \frac{(4\pi/\lambda)\varepsilon}{\sqrt{M_\infty^2 - 1}} \cos\left[\frac{2\pi}{\lambda}\left(x - \sqrt{M_\infty^2 - 1}\,y\right)\right] \qquad (13.7b)$$

The solution which is represented by Eqs. (13.7a) and (13.7b) shows that the velocity components and the pressure are constant along the lines

$$x - \sqrt{M_\infty^2 - 1}\,y = \text{constant}$$

The slope of these lines is given by

$$\frac{dy}{dx} = \frac{1}{\sqrt{M_\infty^2 - 1}} = \tan\theta$$

where $\theta$ is the inclination of the lines with respect to the $x$ axis. Hence

$$\theta = \sin^{-1}\left(\frac{1}{M_\infty}\right)$$

This result shows that the lines along which the flow parameters are constant are actually the Mach lines. That is, signals are propagated along the Mach lines undisturbed. The resulting flow field is illustrated in Fig. 13.1c.

Equation (13.7b) shows that the pressure on the wall is proportional to $\cos(2\pi x/\lambda)$, which means that the pressure peaks are 90° out of phase with the geometric peaks of the wall. It follows, then, that a drag force will exist on the wall for the case of supersonic flow. This is quite different from the result which was obtained for subsonic flow. Figure 13.1a shows a section of the wall, while Fig. 13.1b and c show, respectively, the pressure distribution on the wall for subsonic flow and for supersonic flow. In this figure the value of the pressure coefficient $C_p$ evaluated on the wall is denoted by $\overline{C_p}$. Because of the symmetrical pressure distribution about each geometric peak, there is no drag force in subsonic flow. However, the lack of symmetry in supersonic flow leads to a drag which is called the *wave drag*. Thus the linearized theoretical drag on bodies in compressible flow is as shown in Fig. 13.2d. The theoretical drag becomes infinite for Mach numbers close to, but greater than, unity because the linearized theory breaks down in sonic flow. A transonic theory exists which shows, as it should, that retaining some important terms which the linear theory neglects results in a finite drag coefficient. The actual drag indicated in Fig. 13.2d illustrates this result and also shows that, owing to viscous effects, a drag force exists even for subsonic flows. However, this viscous drag is relatively small for slender bodies when it is compared with the wave drag.

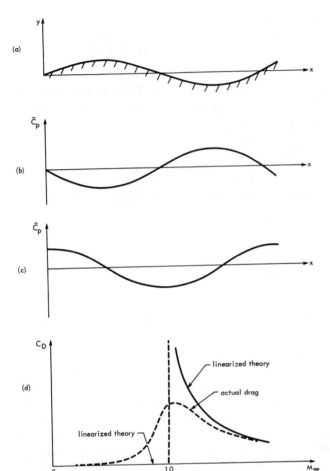

**FIGURE 13.2**
(*a*) Wave-shaped wall, (*b*) surface pressure coefficient for subsonic flow, and (*c*) for supersonic flow, (*d*) drag coefficient versus Mach number.

The foregoing solution for a wave-shaped wall is significant in its own right, and it illustrates some important features of subsonic and supersonic flows. Also since the theory being used is linear, superposition is valid. Thus, by use of Fourier integrals, the solution obtained here may be extended to obtain solutions for compressible flow over arbitrarily shaped surfaces.

## 13.6  PRANDTL-GLAUERT RULE FOR SUBSONIC FLOW

Using small-perturbation theory it is possible, by means of a simple transformation, to reduce all subsonic-flow problems to equivalent incompressible-flow problems. The rule which results from such a transformation is called the Prandtl-Glauert rule.

For subsonic flow over a body whose surface is defined by $y = f(x)$, the perturbation velocity potential must satisfy the following problem:

$$\frac{\partial^2 \Phi}{\partial x^2} + \frac{1}{1 - M_\infty^2} \frac{\partial^2 \Phi}{\partial y^2} = 0$$

$$\frac{\partial \Phi}{\partial y}(x, 0) = U \frac{df}{dx}(x)$$

$$\frac{\partial \Phi}{\partial x}(x, y) = \text{finite} \qquad \text{as } y \to \infty$$

Introduce a new velocity potential $\Phi'$ and a new vertical coordinate $\eta$ which are defined as follows:

$$\Phi = \frac{1}{\sqrt{1 - M_\infty^2}} \Phi'$$

$$\eta = \sqrt{1 - M_\infty^2}\, y$$

Then the problem to be solved for $\Phi'(x, \eta)$ is the following:

$$\frac{\partial^2 \Phi'}{\partial x^2} + \frac{\partial^2 \Phi'}{\partial \eta^2} = 0$$

$$\frac{\partial \Phi'}{\partial \eta}(x, 0) = U \frac{df}{dx}(x)$$

$$\frac{\partial \Phi'}{\partial x}(x, \eta) = \text{finite} \qquad \text{as } \eta \to \infty$$

That is, in the $x\eta$ plane the problem to be solved is that of irrotational motion of an incompressible fluid about a body whose surface is defined by $\eta = f(x)$. Assuming that the ideal-fluid flow problem can be solved, the corresponding pressure coefficient may be evaluated from Eq. (13.4b). Thus, denoting the incompressible pressure coefficient by $C_p'$, it follows that

$$C_p' = -\frac{2}{U} \frac{\partial \Phi'}{\partial x}$$

But the compressible pressure coefficient is given by

$$C_p = -\frac{2}{U} \frac{\partial \Phi}{\partial x}$$

$$= -\frac{1}{\sqrt{1 - M_\infty^2}} \frac{2}{U} \frac{\partial \Phi'}{\partial x}$$

that is,

$$C_p(x, y) = \frac{C_p'(x, y)}{\sqrt{1 - M_\infty^2}} \tag{13.8}$$

That is, the pressure distribution around a body which is in subsonic compressible flow may be obtained from the corresponding incompressible pressure distribution. The rule which connects these two pressure distributions [Eq. (13.8)] is known as the Prandtl-Glauert rule. It establishes the effects of compressibility for subsonic flows and illustrates that, within the linear theory, any subsonic compressible-flow problem may be solved provided that the corresponding incompressible-flow problem may be solved.

## 13.7   ACKERT'S THEORY FOR SUPERSONIC FLOWS

Small-perturbation theory may also be used to establish a theory for supersonic flows. The resulting theory is known as Ackeret's theory. The situation to which this theory applies is shown in Fig. 13.3$a$. Supersonic flow approaches a thin, cambered airfoil which is at an angle of attack $\alpha$ to the free stream whose Mach number is $M_\infty$. The chord of the airfoil is denoted by $c$, $t$ is the maximum thickness, and $h$ is the maximum camber of the airfoil. The equation of the upper surface of the airfoil is $y = \eta_u(x)$, while that of the lower surface is $\eta_l(x)$.

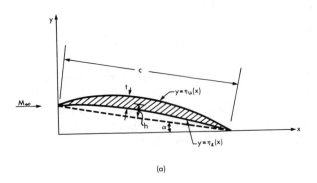

(a)

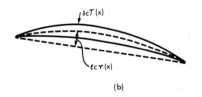

(b)

**FIGURE 13.3**
($a$) Parameters for supersonic airfoil and ($b$) definitions of the half-thickness function $\tau(x)$ and the camber function $\gamma(x)$.

According to the linearized theory, the problem to be solved for the perturbation velocity potential is the following:

$$\frac{\partial^2 \Phi}{\partial x^2} - \frac{1}{M_\infty^2 - 1} \frac{\partial^2 \Phi}{\partial y^2} = 0$$

$$\frac{\partial \Phi}{\partial y}(x, 0) = U \frac{d\eta}{dx}(x)$$

$$\frac{\partial \Phi}{\partial x}(x, y) = \text{finite} \qquad \text{as } y \to \infty$$

Since, in general, the surfaces $y = \eta_u(x)$ and $y = \eta_l(x)$ will be different, the boundary condition on $y = 0$ will be different for the upper and lower surfaces, so that the corresponding values of $\Phi$ will be different. Denoting these solutions by $\Phi_u$ and $\Phi_l$, it therefore follows that the two solutions will be

$$\Phi_u(x, y) = f\left(x - \sqrt{M_\infty^2 - 1}\, y\right)$$

$$\Phi_l(x, y) = g\left(x + \sqrt{M_\infty^2 - 1}\, y\right)$$

Here the left-running solution has been omitted from $\Phi_u$ and the right-running solution has been omitted from $\Phi_l$. This satisfies the condition that signals can travel downstream only in supersonic flow, so that the lines along which signals travel must slope downstream as they move away from the airfoil.

The functions $f$ and $g$ may be evaluated by imposing the boundary conditions at the surface of the airfoil. Thus the boundary conditions at $y = 0+$ and $y = 0-$, together with the corresponding solutions, give

$$f'(x) = -\frac{U}{\sqrt{M_\infty^2 - 1}} \frac{d\eta_u}{dx}(x)$$

$$g'(x) = \frac{U}{\sqrt{M_\infty^2 - 1}} \frac{d\eta_l}{dx}(x)$$

It is not necessary to integrate these expressions in order to evaluate the pressure coefficient. From Eq. (13.4b),

$$C_p = -\frac{2}{U} \frac{\partial \Phi}{\partial x}$$

Thus, denoting the pressure coefficient on the upper surface by $\overline{C_{pu}}$ and that on the lower surface by $\overline{C_{pl}}$, it follows that

$$\overline{C_{pu}} = -\frac{2}{U}f'(x) = \frac{2}{\sqrt{M_\infty^2 - 1}}\frac{d\eta_u}{dx}$$

$$\overline{C_{pl}} = -\frac{2}{U}g'(x) = -\frac{2}{\sqrt{M_\infty^2 - 1}}\frac{d\eta_l}{dx}$$

These results show that the local value of the pressure coefficient is proportional to the local slope of the airfoil surface.

Using these values of the surface pressure coefficient, the lift coefficient of the airfoil may be evaluated as follows:

$$C_L = \frac{1}{\frac{1}{2}\rho_\infty U^2}\frac{\int_0^c (\overline{p_l} - \overline{p_u})\, dx}{c}$$

Here $\rho_\infty$ is the density of the fluid far from the airfoil and $p_l, p_u$ are, respectively, the pressure on the lower surface of the airfoil and the pressure on the upper surface. Then, from the definition of the pressure coefficient it follows that

$$C_L = \frac{1}{c}\int_0^c (\overline{C_{pl}} - \overline{C_{pu}})\, dx$$

$$= -\frac{2}{c\sqrt{M_\infty^2 - 1}}\int_0^c \left(\frac{d\eta_l}{dx} + \frac{d\eta_u}{dx}\right) dx$$

$$= -\frac{2}{c\sqrt{M_\infty^2 - 1}}[\eta_l + \eta_u]_0^c$$

But $\eta_l(c) = \eta_u(c) = 0$ and $\eta_l(0) = \eta_u(0) = \alpha c$. Hence the value of the lift coefficient is

$$C_L = \frac{4\alpha}{\sqrt{M_\infty^2 - 1}} \qquad (13.9a)$$

Equation (13.9a) shows that the lift force which acts on a supersonic airfoil depends only on the Mach number of the flow and on the angle of attack of the airfoil. That is, the lift force is independent of the camber and thickness of the airfoil. This result is quite different from the corresponding result for subsonic flow. Indeed, Eqs. (4.25b) and (13.8) show that the lift coefficient for a Joukowski airfoil at subsonic speeds is

$$C_L = \frac{2\pi}{\sqrt{1 - M_\infty^2}}\left(1 + 0.77\frac{t}{c}\right)\sin\left(\alpha + 2\frac{h}{c}\right)$$

Thus the lift of subsonic airfoils is greatly affected by airfoil thickness and camber, but the lift of supersonic airfoils is not affected by these parameters.

The drag coefficient for the airfoil may be evaluated in a similar manner as follows:

$$C_D = \frac{1}{\frac{1}{2}\rho_\infty U^2} \frac{\int_0^{ac}(\overline{p_l} - \overline{p_u})\,dy}{c}$$

$$= \frac{1}{c}\int_0^{ac}(\overline{C_{pl}} - \overline{C_{pu}})\,dy$$

This integral may be converted to one in $x$ by noting that $dy = (dy/dx)\,dx$, where $dy/dx = d\eta_u/dx$ on the upper surface and $dy/dx = d\eta_l/dx$ on the lower surface of the airfoil. Hence

$$C_D = \frac{1}{c}\int_c^0 \left(\overline{C_{pl}}\frac{d\eta_l}{dx} - \overline{C_{pu}}\frac{d\eta_u}{dx}\right)dx$$

Using the expressions which were derived for the pressure coefficient on the upper and lower surfaces shows that

$$C_D = \frac{2}{c\sqrt{M_\infty^2 - 1}}\int_0^c \left[\left(\frac{d\eta_l}{dx}\right)^2 + \left(\frac{d\eta_u}{dx}\right)^2\right]dx$$

Since the integrand of this integral is positive definite, it is evident that a nonzero drag will exist for nontrivial airfoil shapes, a result which was deduced for the wave-shaped wall in Sec. 13.5.

The manner in which airfoil thickness and camber affect the wave drag may be established by writing the equations of the upper and lower surfaces of the airfoil in terms of the corresponding parameters. Thus, let the thickness parameter and the camber parameter be defined, respectively, by

$$\delta = \frac{t}{c}$$

$$\varepsilon = \frac{h}{c}$$

Here, $t$ and $h$ are the maximum thickness and maximum camber, respectively, as shown in Fig. 13.3a. Then a half-thickness function $\tau(x)$ is defined such that the local value of the airfoil half thickness is $\delta c\tau(x)$, as shown in Fig. 13.3b. Likewise, a camber function $\gamma(x)$ is defined such that the local value of the airfoil camber is $\varepsilon c\gamma(x)$, which is also shown in Fig. 13.3b. Thus, the upper and lower surfaces of the airfoil may be defined in terms of the angle of attack, the half-thickness function, and the camber function. From the definitions of these two functions they must lie in the following range:

$$0 \le \tau(x) \le \tfrac{1}{2}$$

$$0 \le \gamma(x) \le 1$$

In terms of the functions defined above, the equations of the upper and lower surfaces of the airfoil are

$$\eta_u(x) = \alpha(c - x) + \varepsilon c\gamma(x) + \delta c\tau(x)$$

$$\eta_l(x) = \alpha(c - x) + \varepsilon c\gamma(x) - \delta c\tau(x)$$

That is, the upper and lower surfaces are defined by the line through the mean thickness of the airfoil, plus or minus the half thickness, respectively. Thus, the integrand of the integral which defines the drag coefficient may be evaluated as follows:

$$\frac{d\eta_u}{dx} = -\alpha + \varepsilon c\gamma' + \delta c\tau'$$

$$\frac{d\eta_l}{dx} = -\alpha + \varepsilon c\gamma' + \delta c\tau'$$

$$\therefore \quad \left(\frac{d\eta_u}{dx}\right)^2 + \left(\frac{d\eta_l}{dx}\right)^2 = 2\alpha^2 + 2\varepsilon^2 c^2(\gamma')^2 + 2\delta^2 c^2(\tau')^2 - 4\alpha\varepsilon c\gamma'$$

In the foregoing equations the primes denote differentiation with respect to $x$. Substituting the last result into the expression for the drag coefficient yields the following result:

$$C_D = \frac{2}{c\sqrt{M_\infty^2 - 1}} \int_0^c \left[2\alpha^2 + 2\varepsilon^2 c^2(\gamma')^2 + 2\delta^2 c^2(\tau')^2 - 4\alpha\varepsilon c\gamma'\right] dx$$

The first term in the integrand is a constant and may be integrated directly. The last term in the integrand integrates to zero, since $\gamma(0) = \gamma(c) = 0$. Thus the drag coefficient becomes

$$C_D = \frac{4\alpha^2}{\sqrt{M_\infty^2 - 1}} + \frac{4\varepsilon^2 c}{\sqrt{M_\infty^2 - 1}} \int_0^c (\gamma')^2 \, dx + \frac{4\delta^2 c}{\sqrt{M_\infty^2 - 1}} \int_0^c (\tau')^2 \, dx \quad (13.9b)$$

Equation (13.9a) showed that the lift coefficient of supersonic wings is independent of the camber and the thickness of the airfoil. On the other hand, Eq. (13.9b) shows that both camber and thickness increase the drag coefficient of such wings. Hence it may be concluded that supersonic wings should be as straight as possible and as thin as possible. However, it is evident that for structural reasons there is a limit to the minimum thickness to which such wings may be made. Apart from the general guidelines of minimizing the camber and thickness of wing sections, application of the foregoing theory to specific airfoils shows that sharp corners are preferable to rounded corners in supersonic flight.

The investigation of the performance of specific airfoil sections is deferred to the problems at the end of the chapter.

## 13.8  PRANDTL-MEYER FLOW

In this section an exact solution to the equations of two-dimensional flow of a compressible fluid will be derived. The flow situation to which this solution refers consists of supersonic flow approaching a sharp bend in a boundary in which the boundary bends in such a direction that an expansion, rather than a compression, is required to turn the fluid. The resulting flow field is called Prandtl-Meyer flow.

Figure 13.4$a$ shows the flow configuration which is under consideration. Supersonic flow whose Mach number is $M_1$ flows parallel to a boundary which suddenly changes direction as shown. In order to satisfy the boundary condition at the surface, the velocity vector must be deflected in the direction indicated. Since this deflection is opposite in sense to that which was shown to be necessary for shock waves, it may be concluded that an expansion, rather than a compression, is required. Although expansions are continuous processes as opposed to the discrete processes of shock waves, the expansion is illustrated as consisting of a large number of very weak expansion waves. This is known as a *Prandtl-Meyer fan.*

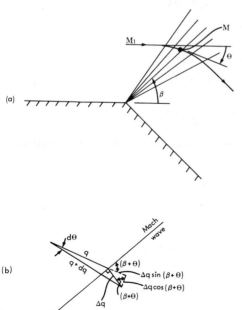

**FIGURE 13.4**

($a$) Prandtl-Meyer fan and ($b$) velocity change through a typical Mach wave.

An arbitrary point in the expansion fan is indicated in Fig. 13.4$a$. The Mach number at this point is $M$, and the deflection of the velocity vector at this point, relative to its original direction, is denoted by $\theta$. The inclination of the Mach wave which passes through the point under consideration is denoted by $\beta$. Then it is known that the leading Mach wave will subtend an angle defined by

$$\beta_1 = \sin^{-1}\left(\frac{1}{M_1}\right)$$

Since the pressure gradient will be normal to each of the Mach lines, the changes in the velocity must also be normal to the Mach lines. Thus if $q$ denotes the magnitude of the velocity vector as it approaches our reference Mach wave and if $\Delta q$ denotes the change in the value of $q$ which is caused by the Mach wave, the velocity diagram will be as shown in Fig. 13.4$b$.

The velocity vector which emerges from the Mach wave will have a magnitude $q + dq$, and it will have been deflected through an angle $d\theta$. Since $\Delta q$ will be infinitesimally small as the limit of an infinite number of Mach waves is approached, the deflection of the velocity vector may be approximated by

$$d\theta = \frac{\Delta q \cos(\beta + \theta)}{q + dq}$$

In the limit all second-order terms will vanish identically, so that this expression may be further reduced to

$$d\theta = \frac{\Delta q}{q} \cos(\beta + \theta)$$

But, from Fig. 13.4$b$,

$$q + dq = q + \Delta q \sin(\beta + \theta)$$

Using this result to eliminate $\Delta q$ yields the following equation for $d\theta$:

$$d\theta = \frac{dq}{q} \cot(\beta + \theta)$$

But it is known that inclination of the Mach wave under consideration is given by

$$\sin(\beta + \theta) = \frac{1}{M}$$

hence

$$\cot(\beta + \theta) = \sqrt{M^2 - 1}$$

Thus the turning angle of the velocity vector becomes

$$d\theta = \sqrt{M^2 - 1}\,\frac{dq}{q}$$

In order to complete the solution, this equation must be integrated to yield an expression for $\theta$ in terms of either $q$ or $M$. Choosing the latter as being more relevant, $q$ must be expressed in terms of the local Mach number $M$. This may be done by use of the definition of the Mach number in the form

$$q = aM$$

hence

$$\frac{dq}{q} = \frac{da}{a} + \frac{dM}{M}$$

To eliminate $a$ from this equation, the energy equation will be employed in the form

$$\tfrac{1}{2}q^2 + \frac{a^2}{\gamma - 1} = \frac{a_0^2}{\gamma - 1}$$

Or, multiplying this equation by $(\gamma - 1)/a^2$ gives

$$\frac{\gamma - 1}{2}M^2 + 1 = \frac{a_0^2}{a^2}$$

$$\therefore \quad a^2 = \frac{a_0^2}{1 + [(\gamma - 1)/2]M^2}$$

and

$$2a\,da = \frac{-a_0^2(\gamma - 1)M\,dM}{\{1 + [(\gamma - 1)/2]M^2\}^2}$$

Thus the expression for $dq$ becomes

$$\frac{dq}{q} = -\frac{\gamma - 1}{2} \frac{M\,dM}{1 + [(\gamma - 1)/2]M^2} + \frac{dM}{M}$$

$$= \frac{1}{1 + [(\gamma - 1)/2]M^2} \frac{dM}{M}$$

Using this result to eliminate $q$ from our expression for the element of turning angle $d\theta$ gives

$$d\theta = \frac{\sqrt{M^2 - 1}}{1 + [(\gamma - 1)/2]M^2} \frac{dM}{M}$$

This equation may now be integrated to give

$$\theta = \nu(M) - \nu(M_1) \tag{13.10a}$$

where

$$\nu(M) = \sqrt{\frac{\gamma + 1}{\gamma - 1}} \tan^{-1}\left(\sqrt{\frac{\gamma - 1}{\gamma + 1}(M^2 - 1)}\right) - \tan^{-1}\left(\sqrt{M^2 - 1}\right)$$

$$\tag{13.10b}$$

Equation (13.10b) defines the so-called Prandtl-Meyer function which is denoted by $\nu(M)$. The solution (13.10a) gives $\theta = 0$ for $M = M_1$ and represents a monotonically increasing function of $M$ for values of $M > M_1$. Then the minimum value of $\theta$ occurs at $M_1 = 1$, the value of $\theta$ being zero, and the maximum value of $\theta$ occurs when $M$ tends to infinity. From Eqs. (13.10a) and (13.10b), this gives the following maximum:

$$\theta_{max} = \frac{\pi}{2}\left(\sqrt{\frac{\gamma + 1}{\gamma - 1}} - 1\right) \tag{13.10c}$$

For $\gamma = 1.4$ this gives a maximum deflection of about $130°$.


## PROBLEMS

**13.1.** Show that the equation to be satisfied by the velocity potential for steady, two-dimensional, irrotational motion of an inviscid fluid is

$$\left(1 - \frac{u^2}{a^2}\right)\frac{\partial^2 \phi}{\partial x^2} - 2\frac{uv}{a^2}\frac{\partial^2 \phi}{\partial x \, \partial y} + \left(1 - \frac{v^2}{a^2}\right)\frac{\partial^2 \phi}{\partial y^2} = 0$$

**13.2.** Introduce a stream function which is defined as follows:

$$\rho u = \rho_0 \frac{\partial \psi}{\partial y}$$

$$\rho v = -\rho_0 \frac{\partial \psi}{\partial x}$$

where $\rho_0$ is a constant reference value of the density. Show that this stream function identically satisfies the continuity equation for steady two-dimensional motion of a compressible fluid and that for irrotational motion the equation to be satisfied by $\psi$ is

$$\left(1 - \frac{u^2}{a^2}\right)\frac{\partial^2 \psi}{\partial x^2} - 2\frac{uv}{a^2}\frac{\partial^2 \psi}{\partial x \, \partial y} - \left(1 - \frac{v^2}{a^2}\right)\frac{\partial^2 \psi}{\partial y^2} = 0$$

**13.3.** The stream function which was defined in Prob. 13.2 may be considered to be a function of the magnitude of the velocity vector, $q$, and its angle $\theta$. That is, we can consider the stream function to be $\psi(q, \theta)$ where $u = q \cos \theta$ and $v = q \sin \theta$. To obtain the differential equation to be satisfied by $\psi(q, \theta)$, proceed as follows.
  (a) Obtain expressions for $d\phi$ and $d\psi$ in terms of $dx$ and $dy$ in which the coefficients are functions of $q$, $\theta$, and the density ratio only. Invert these equations to express $dx$ and $dy$ in terms of $d\phi$ and $d\psi$.
  (b) Use the fact that the velocity potential is $\phi(q, \theta)$ and the stream function is $\psi(q, \theta)$ to eliminate $d\phi$ and $d\psi$ in the expressions obtained above in terms of their derivatives with respect to $q$ and $\theta$.
  (c) Considering both $x$ and $y$ to be functions of $q$ and $\theta$, obtain expressions from differential calculus for $dx$ and $dy$. By equating these expressions with those

obtained in (b) above, obtain expressions for the partial derivatives of $x$ and $y$ with respect to $q$ and $\theta$.

(d) Eliminate both $x$ and $y$ from the results obtained in (c) by forming the second mixed derivatives of $x$ and $y$ with respect to $q$ and $\theta$. In this way obtain expressions for the derivatives

$$\frac{\partial \phi}{\partial \theta} \quad \text{and} \quad \frac{\partial \phi}{\partial q}$$

in terms of derivatives of the stream function $\psi(q, \theta)$. In deriving this result it should be noted that the density $\rho$ is a function of the magnitude of the velocity $q$, but not of its direction $\theta$.

(e) Use the Bernoulli equation for steady flow and the definition of the speed of sound to obtain the result

$$\frac{d}{dq}\left(\frac{\rho_0}{\rho}\right) = \frac{\rho}{\rho_0}\frac{q}{a^2}$$

From this result, and one of the expressions obtained in (d), show that

$$\frac{\partial \phi}{\partial q} = -\frac{\rho_0}{\rho}\frac{1}{q}(1 - M^2)\frac{\partial \psi}{\partial \theta}$$

Finally, eliminate $\phi$ from this expression, using a result obtained in (d), to show that the following equation is to be satisfied by the stream function:

$$q^2\frac{\partial^2 \psi}{\partial q^2} + q(1 + M^2)\frac{\partial \psi}{\partial q} + (1 - M^2)\frac{\partial^2 \psi}{\partial \theta^2} = 0$$

**13.4.** In the Janzen-Rayleigh expansion, find the differential equation to be satisfied by the function $\phi_3$ in the series.

**13.5.** The linearized form of the pressure coefficient is defined by Eq. (13.4b). Find the next correction term in the expression for $C_p$. That is, find the approximate expression for the pressure coefficient which is valid to the second order in small quantities.

**13.6.** Using the results of the linearized theory for compressible flow over a wave-shaped wall, integrate the pressure over one wavelength of the wall and so verify that the drag is zero for subsonic flow. Calculate, also, the drag per wavelength per unit width of the wall for supersonic flow.

**13.7.** An infinitely long cylinder of radius $a + \varepsilon \sin(2\pi x/\lambda)$ is exposed to a uniform axial flow of compressible fluid as shown in Fig. 13.5. If the flow is subsonic and if

$$\frac{\varepsilon}{\lambda} \ll 1 \qquad \frac{\varepsilon}{a} \ll 1$$

the conditions to be satisfied by the perturbation velocity potential are as follows:

$$(1 - M_\infty^2)\frac{\partial^2 \phi}{\partial x^2} + \frac{\partial^2 \phi}{\partial r^2} + \frac{1}{r}\frac{\partial \phi}{\partial r} = 0$$

$$\frac{\partial \phi}{\partial r}(x, a) = U\frac{d\bar{r}}{dx}$$

and subject to the condition that the fluid velocity is everywhere finite. In the foregoing, $\bar{r}$ is the value of $r$ which corresponds to the surface of the cylinder.

Using a linearized theory, find an expression for the pressure coefficient on the surface of the cylinder. Form the ratio of this pressure coefficient to that for a wave-shaped wall, and by expanding this ratio in powers of $\lambda/[a(1 - M_\infty^2)]$, establish the effect of wall curvature.

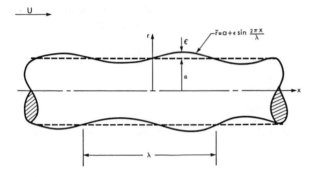

**FIGURE 13.5**
Axial flow over a wave-shaped circular cylinder.

**13.8.** Use Ackeret's theory to find the drag coefficient of the double-wedge airfoil shown in Fig. 13.6 for zero angle of attack.

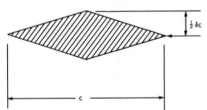

**FIGURE 13.6**
Double-wedge airfoil.

**13.9.** The half-thickness function $\tau(x)$ for the biconvex circular-arc airfoil shown in Fig. 13.7 is given by

$$\tau(x) = \frac{1}{\delta c}\eta(x)$$

and the equation of the upper surface is defined by the equation

$$(\eta + a)^2 + \left(x - \frac{c}{2}\right)^2 = \left(a + \tfrac{1}{2}\delta c\right)^2$$

Using this result and Ackeret's theory, evaluate the drag coefficient of a biconvex circular-arc airfoil at zero angle of attack. Compare this result with that for the double-wedge airfoil treated in the previous section.

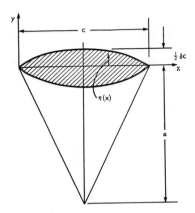

**FIGURE 13.7**
Biconvex circular-arc airfoil.

# FURTHER READING—PART IV

The compressible-flow area of fluid mechanics is not as well endowed with books as the other areas. However, the following material adequately cover and extends the material treated i Part IV of this book.

Ames Research Staff: "Equations, Tables and Charts for Compressible Flow," National Advisory Committee for Aeronautics, Report 1135, 1953.

Liepmann, J. W., and A. Roshko: "Elements of Gasdynamics," John Wiley & Sons, Inc., New York, 1957.

Lighthill, James: "Waves in Fluids," Cambridge University Press, London, 1978.

Shapiro, Ascher H.: "The Dynamics and Thermodynamics of Compressible Fluid Flow," vols. 1 and 2, The Ronald Press Company, New York, 1953.

# APPENDIXES

# APPENDIX A

## VECTOR ANALYSIS

Listed below are some of the vector relations which are particularly useful in the study of fluid mechanics. The derivation of these relationships may be found in most of the books which cover the topic of vector analysis.

### VECTOR IDENTITIES

In the following formulas, $\phi$ is any scalar and $\mathbf{a}$, $\mathbf{b}$, and $\mathbf{c}$ are any vectors.

$$\nabla \times \nabla\phi = 0$$

$$\nabla \cdot (\phi\mathbf{a}) = \phi\nabla \cdot \mathbf{a} + \mathbf{a} \cdot \nabla\phi$$

$$\nabla \times (\phi\mathbf{a}) = \nabla\phi \times \mathbf{a} + \phi(\nabla \times \mathbf{a})$$

$$\nabla \cdot (\nabla \times \mathbf{a}) = 0$$

$$(\mathbf{a} \cdot \nabla)\mathbf{a} = \tfrac{1}{2}\nabla(\mathbf{a} \cdot \mathbf{a}) - \mathbf{a} \times (\nabla \times \mathbf{a})$$

$$\nabla \times (\nabla \times \mathbf{a}) = \nabla(\nabla \cdot \mathbf{a}) - \nabla^2\mathbf{a}$$

$$\nabla \times (\mathbf{a} \times \mathbf{b}) = \mathbf{a}(\nabla \cdot \mathbf{b}) - \mathbf{b}(\nabla \cdot \mathbf{a}) - (\mathbf{a} \cdot \nabla)\mathbf{b} + (\mathbf{b} \cdot \nabla)\mathbf{a}$$

$$\nabla \cdot (\mathbf{a} \times \mathbf{b}) = \mathbf{b} \cdot (\nabla \times \mathbf{a}) - \mathbf{a} \cdot (\nabla \times \mathbf{b})$$

### INTEGRAL THEOREMS

In the following two theorems, which relate surface integrals to volume integrals, $V$ is any volume and $S$ is the surface which encloses $V$, the unit normal on $S$ being denoted by $\mathbf{n}$. $\phi$ is any scalar and $\mathbf{a}$ is any vector.

*Gauss' theorem* (also known as the divergence theorem):

$$\int_s \mathbf{a} \cdot \mathbf{n} \, ds = \int_V \nabla \cdot \mathbf{a} \, dV$$

*Green's theorem:*

$$\int_s \phi \frac{\partial \phi}{\partial n} \, dS = \int_V \left[ \nabla \phi \cdot \nabla \phi + \phi \nabla^2 \phi \right] dV$$

*Stokes' theorem:*

$$\oint \mathbf{a} \cdot \mathbf{dl} = \int_A (\nabla \times \mathbf{a}) \cdot \mathbf{n} \, dA$$

This theorem relates a line integral to an equivalent surface integral. The surface $A$ is arbitrary, but it must terminate on the line **1**.

## ORTHOGONAL CURVILINEAR COORDINATES

Let $x_1, x_2, x_3$ be a set of orthogonal curvilinear coordinates with $\mathbf{e}_1, \mathbf{e}_2, \mathbf{e}_3$ as the corresponding unit base vectors.

*Position vector:*

$$\mathbf{r} = x\mathbf{e}_x + y\mathbf{e}_y + z\mathbf{e}_z$$

where $\mathbf{e}_x$, $\mathbf{e}_y$, and $\mathbf{e}_z$ are fixed in space.

*Base vectors:*

$$\mathbf{e}_i = \frac{\partial \mathbf{r}}{\partial x_i} \bigg/ \left| \frac{\partial \mathbf{r}}{\partial x_i} \right|$$

*Metric-scale factors:*

$$h_i = \left| \frac{\partial \mathbf{r}}{\partial x_i} \right|$$

*Line element:*

$$(d\mathbf{r} \cdot d\mathbf{r}) = h_1^2 (dx_1)^2 + h_2^2 (dx_2)^2 + h_3^2 (dx_3)^2$$

*Cartesian coordinates* (rectangular coordinates):

$$x_1 = x \qquad x_2 = y \qquad x_3 = z$$

$$h_1 = 1 \qquad h_2 = 1 \qquad h_3 = 1$$

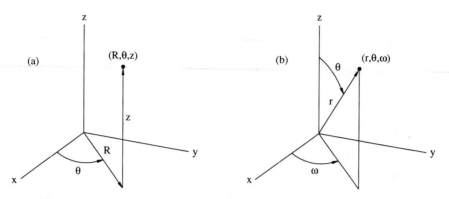

**FIGURE A.1**
Relationship between cartesian coordinates and (*a*) cylindrical, (*b*) spherical coordinates.

*Cylindrical coordinates:*

$$x_1 = R \qquad\qquad x_2 = \theta \qquad\qquad x_3 = z$$

$$h_1 = 1 \qquad\qquad h_2 = R \qquad\qquad h_3 = 1$$

$$x = R \cos \theta \qquad y = R \sin \theta \qquad z = z$$

*Spherical coordinates:*

$$x_1 = r \qquad\qquad x_2 = \theta \qquad\qquad x_3 = \omega$$

$$h_1 = 1 \qquad\qquad h_2 = r \qquad\qquad h_3 = r \sin \theta$$

$$x = r \cos \theta \qquad y = r \sin \theta \cos \omega \qquad z = r \sin \theta \sin \omega$$

*Vector operations:*

In the following let $\phi$ be any scalar and let $\mathbf{a} = a_1\mathbf{e}_1 + a_2\mathbf{e}_2 + a_3\mathbf{e}_3$ be any vector.

**1.** *Gradient:*

$$\nabla\phi = \frac{1}{h_1}\frac{\partial\phi}{\partial x_1}\mathbf{e}_1 + \frac{1}{h_2}\frac{\partial\phi}{\partial x_2}\mathbf{e}_2 + \frac{1}{h_3}\frac{\partial\phi}{\partial x_3}\mathbf{e}_3$$

**2.** *Divergence:*

$$\nabla \cdot \mathbf{a} = \frac{1}{h_1 h_2 h_3}\left[\frac{\partial}{\partial x_1}(h_2 h_3 a_1) + \frac{\partial}{\partial x_2}(h_1 h_3 a_2) + \frac{\partial}{\partial x_3}(h_1 h_2 a_3)\right]$$

### 3. *Curl:*

$$\nabla \times \mathbf{a} = \frac{1}{h_1 h_2 h_3} \begin{vmatrix} h_1 \mathbf{e}_1 & h_2 \mathbf{e}_2 & h_3 \mathbf{e}_3 \\ \dfrac{\partial}{\partial x_1} & \dfrac{\partial}{\partial x_2} & \dfrac{\partial}{\partial x_3} \\ h_1 a_1 & h_2 a_2 & h_3 a_3 \end{vmatrix}$$

### 4. *Laplacian:*

$$\nabla^2 \phi = \frac{1}{h_1 h_2 h_3} \left[ \frac{\partial}{\partial x_1} \left( \frac{h_2 h_3}{h_1} \frac{\partial \phi}{\partial x_1} \right) + \frac{\partial}{\partial x_2} \left( \frac{h_3 h_1}{h_2} \frac{\partial \phi}{\partial x_2} \right) + \frac{\partial}{\partial x_3} \left( \frac{h_1 h_2}{h_3} \frac{\partial \phi}{\partial x_3} \right) \right]$$

$$\nabla^2 \mathbf{a} = \left[ \frac{1}{h_1} \frac{\partial}{\partial x_1} (\nabla \cdot \mathbf{a}) + \frac{1}{h_2 h_3} \left( \frac{\partial}{\partial x_3} \left\{ \frac{h_2}{h_1 h_3} \left[ \frac{\partial}{\partial x_3} (h_1 a_1) - \frac{\partial}{\partial x_1} (h_3 a_3) \right] \right\} \right. \right.$$
$$\left. \left. - \frac{\partial}{\partial x_2} \left\{ \frac{h_3}{h_1 h_2} \left[ \frac{\partial}{\partial x_1} (h_2 a_2) - \frac{\partial}{\partial x_2} (h_1 a_1) \right] \right\} \right) \right] \mathbf{e}_1$$
$$+ \left[ \frac{1}{h_2} \frac{\partial}{\partial x_2} (\nabla \cdot \mathbf{a}) + \frac{1}{h_1 h_3} \left( \frac{\partial}{\partial x_1} \left\{ \frac{h_3}{h_1 h_2} \left[ \frac{\partial}{\partial x_1} (h_2 a_2) - \frac{\partial}{\partial x_2} (h_1 a_1) \right] \right\} \right. \right.$$
$$\left. \left. - \frac{\partial}{\partial x_3} \left\{ \frac{h_1}{h_2 h_3} \left[ \frac{\partial}{\partial x_2} (h_3 a_3) - \frac{\partial}{\partial x_3} (h_2 a_2) \right] \right\} \right) \right] \mathbf{e}_2$$
$$+ \left[ \frac{1}{h_3} \frac{\partial}{\partial x_3} (\nabla \cdot \mathbf{a}) + \frac{1}{h_1 h_2} \left( \frac{\partial}{\partial x_2} \left\{ \frac{h_1}{h_2 h_3} \left[ \frac{\partial}{\partial x_2} (h_3 a_3) - \frac{\partial}{\partial x_3} (h_2 a_2) \right] \right\} \right. \right.$$
$$\left. \left. - \frac{\partial}{\partial x_1} \left\{ \frac{h_2}{h_1 h_3} \left[ \frac{\partial}{\partial x_3} (h_1 a_1) - \frac{\partial}{\partial x_1} (h_3 a_3) \right] \right\} \right) \right] \mathbf{e}_3$$

### 5. *Lagrangian derivative:*

$$(\mathbf{a} \cdot \nabla) \mathbf{a} = \frac{1}{h_1} \left\{ \left( a_1 \frac{\partial a_1}{\partial x_1} + a_2 \frac{\partial a_2}{\partial x_1} + a_3 \frac{\partial a_3}{\partial x_1} \right) - \frac{a_2}{h_2} \left[ \frac{\partial}{\partial x_1} (h_2 a_2) - \frac{\partial}{\partial x_2} (h_1 a_1) \right] \right.$$
$$\left. + \frac{a_3}{h_3} \left[ \frac{\partial}{\partial x_3} (h_1 a_1) - \frac{\partial}{\partial x_1} (h_3 a_3) \right] \right\} \mathbf{e}_1 + \frac{1}{h_2} \left\{ \left( a_1 \frac{\partial a_1}{\partial x_2} + a_2 \frac{\partial a_2}{\partial x_2} + a_3 \frac{\partial a_3}{\partial x_2} \right) \right.$$
$$\left. - \frac{a_3}{h_3} \left[ \frac{\partial}{\partial x_2} (h_3 a_3) - \frac{\partial}{\partial x_3} (h_2 a_2) \right] + \frac{a_1}{h_1} \left[ \frac{\partial}{\partial x_1} (h_2 a_2) - \frac{\partial}{\partial x_2} (h_1 a_1) \right] \right\} \mathbf{e}_2$$
$$+ \frac{1}{h_3} \left\{ \left( a_1 \frac{\partial a_1}{\partial x_3} + a_2 \frac{\partial a_2}{\partial x_3} + a_3 \frac{\partial a_3}{\partial x_3} \right) - \frac{a_1}{h_1} \left[ \frac{\partial}{\partial x_3} (h_1 a_1) - \frac{\partial}{\partial x_1} (h_3 a_3) \right] \right.$$
$$\left. + \frac{a_2}{h_2} \left[ \frac{\partial}{\partial x_2} (h_3 a_3) - \frac{\partial}{\partial x_3} (h_2 a_2) \right] \right\} \mathbf{e}_3$$

# APPENDIX

# B

## TENSORS

In this appendix some of the basic properties of tensors are reviewed. Although much of the material is general in nature, the discussion is restricted to cartesian tensors, since curvilinear tensors are not used in this book.

### NOTATION AND DEFINITION

*Notation:*

The following rules of notation will be followed throughout:

1. If a given index appears only once in each term of a tensor equation, it is a free index and the equation holds for all possible values of that index.
2. If an index appears twice in any term, it is understood that a summation is to be made over all possible values of that index.
3. No index may appear more than twice in any term.

*Definition:*

A tensor of rank $r$ is a quantity having $n^r$ components in $n$-dimensional space. The components of a tensor quantity expressed in two different coordinate systems are related as follows:

$$T'_{ijk \cdots m} = C_{is}C_{jt}C_{ku} \cdots C_{mv}T_{stu \cdots v}$$

where the quantities $C_{mn}$ are the direction cosines between the axes of the two coordinate systems.

A tensor of rank 2 is sometimes called a *dyadic*, a tensor of rank 1 is a *vector*, and a tensor of rank 0 is a *scalar*.

**431**

## TENSOR ALGEBRA

*Addition:*

Two tensors of equal rank may be added to yield a third tensor of the same rank as follow:

$$C_{ij\cdots k}\,A_{ij\cdots k} + B_{ij\cdots k}$$

*Multiplication:*

If tensor $A$ has rank $a$ and tensor $B$ has rank $b$, the multiplication of these two tensors yields a third one of rank $c$.

$$C_{ij\cdots krs\cdots t} = A_{ij\cdots k}\,B_{r\cdots t}$$

*Contraction:*

If any two indices of a tensor of rank $r \geq 2$ are set equal, a tensor of rank $r - 2$ is obtained. For example, if $C_{ij}$ is defined by

$$C_{ij} = A_i B_j$$

then be setting $i = j$ the tensor $C_{ij}$, which is of rank 2, becomes a tensor of rank 0 (i.e., a scalar}.

$$C_{ii} = A_i B_i$$

Thus contraction is equivalent to taking the scalar product of two vectors in vector algebra.

*Symmetry:*

If the tensor $A$ has the property that

$$A_{i\cdots j\cdots k\cdots l} = A_{i\cdots k\cdots j\cdots l}$$

then the tensor $A$ is said to be *symmetric* in the indices $j$ and $k$. As a consequence of the above relation the tensor has only $\frac{1}{2}n(n + 1)$ independent components.

If the tensor $A$ has the property that

$$A_{i\cdots j\cdots k\cdots l} = -A_{i\cdots k\cdots j\cdots l}$$

then the tensor $A$ is said to be *antisymmetric* in the indices $j$ and $k$. Such tensors have only $\frac{1}{2}n(n - 1)$ independent components.

# TENSOR OPERATIONS

*Gradient:*

The gradient of a tensor of rank $r$ is defined by

$$T_{ij \cdots kl} = \frac{\partial R_{ij \cdots k}}{\partial x_l}$$

and yields a tensor of rank $(r + 1)$.

*Divergence:*

The divergence of a tensor of rank $r$ results in a tensor of rank $(r - 1)$.

$$T_{i \cdots jl \cdots m} = \frac{\partial R_{i \cdots jkl \cdots m}}{\partial x_k}$$

*Curl:*

If $R$ is a tensor of rank $r$, the curl operation will produce an antisymmetric tensor of rank $(r + 1)$. In general, the operation is defined by

$$T_{i \cdots j \cdots kl} = \frac{\partial R_{i \cdots j \cdots k}}{\partial x_l} - \frac{\partial R_{i \cdots l \cdots k}}{\partial x_j}$$

In three dimensions, the curl of a tensor of rank 1 (i.e., a vector) may be written in the form

$$T_i = -\varepsilon_{ijk} \frac{\partial R_j}{\partial x_k}$$

where $\varepsilon_{ijk}$ is a constant pseudoscalar defined by

$$\varepsilon_{123} = \varepsilon_{312} = \varepsilon_{231} = 1$$

$$\varepsilon_{213} = \varepsilon_{321} = \varepsilon_{132} = -1$$

$$\varepsilon_{ijk} = 0 \qquad \text{otherwise}$$

# ISOTROPIC TENSORS

*Definition:*

An isotropic tensor is one whose components are invariant with respect to all possible rotations of the coordinate system. That is, there are no preferred directions, and the quantity represented by the tensor is a function of position only and not of orientation.

*Isotropic tensors of rank 0:*

All tenors of rank 0 (i.e., scalars) are isotropic.

*Isotropic tensors of rank 1:*

There are no isotropic tensors of rank 1. That is, vectors are not isotropic, since there are preferred directions.

*Isotropic tensors of rank 2:*

The only isotropic tensors of rank 2 are of the form $\alpha\delta_{ij}$, where $\alpha$ is a scalar and $\delta_{ij}$ is the *Kronecker delta*, which has the property that

$$\delta_{ij} = \begin{cases} 0 & \text{when } i \neq j \\ 1 & \text{when } i = j \end{cases}$$

*Isotropic tensors of rank 3:*

The isotropic tensors of rank 3 are of the form $\alpha\varepsilon_{ijk}$, where $\alpha$ is a scalar and $\varepsilon_{ijk}$ is a pseudoscalar defined under Tensor Operations.

*Isotropic tensors of rank 4:*

The most general isotropic tensor of rank 4 is of the form

$$\alpha\delta_{ij}\delta_{pq} + \beta\left(\delta_{ip}\delta_{jq} + \delta_{iq}\delta_{jp}\right) + \gamma\left(\delta_{ip}\delta_{jq} - \delta_{iq}\delta_{jp}\right)$$

where $\alpha$, $\beta$, and $\gamma$ are scalars.

## INTEGRAL THEOREMS

The following two theorems were given in vector form in Appendix A, and they are reproduced here in tensor form.

*Gauss' theorem* (divergence theorem):

$$\int_s a_i n_i \, dS = \int_V \frac{\partial a_i}{\partial x_i} \, dV$$

*Stokes' theorem:*

$$\oint a_i \, dl_i = \int_A - \varepsilon_{ijk} \frac{\partial a_j}{\partial x_k} n_i \, dA$$

Listed below are the continuity, Navier-Stokes, and energy equations together with the components of stress in the three most commonly used coordinate systems: cartesian, cylindrical, and spherical coordinates. The equations are valid for calorically perfect newtonian fluids in which $\rho$, $\mu$, and $k$ are all constants.

## CARTESIAN COORDINATES

$$\text{Coordinates } \mathbf{r} = (x, y, z)$$

$$\text{Velocity } \mathbf{u} = (u, v, w)$$

$$\frac{\partial u}{\partial x} + \frac{\partial v}{\partial y} + \frac{\partial w}{\partial z} = 0$$

$$\rho \left( \frac{\partial u}{\partial t} + u \frac{\partial u}{\partial x} + v \frac{\partial u}{\partial y} + w \frac{\partial u}{\partial z} \right) = -\frac{\partial p}{\partial x} + \mu \nabla^2 u + \rho f_x$$

$$\rho \left( \frac{\partial v}{\partial t} + u \frac{\partial v}{\partial x} + v \frac{\partial v}{\partial y} + w \frac{\partial v}{\partial z} \right) = -\frac{\partial p}{\partial y} + \mu \nabla^2 v + \rho f_y$$

$$\rho \left( \frac{\partial w}{\partial t} + u \frac{\partial w}{\partial x} + v \frac{\partial w}{\partial y} + w \frac{\partial w}{\partial z} \right) = -\frac{\partial p}{\partial z} + \mu \nabla^2 w + \rho f_z$$

$$\rho C_v \left( \frac{\partial T}{\partial t} + u \frac{\partial T}{\partial x} + v \frac{\partial T}{\partial y} + w \frac{\partial T}{\partial z} \right) = k \nabla^2 T + 2\mu \left[ \left( \frac{\partial u}{\partial x} \right)^2 + \left( \frac{\partial v}{\partial y} \right)^2 + \left( \frac{\partial w}{\partial z} \right)^2 \right]$$

$$+ \mu \left[ \left( \frac{\partial u}{\partial y} + \frac{\partial v}{\partial x} \right)^2 + \left( \frac{\partial u}{\partial z} + \frac{\partial w}{\partial x} \right)^2 + \left( \frac{\partial v}{\partial z} + \frac{\partial w}{\partial y} \right)^2 \right]$$

where
$$\nabla^2 = \frac{\partial^2}{\partial x^2} + \frac{\partial^2}{\partial y^2} + \frac{\partial^2}{\partial z^2}$$

$$\tau_{xx} = 2\mu \frac{\partial u}{\partial x} \qquad \tau_{xy} = \tau_{yx} = \mu \left[ \frac{\partial u}{\partial y} + \frac{\partial v}{\partial x} \right]$$

$$\tau_{yy} = 2\mu \frac{\partial v}{\partial y} \qquad \tau_{yz} = \tau_{zy} = \mu \left[ \frac{\partial v}{\partial z} + \frac{\partial w}{\partial y} \right]$$

$$\tau_{zz} = 2\mu \frac{\partial w}{\partial z} \qquad \tau_{zx} = \tau_{xz} = \mu \left[ \frac{\partial w}{\partial x} + \frac{\partial u}{\partial z} \right]$$

## CYLINDRICAL COORDINATES

$$\text{Coordinates } \mathbf{r} = (R, \theta, z)$$

$$\text{Velocity } \mathbf{u} = (u_R, u_\theta, u_z)$$

$$\frac{1}{R}\frac{\partial}{\partial R}(Ru_R) + \frac{1}{R}\frac{\partial u_\theta}{\partial \theta} + \frac{\partial u_z}{\partial z} = 0$$

$$\rho \left( \frac{\partial u_R}{\partial t} + u_R \frac{\partial u_R}{\partial R} + \frac{u_\theta}{R}\frac{\partial u_R}{\partial \theta} - \frac{u_\theta^2}{R} + u_z \frac{\partial u_R}{\partial z} \right)$$

$$= -\frac{\partial p}{\partial R} + \mu \left( \nabla^2 u_R - \frac{u_R}{R^2} - \frac{2}{R^2}\frac{\partial u_\theta}{\partial \theta} \right) + \rho f_R$$

$$\rho \left( \frac{\partial u_\theta}{\partial t} + u_R \frac{\partial u_\theta}{\partial R} + \frac{u_\theta}{R}\frac{\partial u_\theta}{\partial \theta} + \frac{u_R u_\theta}{R} + u_z \frac{\partial u_\theta}{\partial z} \right)$$

$$= -\frac{1}{R}\frac{\partial p}{\partial \theta} + \mu \left( \nabla^2 u_\theta - \frac{u_\theta}{R^2} + \frac{2}{R^2}\frac{\partial u_R}{\partial \theta} \right) + \rho f_\theta$$

$$\rho \left( \frac{\partial u_z}{\partial t} + u_R \frac{\partial u_z}{\partial R} + \frac{u_\theta}{R}\frac{\partial u_z}{\partial \theta} + u_z \frac{\partial u_z}{\partial z} \right) = -\frac{\partial p}{\partial z} + \mu \nabla^2 u_z + \rho f_z$$

$$\rho C_v \left( \frac{\partial T}{\partial t} + u_R \frac{\partial T}{\partial R} + \frac{u_\theta}{R}\frac{\partial T}{\partial \theta} + u_z \frac{\partial T}{\partial z} \right)$$

$$= k\nabla^2 T + 2\mu \left\{ \left( \frac{\partial u_R}{\partial R} \right)^2 + \left[ \frac{1}{R}\left( \frac{\partial u_\theta}{\partial \theta} + u_R \right) \right]^2 + \left( \frac{\partial u_z}{\partial z} \right)^2 \right\}$$

$$+ \mu \left\{ \left( \frac{\partial u_\theta}{\partial z} + \frac{1}{R}\frac{\partial u_z}{\partial \theta} \right)^2 + \left( \frac{\partial u_z}{\partial R} + \frac{\partial u_R}{S\partial z} \right)^2 + \left[ \frac{1}{R}\frac{\partial u_R}{\partial \theta} + R\frac{\partial}{\partial R}\left( \frac{u_\theta}{R} \right) \right]^2 \right\}$$

where
$$\nabla^2 = \frac{1}{R}\frac{\partial}{\partial R}\left(R\frac{\partial}{\partial R}\right) + \frac{1}{R^2}\frac{\partial^2}{\partial\theta^2} + \frac{\partial^2}{\partial z^2}$$

$$\tau_{RR} = 2\mu\frac{\partial u_R}{\partial R} \qquad\qquad \tau_{R\theta} = \tau_{\theta R} = \mu\left[R\frac{\partial}{\partial R}\left(\frac{u_\theta}{R}\right) + \frac{1}{R}\frac{\partial u_R}{\partial\theta}\right]$$

$$\tau_{\theta\theta} = 2\mu\left(\frac{1}{R}\frac{\partial u_\theta}{\partial\theta} + \frac{u_R}{R}\right) \qquad \tau_{\theta z} = \tau_{z\theta} = \mu\left[\frac{\partial u_\theta}{\partial z} + \frac{1}{R}\frac{\partial u_z}{\partial\theta}\right]$$

$$\tau_{zz} = 2\mu\frac{\partial u_z}{\partial z} \qquad\qquad \tau_{zR} = \tau_{Rz} = \mu\left[\frac{\partial u_z}{\partial R} + \frac{\partial u_R}{\partial z}\right]$$

## SPHERICAL COORDINATES

$$\text{Coordinates } \mathbf{r} = (r, \theta, \omega)$$

$$\text{Velocity } \mathbf{u} = (u_r, u_\theta, u_\omega)$$

$$\frac{1}{r^2}\frac{\partial}{\partial r}\left(r^2 u_r\right) + \frac{1}{r\sin\theta}\frac{\partial}{\partial\theta}\left(u_\theta\sin\theta\right) + \frac{1}{r\sin\theta}\frac{\partial u_\omega}{\partial\omega} = 0$$

$$\rho\left(\frac{\partial u_r}{\partial t} + u_r\frac{\partial u_r}{\partial r} + \frac{u_\theta}{r}\frac{\partial u_r}{\partial\theta} + \frac{u_\omega}{r\sin\theta}\frac{\partial u_r}{\partial\omega} - \frac{u_\theta^2 + u_\omega^2}{r}\right)$$

$$= -\frac{\partial p}{\partial r} + \mu\left(\nabla^2 u_r - \frac{2}{r^2}u_r - \frac{2}{r^2}\frac{\partial u_\theta}{\partial\theta} - \frac{2}{r^2}u_\theta\cot\theta - \frac{2}{r^2\sin\theta}\frac{\partial u_\omega}{\partial\omega}\right) + \rho f_r$$

$$\rho\left(\frac{\partial u_\theta}{\partial t} + u_r\frac{\partial u_\theta}{\partial r} + \frac{u_\theta}{r}\frac{\partial u_\theta}{\partial\theta} + \frac{u_\omega}{r\sin\theta}\frac{\partial u_\theta}{\partial\omega} + \frac{u_r u_\theta}{r} - \frac{u_\omega^2}{r}\cot\theta\right)$$

$$= -\frac{1}{r}\frac{\partial p}{\partial\theta} + \mu\left(\nabla^2 u_\theta + \frac{2}{r^2}\frac{\partial u_r}{\partial\omega} - \frac{u_\theta}{r^2\sin^2\theta} - \frac{2}{r^2}\frac{\cos\theta}{\cos^2\theta}\frac{\partial u_\omega}{\partial\omega}\right) + \rho f_\theta$$

$$\rho\left(\frac{\partial u_\omega}{\partial t} + u_r\frac{\partial u_\omega}{\partial r} + \frac{u_\theta}{r}\frac{\partial u_\omega}{\partial\theta} + \frac{u_\omega}{r\sin\theta}\frac{\partial u_\omega}{\partial\omega} + \frac{u_r u_\omega}{r} + \frac{u_\theta u_\omega}{r}\cot\theta\right)$$

$$= -\frac{1}{r\sin\theta}\frac{\partial p}{\partial\omega} + \mu\left(\nabla^2 u_\omega + \frac{2}{r^2\sin\theta}\frac{\partial u_r}{\partial\omega}\right.$$

$$\left. + \frac{2\cos\theta}{r^2\sin^2\theta}\frac{\partial u_\theta}{\partial\omega} - \frac{u_\omega}{r^2\sin^2\theta}\right) + \rho f_\omega$$

$$\rho C_v \left( \frac{\partial T}{\partial t} + u_r \frac{\partial T}{\partial r} + \frac{u_\theta}{r} \frac{\partial T}{\partial \theta} + \frac{u_\omega}{r \sin \theta} \frac{\partial T}{\partial \omega} \right)$$

$$= k \nabla^2 T + 2\mu \left[ \left( \frac{\partial u_r}{\partial r} \right)^2 + \left( \frac{1}{r} \frac{\partial u_\theta}{\partial \theta} + \frac{u_r}{r} \right)^2 \right.$$

$$\left. + \left( \frac{1}{r \sin \theta} \frac{\partial u_\omega}{\partial \omega} + \frac{u_r}{r} + \frac{u_\theta}{r} \cot \theta \right)^2 \right]$$

$$+ \mu \left\{ \left[ \frac{1}{r} \frac{\partial u_r}{\partial \theta} + r \frac{\partial}{\partial r} \left( \frac{u_\theta}{r} \right) \right]^2 + \left[ \frac{1}{r \sin \theta} \frac{\partial u_r}{\partial \omega} + r \frac{\partial}{\partial r} \left( \frac{u_\omega}{r} \right) \right]^2 \right.$$

$$\left. + \left[ \frac{1}{r \sin \theta} \frac{\partial u_\theta}{\partial \omega} + \frac{\sin \theta}{r} \frac{\partial}{\partial \theta} \left( \frac{u_\omega}{\sin \theta} \right) \right]^2 \right\}$$

where $\quad \nabla^2 = \dfrac{1}{r^2} \dfrac{\partial}{\partial r} \left( r^2 \dfrac{\partial}{\partial r} \right) + \dfrac{1}{r^2 \sin \theta} \dfrac{\partial}{\partial \theta} \left( \sin \theta \dfrac{\partial}{\partial \theta} \right) + \dfrac{1}{r^2 \sin^2 \theta} \dfrac{\partial^2}{\partial \omega^2}$

$$\tau_{rr} = 2\mu \frac{\partial u_r}{\partial r} \qquad\qquad\qquad \tau_{r\theta} = \tau_{\theta r} = \mu \left[ r \frac{\partial}{\partial r} \left( \frac{u_\theta}{r} \right) + \frac{1}{r} \frac{\partial u_r}{\partial \theta} \right]$$

$$\tau_{\theta\theta} = 2\mu \left( \frac{1}{r} \frac{\partial u_\theta}{\partial \theta} + \frac{u_r}{r} \right) \qquad\qquad \tau_{\theta\omega} = \tau_{\omega\theta} = \mu \left[ \frac{\sin \theta}{r} \frac{\partial}{\partial \theta} \left( \frac{u_\omega}{\sin \theta} \right) + \frac{1}{r \sin \theta} \frac{\partial u_\theta}{\partial \omega} \right]$$

$$\tau_{\omega\omega} = 2\mu \left( \frac{1}{r \sin \theta} \frac{\partial u_\omega}{\partial \omega} + \frac{u_r}{r} + \frac{u_\theta \cot \theta}{r} \right) \qquad \tau_{\omega r} = \tau_{r\omega} = \mu \left[ r \frac{\partial}{\partial r} \left( \frac{u_\omega}{r} \right) + \frac{1}{r \sin \theta} \frac{\partial u_r}{\partial \omega} \right]$$

# COMPLEX
# VARIABLES

Summarized below are some of the results of complex-variable theory which are particularly useful in the study of field mechanics.

## ANALYTIC FUNCTION

A function $F(z)$ of the complex variable $z = x + iy$ is said to be analytic if the derivative $dF/dz$ exists at a point $z_0$ and in some neighborhood of $z_0$ and if the value of $dF/dz$ is independent of the direction in which it is calculated.

## SINGULAR POINTS

A singular point of the function $F(z)$ is any point at which $F(z)$ is not analytic. If $F(z)$ is analytic in some neighborhood of the point $z_0$, but not at $z_0$ itself, then $z_0$ is called an *isolated singular point* of $F(z)$.

## DERIVATIVE OF AN ANALYTIC FUNCTION

If $F(z)$ is analytic, then $dF/dz$ will exist and may be calculated in any direction, so that

$$\frac{dF}{dz} = \frac{\partial F}{\partial x} = -i\frac{\partial F}{\partial y}$$

## CAUCHY-RIEMANN EQUATIONS

If $F(z) = \phi(x, y) + i\psi(x, y)$ is an analytic function, the real part and the imaginary part of $F(z)$ must satisfy the Cauchy-Riemann equations.

$$\frac{\partial \phi}{\partial x} = \frac{\partial \psi}{\partial y}$$

$$\frac{\partial \phi}{\partial y} = -\frac{\partial \psi}{\partial x}$$

The Cauchy-Riemann equations are necessary, but not sufficient, conditions for an analytic function. Eliminating first $\phi$, then $\psi$, from the Cauchy-Riemann equations shows that both $\phi$ and $\psi$ are harmonic functions; that is, they must satisfy Laplace's equation.

## MULTIPLE-VALUED FUNCTIONS

Many functions are analytic but assume more than one value at any point $z = Re^{i\theta}$ on the complex plane as $\theta$ increases by multiples of $2\pi$. This difficulty is overcome by replacing the single complex plane, which is valid for all $\theta$, by a series of *Riemann sheets* which are connected to each other along a *branch cut* which runs (usually along the negative real axis) between two *branch points* (frequently $z = 0$ and $z = \infty$) which are singular points of the function.

## CAUCHY-GOURSAT THEOREM

If $F(z)$ is analytic at all points inside and on a closed contour $c$, then

$$\int_c F(z)\, dz = 0$$

## CAUCHY INTEGRAL FORMULA

If $F(z)$ is analytic at all points inside and on a closed contour $c$ and if $z_0$ is any point inside $c$, then

$$\frac{d^n F}{dz^n}(z_0) = \frac{n!}{2\pi i} \int_c \frac{F(z)}{(z - z_0)^{n+1}}\, dz \qquad \text{for } n \geq 1$$

and

$$F(z_0) = \frac{1}{2\pi i} \int_c \frac{F(z)}{(z - z_0)}\, dz$$

## TAYLOR SERIES

If $F(z)$ is analytic at all points within a circle $r < r_0$ whose center is at $z_0$, then $F(z)$ may be represented by the following series:

$$F(z) = F(z_0) + (z - z_0)\frac{dF}{dz}(z_0) + \frac{(z - z_0)^2}{2!}\frac{d^2F}{dz^2}(z_0) + \cdots$$

where the radius of convergence $r_0$ is the distance from the point $z_0$ to the nearest singularity. The general form of this series, as given above, is known as the Taylor series, and the special case $z_0 = 0$ is known as Maclaurin's series.

## LAURENT SERIES

If $F(z)$ is analytic at all points within the annular region $r_0 < r < r_1$ whose center is at $z_0$, then $F(z)$ may be represented by the following series:

$$F(z) = \cdots + \frac{b_2}{(z - z_0)^2} + \frac{b_1}{z - z_0} + a_0 + a_1(z - z_0) + a_2(z - z_0)^2 + \cdots$$

where

$$a_n = \frac{1}{2\pi i}\int_c \frac{F(\xi)}{(\xi - z_0)^{n+1}}\,d\xi \qquad n = 0, 1, 2, \ldots$$

and

$$b_n = \frac{1}{2\pi i}\int_{c_0} \frac{F(\xi)}{(\xi - z_0)^{-n+1}}\,d\xi \qquad n = 0, 1, 2, \ldots$$

Here the contour $c$ corresponds to $r = r_0$ and the contour $c_0$ corresponds to $r = r_1$. The series is convergent for the smallest radius $r_0$ and the largest radius $r_1$ such that there is no singularity in the region $r_0 < r < r_1$. The part of the series which contains the $b_n$ coefficients is known as the *principal part*. The general form of the series is known as the Laurent series, and in the special case in which $r_0$ may be extended to zero, that is, there is no principal part, the series becomes the Taylor series.

## RESIDUES

The residue of a function at a point $z_0$ is defined as the coefficient $b_1$ in its Laurent series about the point $z_0$. That is, the residue at $z_0$ is the coefficient of the $1/z$ term in the Laurent series of the function written about the point $z_0$.

## RESIDUE THEOREM

If $F(z)$ is analytic within and on a closed curve $c$, except for a finite number of singular points $z_1, z_2, \ldots, z_n$, then

$$\int_c F(z)\,dz = 2\pi i(R_1 + R_2 + \cdots + R_n)$$

where $R_1$ is the residue of $F(z)$ at $z_1$, $R_2$ is the residue at $z_2$, and $R_n$ is the residue at $z_n$.

## TYPES OF SINGULAR POINTS

In order to evaluate the residue of a function at some point, it is useful to know the type of singularity which exists at that point.

*Branch points:*

These are the singular points at the end of each branch cut of a multiple-valued function. The residue theorem does not apply at branch points.

*Essential singular points:*

If the principal part of the Laurent series of the expansion of a function about some point contains an infinite number of terms, that point is an essential singular point.

*Pole of order m:*

If the principal part of the Laurent series of the expansion of a function about some point contains only terms up to $(z - z_0)^m$, that point is a pole of order $m$. That is, if $F(z_0)$ is a pole of order $m$, then $(z - z_0)^m F(z_0)$ will be analytic.

*Simple pole:*

If the principal part of the Laurent series of the expansion of a function about some point contains only a term proportional to $z - z_0$, that point is a simple pole. That is, if $F(z_0)$ is a simple pole, then $(z - z_0)F(z_0)$ will be analytic.

## CALCULATION OF RESIDUES

The following methods may be used to calculate the residue of a function $F(z)$ at a singular point $z_0$:

1. Expand $F(z)$ in a series about $z_0$, and so obtain the coefficient of the term $1/(z - z_0)$. This fundamental method uses the definition of the residue and is valid for all types of singularities.
2. If the point $z_0$ is a pole of order $m$, the residue may be calculated by taking the following limit:

$$R = \lim_{z \to z_0} \frac{1}{(m-1)!} \frac{d^{m-1}}{dz^{m-1}} \left[ (z - z_0)^m F(z) \right]$$

**3.** If the point $z_0$ is a simple pole, the residue may be calculated by taking the following limit:

$$R = \lim_{z \to z_0} (z - z_0)F(z)$$

**4.** If $F(z)$ may be put in the form $F(z) = p(z)/q(z)$ where $q(z_0) = 0$ but $dq/dz(z_0) \neq 0$, and where $p(z_0) \neq 0$, the residue may be calculated by taking the following limit:

$$R = \lim_{z \to z_0} \frac{p}{dq/dz}$$

## CONFORMAL TRANSFORMATIONS

A conformal transformation is a mapping from the $z$ plane to the $\zeta$ plane of the form $z = f(\zeta)$, where $f$ is an analytic function of $\zeta$. Conformal transformations preserve angles between small arcs except at points where $df/d\zeta = 0$. Such points are called *critical points* of the transformation, and smooth curves through such points in the $\zeta$ plane may give angular corners in the $z$ plane.

## SCHWARZ-CHRISTOFFEL TRANSFORMATION

Apart from the elementary functions, the Schwarz-Christoffel transformation is one of the most common and useful forms of mapping. It maps the interior of a closed polygon in the $z$ plane onto the upper half of the $\zeta$ plane while the boundary of the polygon maps onto the real axis of the $\zeta$ plane. The transfor-

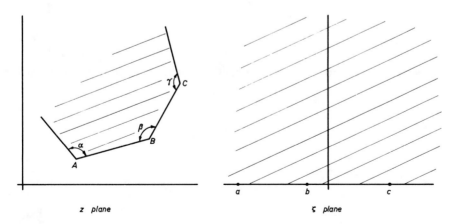

z plane    ζ plane

**FIGURE D.1**
Corresponding regions in the original ($z$ plane) and the mapping plane ($\zeta$ plane) for the Schwarz-Christoffel transformation.

mation is of the form

$$\frac{dz}{d\zeta} = K(\zeta - a)^{\alpha/\pi - 1}(\zeta - b)^{\beta/\pi - 1}(\zeta - c)^{\gamma/\pi - 1} \cdots$$

Here the vertices $A$, $B$, $C$, etc., in the $z$ plane subtend the interior angles $\alpha$, $\beta$, $\gamma$, etc., as indicated in Fig. D.1, and $a$, $b$, $c$, etc., are the points on the real axis of the $\zeta$ plane corresponding to the vertices $A$, $B$, $C$, etc. Since the polygon in the $z$ plane is closed, the angles $\alpha$, $\beta$, $\gamma$, etc., must satisfy the relation

$$\alpha + \beta + \gamma + \cdots = (n - 2)\pi$$

where $n$ is the number of vertices in the polygon. The constant $K$ determines the scale of the polygon and its orientation, while the constant of integration determines the location of the origin in the $z$ plane. Of the constants $a$, $b$, $c$, etc., any three may be chosen arbitrarily (typically $-1$, $0$, $1$) and any remaining ones will be determined by the shape of the polygon.

# APPENDIX
# E

# THERMODYNAMICS

Thermodynamics is a complete subject in itself. However, in the study of fluid mechanics only a few fundamental relationships from thermodynamics are needed, and these arise mainly in the study of compressible flow. The following summary contains all the results from thermodynamics which are required in this book.

## ZEROTH LAW

The zeroth law of thermodynamics states that there exists a variable of state, the temperature $T$, and that two systems which are in thermal contact are in equilibrium only if their temperatures are equal.

## FIRST LAW

The first law of thermodynamics states that there exists a variable of state, the internal energy $e$. If an amount of work $\delta w$ is done on a thermodynamic system and an amount of heat $\delta q$ is added to it, the equilibrium states before and after the process are related by

$$\delta e = \delta w + \delta q$$

That is, the change in the internal energy equals the work done on the system plus the heat added to it during the process or event.

## EQUATIONS OF STATE

There are two thermal equations of state which are commonly used, one being for ideal gases and the other for real gases.

*Thermally perfect gas:*

The equation of state for a thermally perfect gas is

$$p = \rho RT$$

where $R$ is the gas constant for that particular gas. Frequently this equation is used to define a perfect gas. That is, gases which obey the above equation of state are defined as perfect gases.

*Van der Waals equation:*

An approximate equation of state for real gases is given by the van der Waals equation, which is

$$p = \rho RT \left[ \frac{1}{1 - \beta\rho} - \frac{\alpha\rho}{RT} \right]$$

where

$$\frac{\alpha}{\beta} = \tfrac{27}{8} RT_c$$

and

$$\frac{\alpha}{\beta^2} = 27 p_c$$

Here $p_c$ and $T_c$ are, respectively, the critical pressure and temperature of the gas.

## ENTHALPY

The enthalpy $h$ of a gas is defined by the following equation:

$$h = e + \frac{p}{\rho} = e + pv$$

where $v$ is the volume of the gas per unit mass.

## SPECIFIC HEATS

There are two specific heats in common usage, that at constant volume and that at constant pressure.

*Constant volume:*

The specific heat at constant volume $C_v$ is defined by

$$C_v \equiv \left( \frac{dq}{dT} \right)_v = \frac{\partial e}{\partial T} = \frac{\partial h}{\partial T} + \left( \frac{\partial h}{\partial p} - v \right) \left( \frac{\partial p}{\partial T} \right)_v$$

From the defining identity, the other relations follow without approximation.

*Constant pressure:*

The specific heat at constant pressure $C_p$ is defined by

$$C_p \equiv \left(\frac{dq}{dT}\right)_p = \frac{\partial e}{\partial T} + \left(\frac{\partial e}{\partial v} + p\right)\left(\frac{\partial v}{\partial T}\right)_p = \frac{\partial h}{\partial T}$$

*Perfect gas:*

Using the above definitions of the specific heats and considering a perfect gas, it follows that

$$C_p - C_v = R$$

Under these circumstances it can be shown that $e$ and $h$ are functions of the temperature $T$ only and may be expressed in the form

$$e(T) = \int C_v \, dT + \text{constant}$$

$$h(T) = \int C_p \, dT + \text{constant}$$

If $C_v$ and $C_p$ are constants, independent of $T$, the gas is called *calorically perfect*, and it follows that

$$e = C_v T + \text{constant}$$
$$h = C_p T + \text{constant}$$

## ADIABATIC, REVERSIBLE PROCESSES

The following relations are valid for adiabatic, reversible processes:

$$\frac{\rho}{\rho_0} = \left(\frac{T}{T_0}\right)^{1/(\gamma-1)}$$

$$\frac{p}{p_0} = \left(\frac{T}{T_0}\right)^{\gamma/(\gamma-1)} = \left(\frac{\rho_0}{\rho}\right)^{\gamma}$$

where $\gamma = C_p/C_v$ and $\rho_0, T_0, p_0$ are constants.

## ENTROPY

There exists a variable of state, the entropy $s$. If heat is added to a system, the change in entropy between the initial and final equilibrium states will be given by

$$s_B - s_A = \int_A^B \frac{dq}{T}$$

where the integral is evaluated for a reversible process.

## SECOND LAW

The second law of thermodynamics states that for any spontaneous process the entropy change is positive or zero. That is,

$$s_B - s_A \geq \int_A^B \frac{dq}{T}$$

For a calorically perfect gas it follows that

$$s - s_0 = C_p \log \frac{T}{T_0} - R \log \frac{p}{p_0}$$

$$= C_v \log \frac{T}{T_0} + R \log \frac{\rho_0}{\rho}$$

## CANONICAL EQUATIONS OF STATE

The heat-addition term in the first law may be eliminated in favor of the entropy, yielding an equation which involves variables of state only. From this the following identities may be established:

$$\left(\frac{\partial e}{\partial s}\right)_v = T \qquad \left(\frac{\partial e}{\partial v}\right)_s = -p$$

$$\left(\frac{\partial h}{\partial s}\right)_p = T \qquad \left(\frac{\partial h}{\partial p}\right)_x = v$$

## RECIPROCITY RELATIONS

By considering $s$ to be a function of $p$ and $T$, the following reciprocity relations follow:

$$\frac{\partial s}{\partial T} = \frac{1}{T}\frac{\partial h}{\partial T} \qquad \frac{\partial s}{\partial p} = \frac{1}{T}\left(\frac{\partial h}{\partial p} - v\right)$$

From these reciprocity relations the following equation is obtained which relates the caloric and thermal equations of state:

$$\frac{\partial h}{\partial p} = v - T\frac{\partial v}{\partial T}$$

Similarly, by considering $s$ to be a function of $v$ and $T$, the following reciprocity relations are obtained:

$$\frac{\partial s}{\partial T} = \frac{1}{T}\frac{\partial e}{\partial T} \qquad \frac{\partial s}{\partial v} = \frac{1}{T}\left(\frac{\partial e}{\partial v} + p\right)$$

From these, the relation between the caloric and thermal equations of state is found to be

$$\frac{\partial e}{\partial v} = -p + T\frac{\partial p}{\partial T}$$

# INDEX